Smart Grids as Cyber Physical Systems Volume 1

Scrivener Publishing
100 Cummings Center, Suite 541J
Beverly, MA 01915-6106

Publishers at Scrivener
Martin Scrivener (martin@scrivenerpublishing.com)
Phillip Carmical (pcarmical@scrivenerpublishing.com)

Smart Grids as Cyber Physical Systems Volume 1

Artificial Intelligence, Cybersecurity, and Clean Energy for Next Generation Smart Grids

Edited by

O.V. Gnana Swathika

K. Karthikeyan

and

Sanjeevikumar Padmanaban

Scrivener Publishing

WILEY

This edition first published 2024 by John Wiley & Sons, Inc., 111 River Street, Hoboken, NJ 07030, USA and Scrivener Publishing LLC, 100 Cummings Center, Suite 541J, Beverly, MA 01915, USA

For more information about Scrivener publications please visit www.scrivenerpublishing.com.

Wiley Global Headquarters
111 River Street, Hoboken, NJ 07030, USA

For details of our global editorial offices, customer services, and more information about Wiley products visit us at www.wiley.com.

Library of Congress Cataloging-in-Publication Data

ISBN: 9781394261697

Front cover images supplied by Pixabay.com
Cover design by Russell Richardson

Set in size of 11pt and Minion Pro by Manila Typesetting Company, Makati, Philippines

Printed in the USA

10 9 8 7 6 5 4 3 2 1

Contents

Preface

Integrating artificial intelligence into a smart grid system contributes to the trajectory of growth in a knowledge-based economy suitable for the energy transition. They play a key role in smart cities, demanding intelligent energy transition that keeps the importance of energy-saving and energy management technologies. Hence, it eventually paves the way for drastic changes like smart buildings, smart transport, and smart health care systems. Information and communication technologies are a feature of any smart grid system. *Smart Grids as Cyber Physical Systems* is a next-generation solution with significant modules that facilitate the integration of the grid and cyberspace. It operates in a closed-loop mechanism wherein the smart grid system adapts itself based on real-time operating conditions. Smart Grid systems are designed and controlled in a framework that concentrates on clean energy source integration. This integration is done with conscious monitoring of parameters like voltage and frequency control, power quality, and storage features. The work focuses on realizing proper energy management and protection systems operating under uncertainty and prioritizes robustness and sustainable solutions. This book further offers clarity on specific operating and security issues that may arise that deteriorate the overall operation and efficiency of Smart Grid Systems. It also provides techniques to monitor and protect the system. It also aids in designing a threat-free system. It also discusses how Artificial Intelligence (AI) may be extensively deployed in prediction for energy generation, electric grid-related line loss prediction, load forecasting, and for predicting equipment failure prevention. Building CPS with effective communication, control, intelligence, and security is discussed from societal and research perspectives.

1

Grid Independent Dynamic Charging of EV Batteries Using Solar Energy

P. Balamurugan[1*], Tekumalla Lakshmi Sowjanya[2], Manas Goyan[2], J.L. Febin Daya[1] and V. Ananthakrishnan[3]

[1]*eVITRC, Vellore Institute of Technology - Chennai, Tamil Nadu, India*
[2]*VIT-Bhopal University, Madhya Pradesh, India*
[3]*School of Electrical Engineering, Vellore Institute of Technology - Chennai, Tamil Nadu, India*

Abstract

This research aims to utilize energy emitted by the sun as a source to charge an electric car and to eliminate the need for installing more charging stations to save electricity. The proposed concept of dynamic charging from solar source minimizes carbon footprint. The concept can be applied in a wide range by installing flexible solar panels that are available commercially in the market. This concept enables embedding the flexible panels during the manufacturing stage and widespread utilization of available solar energy surplus. This will reduce the consumption of electricity generated from conventional energy sources to charge the electric vehicles. The dynamic charging will also extend the range provided by the battery per charge cycle improving the reliability of the electric vehicle.

Keywords: EV charging, dynamic charging, flexible PV panels, boost converter

1.1 Introduction

Climate changes due to emission of CO_2 from remnant fuels is a major concern in recent years. This enabled electrical engineers and researchers to look out for alternate sources of energy like solar farms, wind farms, and other sources like biomass. Vehicles are the primary sources where the CO_2 emission is present everywhere compared to large industries where it

**Corresponding author*: balamurugan.p@vit.ac.on

O.V. Gnana Swathika, K. Karthikeyan and Sanjeevikumar Padmanaban (eds.) Smart Grids as Cyber Physical Systems Volume 1: Artificial Intelligence, Cybersecurity, and Clean Energy for Next Generation Smart Grids, (1–12) © 2024 Scrivener Publishing LLC

is localized and regulated. Hence, electric vehicles (EV) came into picture, resulting in rapid manufacturing and production of EVs. The use of additional battery storage in automotive systems for supplying the power train and internal utilities extending the range of operation influencing economic challenges is the next challenge. As a promising technology and preference, harvesting renewable energy resources is a boon for both EV industries and the transportation sector. Charging batteries and similar storage systems directly from renewable resources increases the reliability of the battery-operated vehicles. Utilizing energy from alternative energy resources as an alternative to conventional electric network to run the EV is perceived to expand the complete system efficacy and diminish carbon footprint.

Energy from photovoltaic (PV) panels are progressively competing with other forms of renewable sources of energy due to its splendid nature and is easily extractable. Hence, the adaptability of e-transportation is facilitated much by PV sources. In recent days, utilizing PV as a primary source of energy is adapted from small scale to large industries, airports, and residential applications. Since it exhibits lesser maneuver cost and maintenance, low greenhouse gas emanations, and self-governing capability, PV is adopted everywhere. Vast exploration of technology has been adopted lately to charge EV with energy from solar. Many countries have initiated design standards for PV systems.

Analyzing the current scenario and forthcoming challenges in the implementation of EV and charging systems, it is essential to adopt operation of EVs independent on conventional grid for charging. Several agencies analyzed the challenges and deployed several universal standards and codes for charging EVs. Considering wide-spread applications and situations, batteries are charged by PVs which are installed on the vehicles, causing the reduction in the installation of charging stations.

This work, "Solar Powered 4-Wheeler" is a dynamic system, wherein the energy from the sun is transformed into the electrical energy using the required number of solar panels of sufficient area mounted on the EV, to power up the lead–acid battery, which is used in the 4-wheeler [1].

As this is a dynamic system, due to weather conditions, the charging is influenced by parameters like irradiance and temperature. The unregulated output of the PV panel is regulated using DC-DC charge controller that is capable of boosting or stepping down the PV output with simultaneous tracking of maximum power output from the panel. The maximum power point tracking (MPPT) controller is better for the high-power range applications rather than the pulse width modulation (PWM) controller to measure the maximum power point (MPP) and keep the battery in charging mode whenever the variable parameters get changed [2]. Solar cells are implemented on the surface of the scooter by calculating the parameters of the single cell and connect those cells in the series and parallel to get the output that is helpful to charge the battery [3]. The battery plays a major role in electric vehicles. Knowing the total energy left in a battery and the driving conditions, the control system notifies the user, and the range for which the battery will perform until the next recharge, which is a measure of the intermittent ability of the battery [4]. The state-of-charge (SOC) of the battery can be resolute from the discharging characteristics of the battery. An SOC of 100% is a measure of area under the discharge characteristics of a battery from 100% charge to 0% charge in the battery [5]. The battery charging is done using three stages of charging of the battery to maintain the standard and charging takes place with respect to the SOC of battery by constant-current and constant-voltage methods [6].

1.2 Proposed Methodology

In this proposed methodology, flexible solar panels will be mounted on car surfaces, where maximum light will fall. Since the PV panels are flexible, their size and power ratings are not regular. Hence, the power rating of the panel must be calculated using the number of cells present in the section and the data sheet of the PV panel. Based on the efficiency of solar panel, the size and power are finalized. From the data sheet, the power output of a single cell is computed and based on the area of the panel, the total power of the PV is calculated. To get the desired amount of power generation, many such cells will be connected in series and parallel. Once the PV panels are designed, then it requires a DC-DC converter to extract power from the PV panel. In this work, a boost converter is chosen to extract power from the PV. The boost converter provides necessary charging current at desired voltage to the battery present in EV. The power output of PV varies depending on cell temperature and irradiance falling on the panel. Hence, it is necessary to regulate and extract maximum power output from PV with suitable control technique using boost converter. Maximum power point tracking (MPPT) is one of the classical approaches that aims at tracking and extracting the maximum power output from the PV panel and feeds the connected loads. For tracking the maximum power, the perturb and observe (P&O) algorithm is applied. The energy output of the boost converter will be given to the battery for storage or an inverter to convert the DC to AC. The algorithm for the charge controller will adjust the current and voltage with respect to the SOC of the battery. Based on the current from the boost converter, the charging time of the battery can be computed. The functional illustration of the planned system is depicted in Figure 1.1.

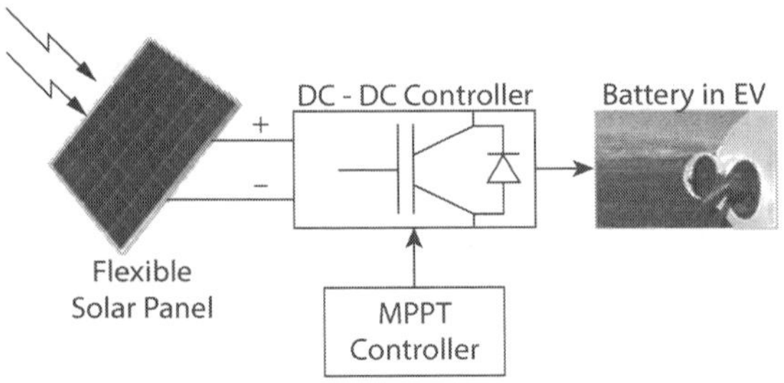

Figure 1.1 Functional illustration of the proposed system.

The specifications of Renogy UK monocrystalline flexible solar panels are provided in Table 1.1. The image of Renogy UK monocrystalline flexible PV panel is shown in Figure 1.2 for reference along with its dimensions specified.

Single solar cell parameters from the selected manufacturer solar panel

- Cell area = 6.5 * 3.3 in = 16.51 * 8.382 cm^2 = 138.39 cm^2 = 13,839 mm^2
- Open circuit voltage from the single cell = V_{oc} of solar panel/ No. of cells in panel
- No. of cells connected in series = PV module voltage / voltage at the operating condition
- Here, PV module voltage is 52 because we need 52 V from the source to charge 48 V

Table 1.1 Renogy UK monocrystalline flexible PV panel specifications.

Parameter	**Values**
Maximum power	100 W
Open circuit voltage (V_{oc})	23.5 V
Voltage at maximum power point	19.4 V
Temperature coefficient (V_{oc}, I_{sc}, P_{max})	−0.29%, 0.05%, −0.38%
Short circuit current (I_{SC})	5.51 A
Current at maximum power point	5.2 A
Irradiance and nominal temperature	1000 W/m^2 and 25°C
Panel dimensions in mm	1093 x 582
No. of cells	36

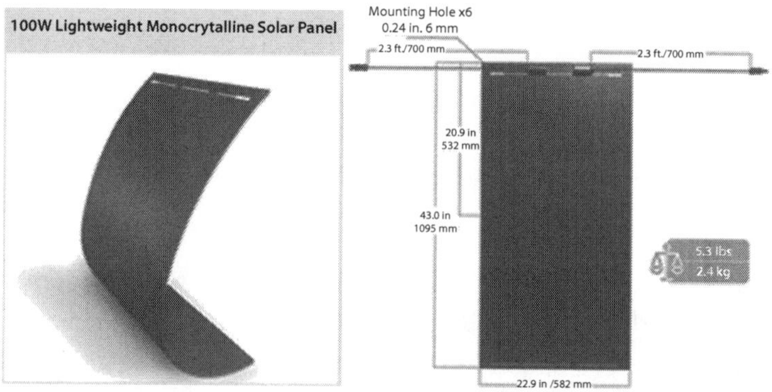

Figure 1.2 Renogy UK monocrystalline flexible PV panel.

Light generated current (I_L) = 5.5453 A
Diode saturation current (I_0) = 12.358 pA
Diode ideality factor = 0.94781
Shunt resistance = 185.7255 Ω
Series resistance = 0.26886 Ω
Series connected modules per string = 2

With all the specifications of the solar PV source, the simulation is performed in MATLAB/Simulink environment. The characteristics of the PV panel at two different temperatures are shown in Figure 1.3.

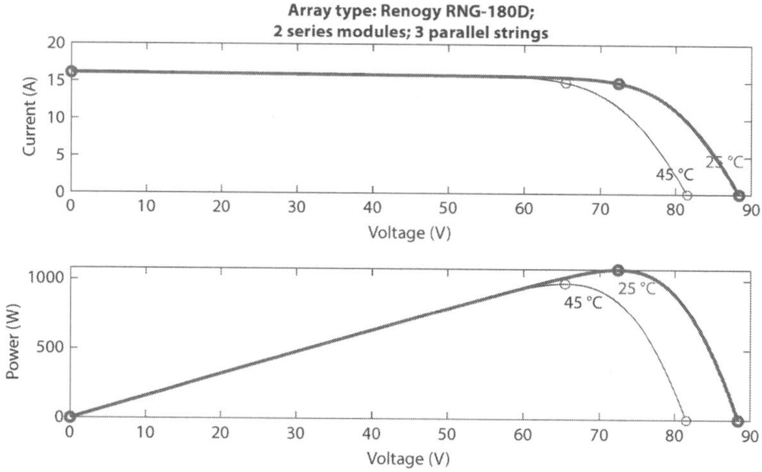

Figure 1.3 I-V and P-V characteristics of PV panel.

1.3 Design of Boost Converter

The power circuit of the boost converter is illustrated in Figure 1.4. The tendency of an inductor to store energy in the form of magnetic field in it, resulting in resisting the changes in current flow through it, is the vital principle behind the boost converter. The energy stored in the inductor in series to the voltage source retains the output voltage, exceeding the input voltage of the boost converter. The working of a boost converter is understood based on the semiconductor switch S in the circuit.

The switch 'S' is operated with a period T, in which it is closed for T_{ON} sec and open for T_{OFF} sec. When the switch is closed, the energy stored in the inductor increases because of the increase in current flowing through it. The polarity of voltage is opposite to that of the supply. When opened, the stored energy in the inductor is discharged to the capacitor, with the polarity aiding the supply voltage. Hence, the voltage across the load is greater than the supply, V_{in}. The design specifications of boost converter are listed in Table 1.2.

The simulation model of the proposed system is shown in Figure 1.5 for reference. In this model, two PV panels are connected in parallel to

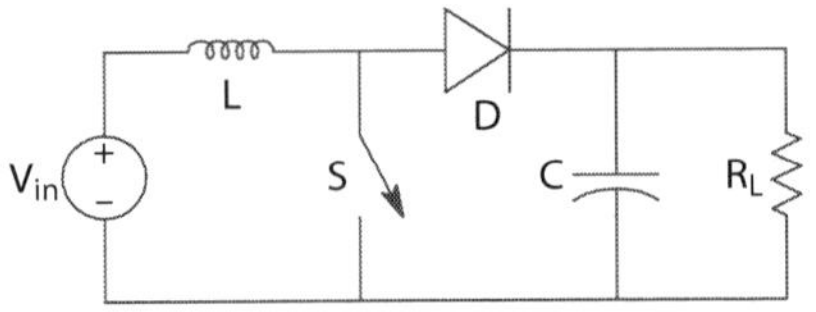

Figure 1.4 Boost converter.

Table 1.2 Boost converter specifications.

Parameter	Values
Input voltage	24 V
Output voltage	52 V
Rated power	150 W
Switching frequency	50 kHz
Output current	5.49 A
Inductance (L)	414 μH
Capacitance (C)	470 μF

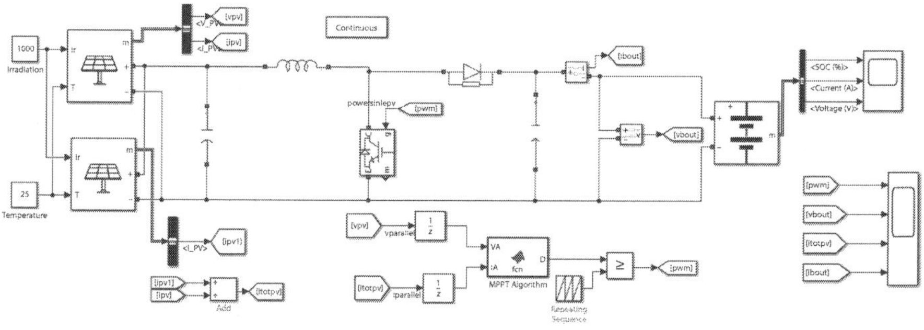

Figure 1.5 Simulink model of the proposed open-loop type battery charger.

achieve more current at constant voltage. The output from the PV panel is tracked using MPPT algorithm embedded in the simulation as a MATLAB function block. The output duty ration is converted into a PWM signal at 50-kHz switching frequency and is applied to the insulated-gate bipolar transistor (IGBT) switch of the boost converter. A battery of 48 V, 50 AH, with an initial SOC of 80% is taken to provide a proof of concept.

Since the panels are considered to be mounted on the top of the car, the same temperature and irradiation levels are considered for the simulation. In practice, if more such panels are provided on the sides, bonnet, etc., the net power output may vary based on partial shadow conditions. But the objective of extending the range of driving an EV using self-charging capability is satisfied with the proposed model.

1.4 Perturb and Observe Algorithm for Tracking Maximum Power

Perturb and observe (P&O) is the most common algorithm employed to extract maximum available power from the renewable energy resources. Through this algorithm, we generate the duty cycle of the PWM required to operate the switch in boost converter. The input to P&O is the output voltage, V(k), of the panel and its output current, I(k). With this voltage and current, the power at k^{th} instant is calculated and is compared with power, p(k), during previous sampling interval. At every instant, the difference in power, Δp(k), and voltage output voltage, ΔV(k), is computed to arrive at the desired duty ratio for the switch 'S'. The flowchart for P&O algorithm is illustrated in Figure 1.6.

If we take the power *vs* voltage current for solar PV panels and assume there is only an MPP, which is both local and global.

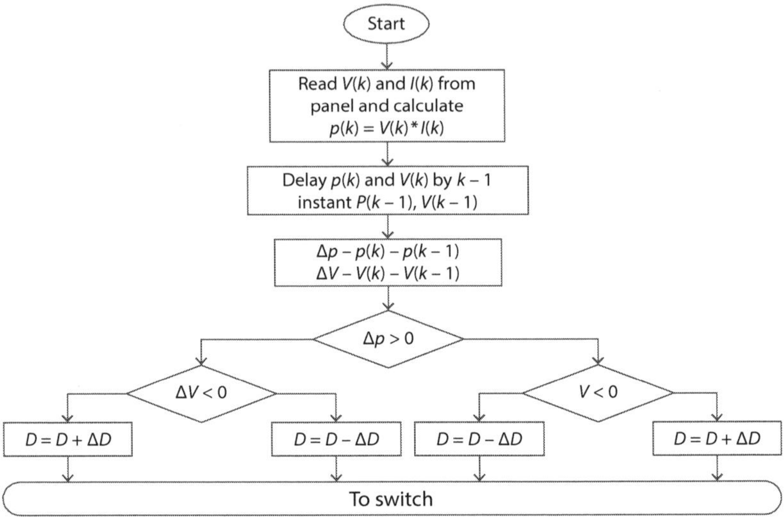

Figure 1.6 Flowchart for P&O algorithm.

There, PV curve is divided into three regions. The MPP is in the peak of the curve, so the regions left-hand side (LHS) and right-hand side (RHS) to the peak point is formed. In the LHS, change in power with respect to voltage is greater than zero, we observed from zero to right when the voltage in the region is increased, and when the voltage is decreased, the power is decreased on the right-hand side and change in power with respect to voltage is less than the zero. As a result, if the voltage increases then power decreases, and if voltage decreases then power increases. At the MPP, it is observed that the change in power is zero, hence, is the condition for maximum. The P&O algorithm examines the power-to-voltage slope. If the slope is positive, it indicates that we are in the LHS of MPPT. If the slope is negative, we are in RHS and we are at MPP if slope = 0. Hence, this logic is employed to construct the flowchart in which perturb the panel voltage and observe the change in power, determining that operating point is on LHS or RHS and proceeding to MPP. The first step is to sample the PV module instantaneous voltage and current referred to as v(k) and I(k), respectively. After that, they are multiplied to get the instantaneous power, p(k). The p(k) is then compared to the power, p(k)-p(k-1), of the previous case. We will refer to the difference as Δp. Similarly, ΔV is formed by subtracting the previous voltage from the present voltage. The numerator and denominator slope of the PV curve are Δp and ΔV, respectively.

If both Δp and ΔV are positive, the slope is positive, that means we are at the left side of MPP and approaching MPP. As a result, to get closer to MPP, we must increase the voltage of the panel and refer to the V_{ref} of panel. In conclusion, perturbation refers to power increase, the perturbation repeats in the same direction. Otherwise, it repeats in the opposite direction, and this is applied to each sampling period until the slope is zero, and until the MPP is reached. Finally, save the values of V(k) and p(k) as Δp and ΔV, respectively, and these will be used to perturb the next sampling period. The flowchart output is V_{ref} which is compared to the panel voltage to calculate the error. And the controller receives the error, and it converts the duty ratio, and the PWM generator receives the duty ratio and generates the PWM signal.

1.5 Charge Controller

In EV, Li-ion batteries are used, and the charging of these batteries is done based on its SOC. A typical charging cycle of battery comprises of three stages: bulk stage, absorption stage, and floating stage. The bulk stage is also called the constant current stage. In this stage, the battery is charged typically from 70% to 90% of SOC by maintaining a high and constant current. While the battery current is constant, the charge in the battery builds up with significant rise in its voltage. The absorption stage is also called constant voltage stage, where the voltage of the battery is kept constant. Here, varied battery current is needed to maintain the constant battery voltage. Here, the battery current decreases, and the SOC increases. When the battery current decreases to a specific time, the charge controller switches to float stage. This specified point corresponds to 98% of the SOC. In the float charge stage, also called trickle charging, the battery current is limited to a very low value called trickle current. This stage serves two purposes: it brings the battery from 98% SOC to 100% SOC; it maintains the battery at 100% SOC. So, with respect to the SOC of the battery, we need to control the current and voltage given to the battery. The algorithm with respect to the MPPT P&O algorithm will control the output of the boost converter. The flowchart describing the charging algorithm is shown in Figure 1.7. In the proposed system, the initial SOC is considered to be at 80%, hence, the battery needs to be charged at a constant current mode. The waveforms under constant current charging, the battery voltage, charging current of the battery, and SOC are shown in Figure 1.8.

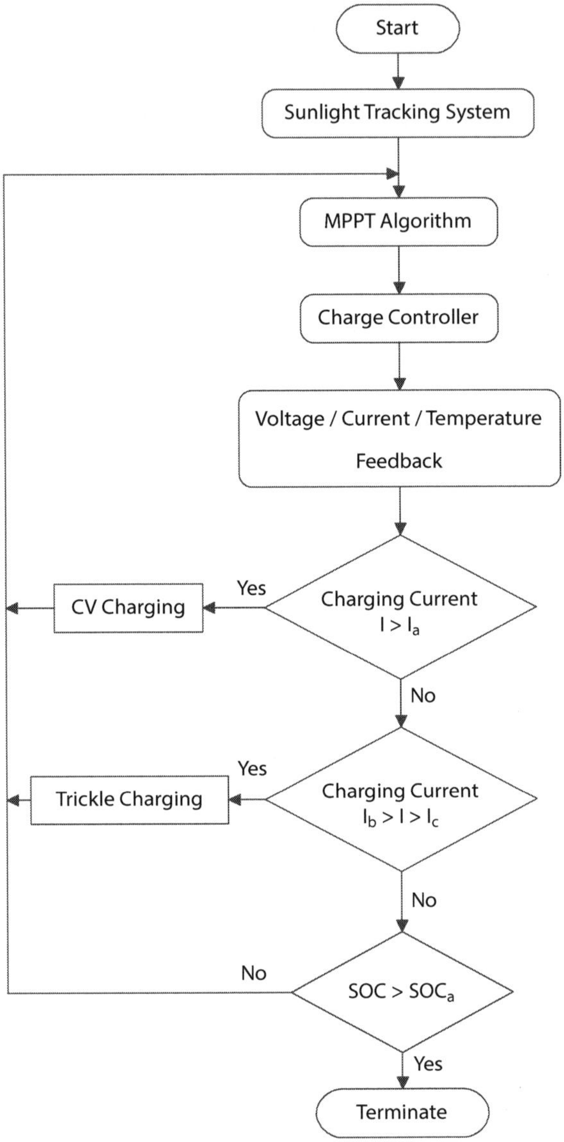

Figure 1.7 Flowchart of charge controller.

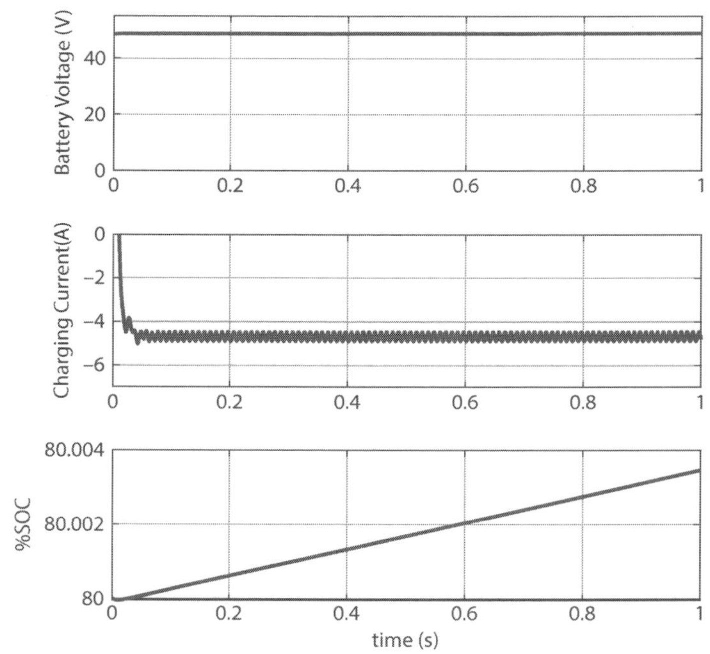

Figure 1.8 Waveform of battery voltage, charging current, and %SOC.

1.6 Conclusion

This chapter intended to design a "solar-powered charging system for 4-wheeler" and considered a flexible PV panel mounted on the surface of a car to provide dynamic charging. Solar is a renewable energy available throughout the year in our country and in tropical regions. When harvested appropriately, tremendous energy can be extracted and utilized for several applications in future. Utilizing solar energy for dynamic charging of battery-operated vehicles greatly improves the reliability and range of electric vehicles. The only way to expand the energy harvesting is locating PV panels on the vehicle surfaces without increasing the overall weight and volume of the vehicle. Research is underway on the materials to reduce weight and volume with simultaneous improvement in the reliable operation of the solar panels in real-time applications. In this work, a step to improve the applicability of flexible PV panels for automobile applications was developed, which reduces the overall charging requirements from the conventional sources of electricity. This enables the reduction in carbon footprints and other problems associated with charging of batteries.

The benefit of extended range of EVs is triggering people's focus towards it. The system is very simple to adapt, use, and manage with less maintenance and no hidden fuel costs. When realized, the proposed system results in the most affordable EVs and reduced per unit electricity consumption. Despite the absence of sunlight, the energy ensnared by the solar panel can be proficiently transformed into electrical energy and stockpiled in a battery. The site of these kinds of applications is progressively growing with the fading fossil fuel resources.

References

1. Asrori Asrori, Fathkur Rohman, Elka Faizal, Muhamad Karis, "The Design and Performance Investigation of Solar E-Bike using Flexible Solar Panel by Different Battery Charging Controller", *Int. J. Mech. Prod. Eng. Res. Dev. (IJMPERD)*, Vol.10 (3), pp 14431-14442, June 2020.
2. Jayesh S. Barad, Mahesh S. Chauhan, Dharmesh S. Barad, Hitesh Parmar, "Calculation Methodology and Development of Solar Power Generating System for Household Appliances," *Int. J. Novel Res. Dev. (IJNRD)*, vol. 2(4), pp142-147, April 2017.
3. Kattimani H.D, Sayalee Sanjay Kahandal, Mohammad Adil Ansari, Nikita Chuhan, Rupansha Khare "Solar Powered Portable Electrical Vehicle Charging Station", *Int. Res. J. Eng. Tech.*, Vol.07 (07), pp541-547, July 2020.
4. A. Hariprasad, I. Priyanka, R. Sandeep, V. Ravi, O. Shekar, 2020, Battery Management System in Electric Vehicles, *Int. J. Eng. Res. Technol. (IJERT)*, Vol.09 (05), (May 2020),
5. Ng, Kong Soon, Chin-Sien Moo, Yi-ping Chen and Yao-Ching Hsieh. "Enhanced coulomb counting method for estimating state-of-charge and state-of-health of lithium-ion batteries." *Appl. Energy* 86 (2009): 1506-1511.
6. Ahmad F. Tazay, 2021, Design and Implementation of Solar Powered Electric Vehicle for On-Campus University Applications, *Int. J. Eng. Res. Technol. (IJERT)*, Vol 10,(10), October 2021.

2

RS-11-I Design and Control of Solar-Battery-Based Microgrid System

Buddhadeva Sahoo[1]*, Subhransu Ranjan Samantaray[1], Pravat Kumar Rout[2] and Pritam Bhowmik[3]

[1]*School of Electrical Sciences, IIT Bhubaneswar, Bhubaneswar, India*
[2]*Department of Electrical and Electronics Engineering, SOADU, Bhubaneswar, India*
[3]*School of Electrical Engineering, Vellore Institute of Technology, Chennai, India*

Abstract

This study proposes a novel reduced switch 11-level inverter (RS-11-I) design for a single stage, grid-connected, solar battery-based microgrid system. To achieve excellent microgrid performance, a reduced boost converter requirement and reduced phase-locked loop (PLL) control are proposed for single-stage operation. In addition to that, a battery energy storage device is integrated to improve the reliability of power flow. Moreover, a novel PLL-less improved instantaneous power theory-based control approach is suggested for inverter pulse generation and maximum power extraction. The inherent characteristics of voltage source-based shunt active filters are used to reduce the total harmonic distortions of the system during non-linear load conditions. To guarantee the significance of the proposed technique, the MATLAB-developed system is tested during steady-state and transient-state conditions. This study is a basis for the recent power market, where renewable power can meet the power demand and maximize its profit by distributing the power to the grid.

Keywords: Reduced switch 11-level inverter, maximum power point tracking, harmonic distortions, voltage source converter

**Corresponding author*: buddhadeva14@gmail.com/bs36@iitbbs.ac.in

O.V. Gnana Swathika, K. Karthikeyan and Sanjeevikumar Padmanaban (eds.) Smart Grids as Cyber Physical Systems Volume 1: Artificial Intelligence, Cybersecurity, and Clean Energy for Next Generation Smart Grids, (13–28) © 2024 Scrivener Publishing LLC

2.1 Introduction

Recently, high population and industrialization has increased the energy demands [1]. To solve these issues, researchers and engineers are emphasizing renewable energy-based power generation. From most renewable systems, solar-based power stations are gaining interest because of their easy availability, simpler design, and control action [2]. However, solar energy generation mostly depends on variable outer conditions such as irradiance and temperature. To extract maximum power and smooth power generation, numerous maximum power extraction algorithm is suggested in the literature [3, 4]. From the literature, it is found that the modified perturb and observer-based (MP&O) control algorithm shows excellent performances in most of the applications [5, 6]. This finding motivates the researcher to implement the most effective MP&O method for microgrid applications.

In microgrid applications, a voltage source converter plays an important role by providing grid synchronization, facilitating renewable energy integration, and harmonic elimination [7, 8]. However, traditional inverters are incapable to provide excellent operation during variable and non-linear load applications [9, 10]. To circumvent the above demerits, multi-level inverters (MLIs) become the first choice of researchers. The main and very important merits of MLIs are reduced harmonic elimination, applicability for different voltage applications, provide linear operation, frequency synchronization, decrease power loss, reduced Electromagnetic interference (EMI) at the inverter output, etc. In literature [11, 12], mostly three types of MLIs such as diode clamped, flying capacitor, and cascaded H-bridge inverters are discussed. However, these inverters' suitability is decreased due to the excess use of power electronic components, complexity in capacitor voltage balancing, and absence of voltage boosting capability. To provide excellent performance, reduced switch multi-level inverters (RSMLI) are developed and implemented during complex system applications [13, 14]. This study motivates the development of a suitable RSMLI topology for excellent microgrid operation.

Looking at the design perspective, two techniques such as single and two stages are selected for microgrid applications. However, in Reference [15], it is identified that for excess input voltage (greater than 350 V), the two-stage-based system efficiency is decreased. In a single-stage operation, the inverter is placed between solar Photovoltaic (PV) and the grid. This not only facilitates the interlinking capacity but also shows the inherent characteristics of shunt active filter operation [16]. To operate the inverter efficiently, the role of a controller action is very important [17]. From the literature [18, 19], it is found that neural and fuzzy control, PQ theory, instantaneous power theory, synchronous frame theory, vector learning machine, predictive approach, and vector quantization approach-based approaches are suggested for unity power factor microgrid applications. However, due to the lack of appropriate mathematical formulations, complexity in design, and additional computation burden, these control approaches lag in their suitability in real applications [20]. Secondly, the uncertain solar power generation also needs an energy storage device in the system device to supply continuous power supply [21, 22]. Therefore, for smoother power management, a better coordination-based controller is needed for both inverter and battery converter applications. The major contribution of the study is presented below.

1. A novel reduced switch 11-level inverter (RS-11-I) is designed and implemented for a single-stage, grid-connected, solar battery-based microgrid system.
2. The single-stage approach reduces the additional boost converter requirement and switching losses.
3. To reduce the controller complexity and provide faster action, a reduced PLL-based improved instantaneouspower theory-based controller is proposed.
4. A battery energy storage device and related coordinated control strategy is proposed to offer excellent power management operation.

The detailed explanation related to the proposed approach is highlighted in the following sections such as modelling, inverter action, control, results, and conclusion sections.

2.2 Solar Battery System Modelling

Figure 2.1 illustrates the complete structure of a solar battery-based system to feed the non-linear load and utility grid. The output of the solar system is directly attached to the dc-link of the inverter. In this study, an RSMLI is used to enhance the system's performance. In addition to that, for improving power reliability, a bidirectional converter is selected for battery energy integration. As illustrated in Figure 2.2, interfacing inductors are used to reduce the current harmonics obtained from the inverter. In this system, the battery charging and discharging conditions at different grid and load

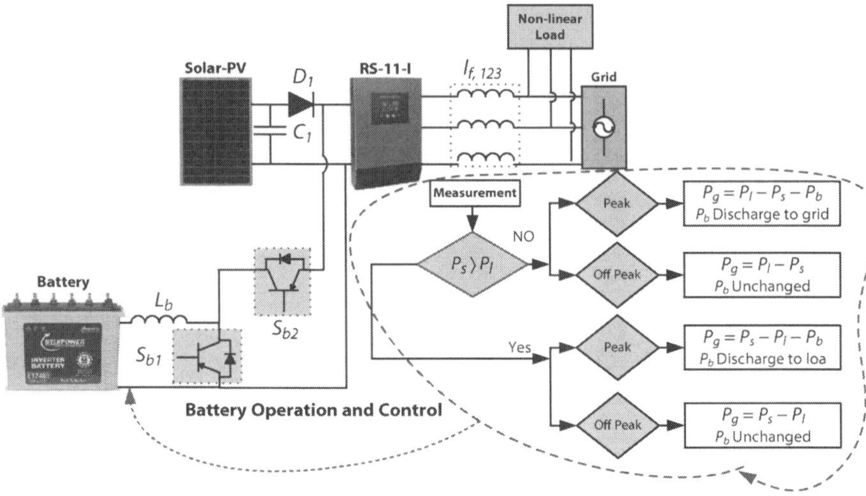

Figure 2.1 Solar battery-based system modelling.

situations are also presented. The detailed explanations related to the proposed RS-11-I are presented in the following sections.

2.2.1 Reduced Switch 11-Level Inverter (RS-11-I)

The proposed RS-11-I diagram is illustrated in Figure 2.2. Reduced switch 11-level inverter modelling includes 10 unidirectional switches (T_1, T_2, T_3, T_4, T_5, T_9, T_{10}, T_{11}, and T_{12}, two bidirectional switches (T_7 and T_8), three capacitors (C_1, C_2 and C_3), two diodes D_1 and D_2 and one DC voltage source

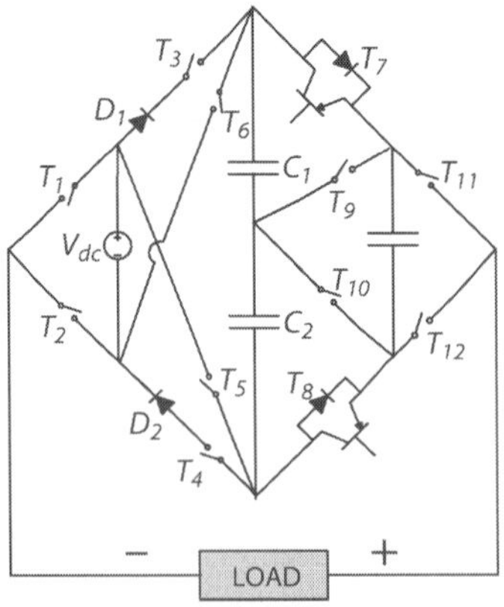

Figure 2.2 Proposed RS-11-I diagram.

V_{dc}. This DC voltage value is dependent upon the integrated energy system. In this proposed system, solar PV is used for energy production. Reduced switch 11-level inverter boosts the dc voltage generation to 2.5 of the total voltage. The switch turned on and turned off conditions decide the voltage level. In RS-11-I, the capacitors are placed in the circuit in such a way that they are capable of regulating the charging and discharging activities as per the desired output voltage. Charging capacity of C_1 and C_2 are set to 0.5 V_{dc} and charging capacity of C_3 is set to V_{dc}, to offer 11-level output voltage as 0, $\pm 0.5V_{dc}$, $\pm V_{dc}$, $\pm 1.5V_{dc}$, $\pm 2V_{dc}$, and $\pm 2.5V_{dc}$. The detailed switching sequences of the proposed inverter are presented in Table 2.1.

2.3 Reduced PLL-Based Control Modelling

The overall control structure of the proposed system is illustrated in Figure 2.3. Figure 2.3 shows the DC-link voltage regulation and inverter pulse generation for the proposed inverter action. The explanations related to the control strategies are presented below.

2.3.1 DC-Link Voltage Regulation

In this study, to extract maximum output (V_m) from solar-based generation system, a modified perturb and observe (MP&O)-based maximum tracking algorithm is used [7]. Generated V_m is equal to the DC-link voltage of RS-11-I. To enhance the inverter operation and required power flow, a reference inverter voltage (V_r^*) is set during the controller design. During inappropriate power generation, the bidirectional converter is used to balance the DC-link voltage. Comparing the reference and maximum solar output voltage, the error in voltage (V_e) is passed through the Proportional and Integral (PI) controller to linearize the error component. At (t+1) instant, the linear signal I_{e1} becomes

$$I_{e1}(t+1) = I_{e1}(t) + k_p\left\{I_{e1}(t+1) - I_{e1}(t)\right\} + k_d I_{e1}(t+1) \quad (2.1)$$

$$I_{e1}(t+1) = V_r^* - V_m \quad (2.2)$$

After getting the appropriate information from the solar system, it is compared with the battery current (I_b) and passed through another PI controller to extract the linear signal (I_{e2}) from the error component.

Table 2.1 Switching table for the proposed inverter RS-11-I.

T_1	T_2	T_3	T_4	T_5	T_6	T_7	T_8	T_9	T_{10}	T_{11}	T_{12}	C_1	C_2	C_3	V_0
✔	*	*	*	✔	*	*	✔	*	*	*	✔	NE	NE	NE	0
*	✔	*	*	*	✔	*	*	✔	*	✔	*	Dis	NE	NE	0.5 V_{dc}
✔	*	✔	✔	*	*	✔	✔	*	*	*	✔	Ch	Ch	Ch	V_{dc}
✔	*	*	*	*	✔	*	*	✔	*	✔	*	Dis	NE	NE	1.5 V_{dc}
✔	*	*	*	*	✔	*	✔	*	*	*	✔	Dis	Dis	NE	2 V_{dc}
✔	*	*	*	*	✔	*	*	✔	*	*	✔	Dis	NE	Dis	2.5 V_{dc}
✔	*	*	*	✔	*	*	✔	*	*	*	✔	NE	NE	NE	0
✔	*	*	*	✔	*	*	*	*	✔	*	✔	NE	Dis	NE	−0.5 V_{dc}
*	✔	✔	✔	*	*	✔	✔	*	*	✔	*	Ch	Ch	Ch	$-V_{dc}$
*	✔	*	*	✔	*	*	*	*	✔	*	✔	NE	Dis	NE	−1.5 V_{dc}
*	✔	*	*	✔	*	✔	*	*	*	✔	*	Dis	Dis	NE	−2 V_{dc}
*	✔	*	*	✔	*	*	*	*	✔	✔	*	NE	Dis	Dis	−2.5 V_{dc}

✔ : On state, *: Off state, NE: No effect state, Dis: Discharging state, Ch: Charging state.

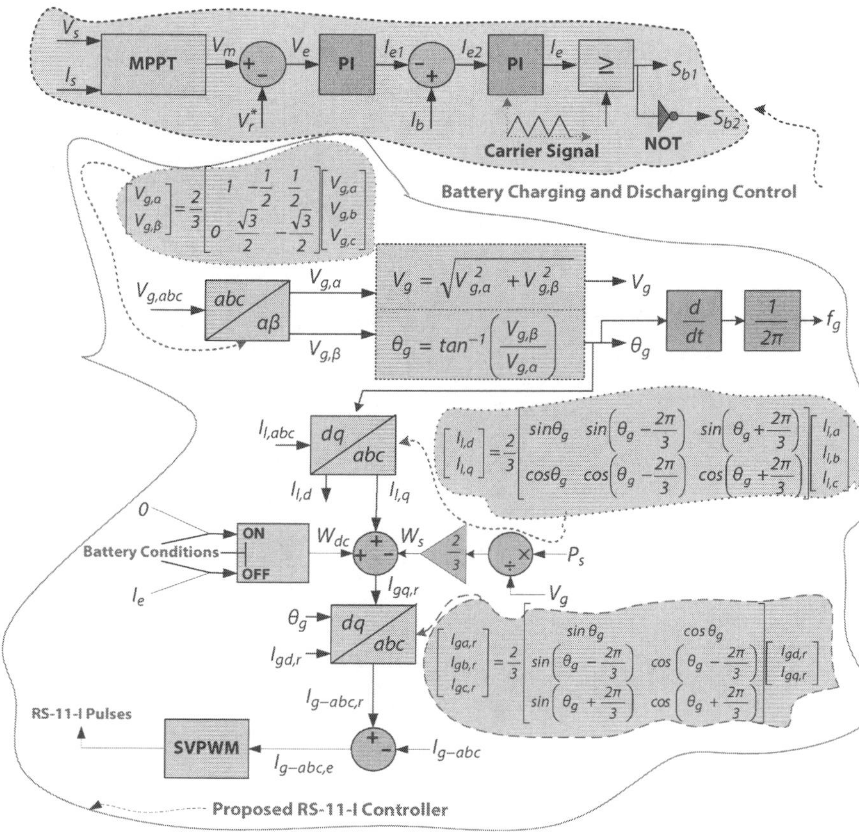

Figure 2.3 Reduced PLL-based improved instantaneous power theory-based control algorithms.

Comparing I_{e2} with the carrier signal, the appropriate switching pulses (S_{b1} and S_{b2}) are generated for the bidirectional converter. Due to the above control structure and appropriate switching pulse generation, the bidirectional converter can balance the DC-link voltage of the inverter.

2.3.2 RS-11-I Control Application

The main objective of the controller is to produce an appropriate current signal for availing a smoother grid synchronization and unit power factor. To offer linear grid current, RS-11-I is needed to show the shunt active filter capability. To compute the voltage angle, in this study, the vector angle of the voltage $\left(\theta_g = tan^{-1}\left(\frac{V_{g,\beta}}{V_{g,\alpha}}\right)\right)$ is considered with the active axis. Specifically, to avoid complexity and avoid tunning, PLL is avoided during the controller action. The complete control algorithm is explained below. Initially, the grid voltage components ($V_{g,a}$, $V_{g,b}$ and $V_{g,c}$) are estimated from two tracked line components ($V_{g,ab}$, $V_{g,bc}$) as discussed in Reference [13].

$$V_{g,a} = \frac{2V_{g,ab} + V_{g,bc}}{3}; V_{g,b} = \frac{-V_{g,ab} + V_{g,bc}}{3}; V_{g,c} = \frac{-V_{g,ab} - 2V_{g,bc}}{3} \quad (2.3)$$

From Equation 2.3, the $\alpha\beta$ voltage components can be computed as

$$V_{g,\alpha} = \begin{bmatrix} \frac{2}{3} & -\frac{1}{3} & -\frac{1}{3} \end{bmatrix} \begin{bmatrix} V_{g,a} \\ V_{g,b} \\ V_{g,c} \end{bmatrix} \quad (2.4)$$

$$V_{g,\beta} = \begin{bmatrix} 0 & \frac{1}{\sqrt{3}} & -\frac{1}{\sqrt{3}} \end{bmatrix} \begin{bmatrix} V_{g,a} \\ V_{g,b} \\ V_{g,c} \end{bmatrix} \quad (2.5)$$

The $\alpha\beta$ voltage components are used to obtain the resultant voltage component ($V_g = \sqrt{V_{g,\alpha}^2 + V_{g,\beta}^2}$) as illustrated in Figure 2.3. From θ_g the grid frequency (f_g) can be computed and represented as:

$$f_g = \frac{1}{2\pi}\frac{d\theta_g}{dt} \quad (2.6)$$

After tracking the load current component ($I_{l,abc}$), the fundamental load current ($I_{l,d}$ and $I_{l,q}$) components are computed as [13]:

$$\begin{bmatrix} I_{l,d} \\ I_{l,q} \end{bmatrix} = \frac{2}{3} \begin{bmatrix} sin\theta_g & sin\left(\theta_g - \frac{2\pi}{3}\right) & sin\left(\theta_g + \frac{2\pi}{3}\right) \\ cos\theta_g & cos\left(\theta_g - \frac{2\pi}{3}\right) & cos\left(\theta_g + \frac{2\pi}{3}\right) \end{bmatrix} \begin{bmatrix} I_{l,a} \\ I_{l,b} \\ I_{l,c} \end{bmatrix} \quad (2.7)$$

As θ_g lies within $-\pi$ to π, the fundamental load component is expressed by $I_{l,q}$. The feedforward solar term can be derived as:

$$W_s = \frac{2P_s}{3V_g} \tag{2.8}$$

where P_s is termed as solar power. The solar battery-based system works in two modes as constant and variable power state conditions. In constant power state conditions, a fixed power is injected to the grid at varying load and generation conditions. For a constant power supply, the most important condition is that the battery should be connected to the system. During the decrease in power generation as compared to the power demand, the battery can fulfill the rest power to the grid. However, if the combined solar and battery power do not meet the constant power demand, then the system cannot operate through constant power state conditions. In variable power state conditions, solar power can be computed by solar voltage and current components. This indicates that solar power varies as per the operating conditions of voltage and current conditions. Therefore, the power exchange condition depends upon the solar irradiance and load variation conditions. In addition to that, a suitable DC-link constant is required to maintain the DC-link constant at its desired value. In this study, the bidirectional converter is used to maintain the DC-link voltage, which indicates that W_{dc} is equal to zero. However, if the energy storage device is disconnected from the system, then W_{dc} is equal to I_{e1} as mentioned in Equations 2.1 and 2.2. The reference quadrature component ($I_{gq,r}$) can be computed as:

$$I_{gq,r} = I_{lq} - W_s + W_d \tag{2.9}$$

In this study, the reference d-axis component ($I_{gd,r}$) is considered as zero. After getting the reference components ($I_{gd,r}$ and $I_{gq,r}$) the three-phase grid reference current component ($I_{gabc,r}$) can be obtained as:

$$\begin{bmatrix} I_{ga,r} \\ I_{gb,r} \\ I_{gc,r} \end{bmatrix} = \frac{2}{3} \begin{bmatrix} sin\theta_g & cos\theta_g \\ sin\left(\theta_g - \frac{2\pi}{3}\right) & cos\left(\theta_g - \frac{2\pi}{3}\right) \\ sin\left(\theta_g + \frac{2\pi}{3}\right) & cos\left(\theta_g + \frac{2\pi}{3}\right) \end{bmatrix} \begin{bmatrix} I_{gd,r} \\ I_{gq,r} \end{bmatrix} \tag{2.10}$$

To generate appropriate switching pulses, the reference component is compared with the tracked grid current components, and the error signal is passed through the space vector pulse width modulation (SVPWM) techniques.

2.4 Result Analysis

The proposed RS-11-I and reduced PLL-based improved instantaneous power theory-based controller application is tested for both constant and variable non-linear load conditions. To justify and validate the controller action, the proposed system findings are compared with the traditional inverter as well as PLL-based control applications.The test results as per the possible conditions are presented below.

Case-a: comparative study: traditional 3-level and proposed 11-level inverter approach.

To test the system dynamics of the microgrid system, the developed model performance is evaluated during variable grid active and reactive power demand and constant load power demand conditions. To justify the contribution, the proposed RS-11-I-based improved instantaneous power theory approach is compared with the traditional three-level inverter-based P-Q approach. The comparative test results are illustrated in Figure 2.4. In Figure 2.4 (a-c), the traditional inverter and control-based results are illustrated. In Figure 2.4 (a), the non-linear load current resultis illustrated and obtained as 2 A. By using the traditional controller, the three-level inverter generated only a three-level output voltage as illustrated in Figure 2.4 (b). However, the design of a three-level inverter contains more switches, diodes, and capacitors, to generate only a three-level inverter voltage. Due to the excess switch and complex structure, the harmonic contained, and losses are also increased. In Figure 2.4 (c), it is visualized that due to the use of the traditional approach, the inverter current contains more harmonics and is not acceptable as per the IEEE-1541 standards.

Looking at the present need and reduced complexity, the developed RS-11-I and control strategy is tested at variable grid codes and load demand. In Figure 2.4 (d-e), the respective solar voltage as 117.5 V and current as 5.516 A results are illustrated at constant irradiance. It illustrates that by using the MP&O technique, the solar system can track its maximum voltage and current, respectively. In Figure 2.4 (f-g), the phase voltage and proposed inverter voltages are illustrated. From Figure 2.4 (g), it is illustrated that the proposed inverter can generate the desired 11-level output voltage. The output of the inverter also validates the existence of the

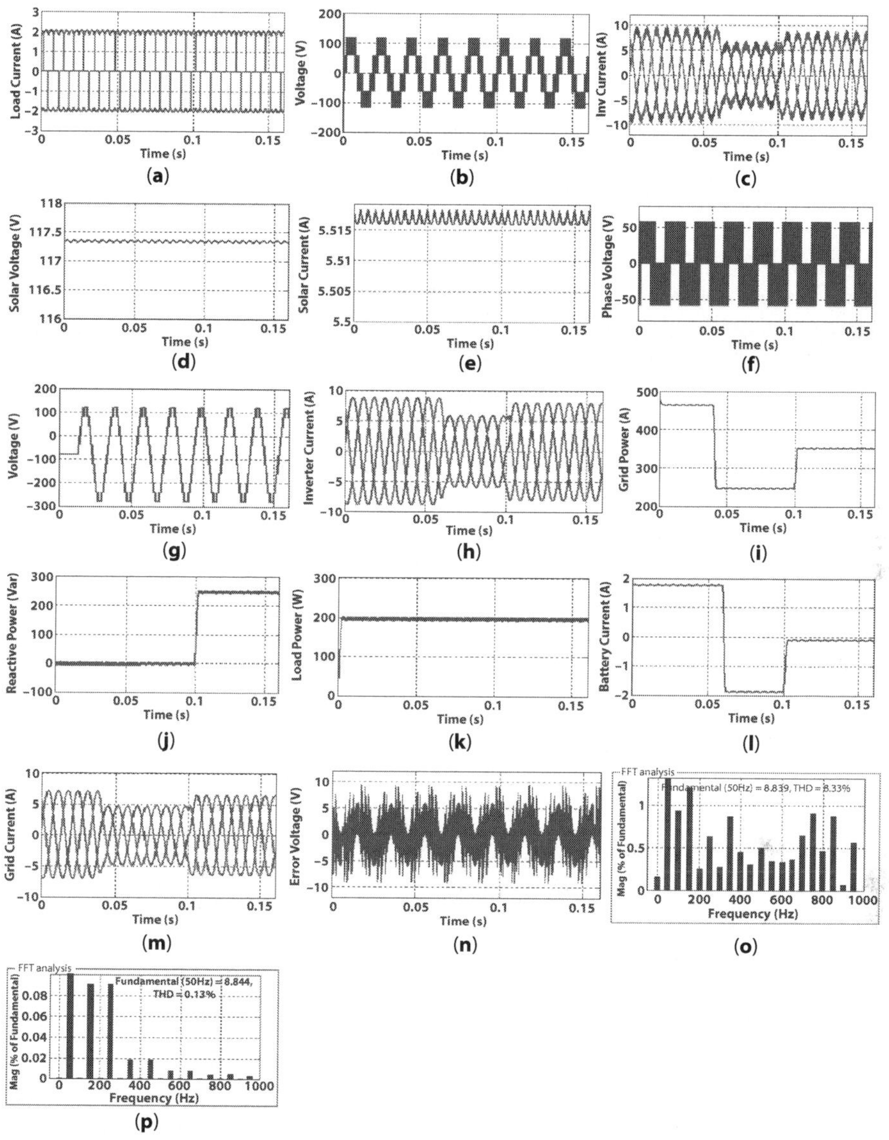

Figure 2.4 Variable grid and constant load demand: (1) three-level inverter approach: (a) non-linear load current, (b) inverter output voltage, (c) inverter current; (2) 11-level inverter approach: (d) solar voltage, (e) solar current, (f) phase voltage, (g) inverter voltage, (h) inverter current, (i) grid current, (j) reactive power, (k) battery current, (l) load power, (m) grid power, (n) error in voltage; (3) harmonics analysis: (o) traditional THD current, (p) proposed THD current results.

controller requirement for real-time applications. In Figure 2.4 (h), it can be observed that due to the use of improved control strategies and inverter design, the inverter generates a linear output current. As compared to the traditional current result illustrated in Figure 2.4 (c),the proposed current results are more sinusoidal and applicable for real-time applications. In this test condition, the grid active and reactive power demand is set and changed at different time intervals as illustrated in Figure 2.4 (i- j), respectively. In addition to that, the undertaken non-linear load also demands 200-W real power, as illustrated in Figure 2.4 (k). The excess grid and load power demand are fulfilled by the energy storage device and during light load demand, the battery charging condition is activated. According to the demand and supply, the battery charging and discharging conditions are decided. The corresponding charging and discharging results are illustrated in Figure 2.4 (l). The grid current and error in voltage component results are illustrated in Figure 2.4 (m-n), respectively. The linear current results guaranteed the system's robustness during variable state conditions. To obtain a clear idea, the Total Harmonic Distortion (THD) values of inverter current results are computed by using both traditional and proposed approaches. It is found that, by using the proposed approach, the THD current value is significantly improved, as illustrated in Figure 2.4 (o-p), respectively. The above findings show the requirement of the proposed controller and inverter for real-time microgrid system applications.

Case-b: test model performance estimation through modified nonlinearity observer-based sliding mode controller (MNOSMC) controller.

Similarly, in this condition, the viability of the test system is evaluated at varying irradiance, constant grid, and load power conditions. The proposed test system results are compared with the traditional results, to provide a clear idea about the improvement and requirements in real-time applications. The corresponding results are illustrated in Figure 2.5 (a-l). Due to the non-linear load applications, the load current result contains non-linearity as illustrated in Figure 2.5 (a). By using the traditional controller, the three-level inverter generates appropriate output voltage levels as illustrated in Figure 2.5 (b). However, these inverter designs require more components and result in more losses and cost, respectively. However, the traditional controller is incapable of eliminating the nonlinearity present in the inverter output voltage. Due to this, the inverter current contains more harmonics, as illustrated in Figure 2.5 (c). These outcomes show the incapability present in the traditional controller and inverter action.

To overcome the traditional approach lag falls, the developed model performance is studied by using the proposed RS-11-I and controller action. By using the MP&O technique, the solar PV can track the maximum voltage

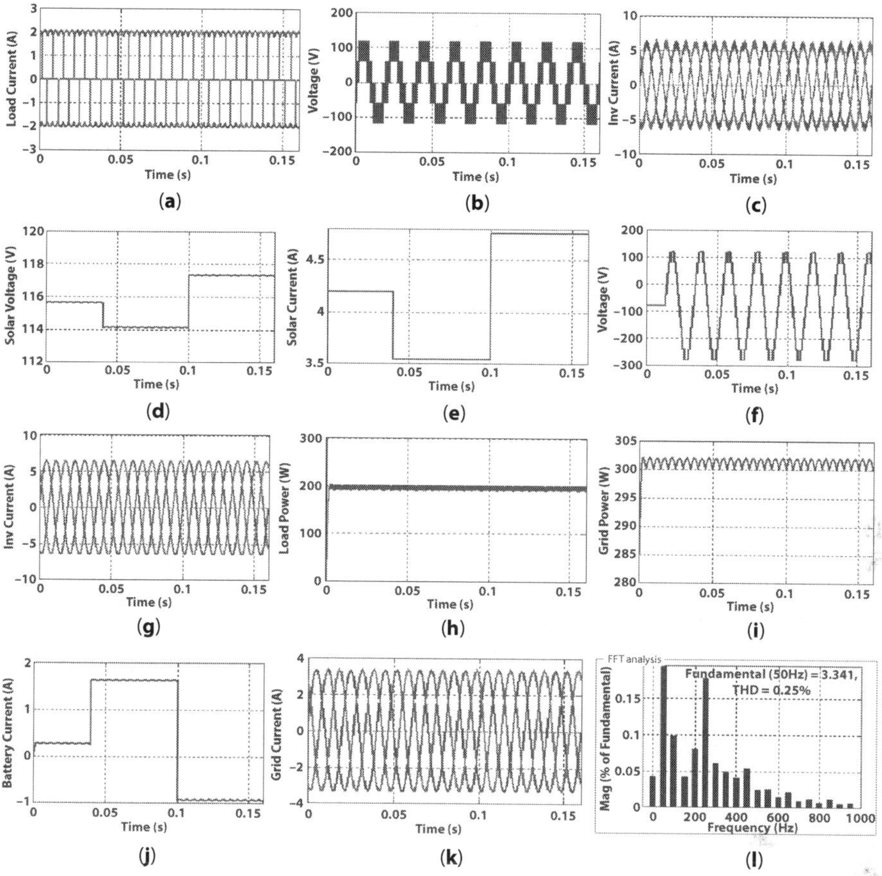

Figure 2.5 Variable irradiance, constant grid and load demand: (1) three-level inverter approach: (a) non-linear load current, (b) inverter output voltage, (c) inverter current; (2) 11-level inverter approach: (d) solar voltage, (e) solar current, (f) inverter voltage, (g) inverter current, (h) battery current, (i) load power (j) grid power, (k) grid current; (3) harmonics analysis: (l) proposed THD current results.

and current at different irradiance conditions, as illustrated in Figure 2.5 (d-e), respectively. In addition to that, the proposed controller is capable of generating appropriate switching pulses for the inverter and the inverter can generate the required 11-level output voltage, as illustrated in Figure 2.5 (f). In the proposed approach, the inverter can generate linear output current, as illustrated in Figure 2.5 (g). By comparing traditional inverter current results, the proposed current results are more linear and acceptable as per the IEEE-1541 standards. Due to the variable environmental condition, the power generation is also varied, as illustrated in Figure 2.5 (d-e). In this test condition, the load and grid power demand is considered as constant value, as illustrated in Figure 2.5 (h-i). According to the power demand and generation, the battery charging and discharging condition is decided as per the proposed control strategy. The charging and discharging condition of the battery is illustrated in Figure 2.5 (j). Due to the appropriate charging and discharging capability of the battery, the grid current becomes linear and constant value, as illustrated in Figure 2.5 (k). To compute the THD percentage, the grid current results are passed through the fast Fourier transform analysis and found lesser THD, as illustrated in Figure 2.5 (l). This indicates the need for the proposed controller and inverter action in a real-time microgrid system to improve dynamic stability. Therefore, the proposed techniques are suggested for recent microgrid system applications.

2.5 Conclusion

The prosed RS-11-I design for a single-stage, grid-connected, solar battery-based microgrid system reduces the component requirement, suitable structure, and cost-effective solution, and reduces power losses. The development of a single-stage operation and reduced PLL requirement also reduces the system as well as controller complexity. The integration of battery energy storage devices also facilitates better power flow through different charging and discharging cycles. By using the novel PLL-less improved instantaneous power theory-based control approach, the proposed RS-11-I can generate the required 11-level voltage output. The increased voltage levels reduce the voltage error and are capable of generating more linear output waveforms for real-time applications. In this study, the inherent characteristics of voltage source-based shunt active filters are also studied to reduce the total harmonic distortions of the system during non-linear load conditions. The MATLAB-developed microgrid system is tested to show the requirement of the proposed technique during both steady-state and transient-state conditions in modern power system applications. This study is a basis for the recent power market, where renewable power can meet the power demand and maximize its profit by distributing the power to the grid.

Acknowledgment

Assistance provided by the Science and Engineering Research Board (SERB) Government of India, under the file number, PDF/2022/000029 and IIT-Bhubaneswar.

Funding Statement

Funding is provided by SERB, Government of India.

References

1. Sahoo, Buddhadeva, Sangram Keshari Routray, and Pravat Kumar Rout. "AC, DC, and hybrid control strategies for smart microgrid application: A review." *Int. Trans. Electr. Energy Syst.* 31.1 (2021): e12683.
2. Kumar, P. Satish, *et al.* "Energy management system for small scale hybrid wind solar battery based microgrid." *IEEE Access* 8 (2020): 8336-8345.
3. Sahoo, Buddhadeva, *et al.* "Power quality and stability assessment of hybrid microgrid and electric vehicle through a novel transformation technique." *Sustainable Energy Technol. Assess.* 51 (2022): 101927.
4. Almadhor, Ahmad, *et al.* "A hybrid algorithm (BAPSO) for capacity configuration optimization in a distributed solar PV based microgrid." *Energy Rep.* 7 (2021): 7906-7912.
5. Chandak, Sheetal, *et al.* "A Brief Analysis on Microgrid Control." Innovation in Electrical Power Engineering, Communication, and Computing Technology. Springer, Singapore, 2022. 541-553.
6. Muhtadi, Abir, *et al.* "Distributed energy resources based microgrid: Review of architecture, control, and reliability." *IEEE Trans. Ind. Appl.* (2021).
7. Sahoo, Buddhadeva, *et al.* "Advanced Adaptive Filter-Based Control Strategy for Active Switch Inverter Operation." Innovation in Electrical Power Engineering, Communication, and Computing Technology. Springer, Singapore, 2022. 183-195.
8. Singh, Bhim, and Sukumar Mishra. "Dual mode operational control of single stage PV-battery based microgrid." 2018 IEEMA Engineer Infinite Conference (eTechNxT). *IEEE*, 2018.
9. Sahoo, Buddhadeva, *et al.* "Neutral Clamped Three-level Inverter based Fractional Order Filter Design for Power Quality Advancement." 2021 International Conference in Advances in Power, Signal, and Information Technology (APSIT). *IEEE*, 2021.
10. Narayanan, Vivek, Seema Kewat, and Bhim Singh. "Solar PV-BES based microgrid system with multifunctional VSC." *IEEE Trans. Ind. Appl.* 56.3 (2020): 2957-2967.
11. Sahoo, Buddhadeva, *et al.* "Advanced Reactive Power Control Technique for Wind Power Application." 2021 International Conferencein Advances in Power, Signal, and Information Technology (APSIT). *IEEE*, 2021.

12. Suman, Gourav Kumar, Josep M. Guerrero, and Om Prakash Roy. "Optimisation of solar/wind/bio-generator/diesel/battery based microgrids for rural areas: A PSO-GWO approach." *Sustainable Cities Soc.* 67 (2021): 102723.
13. Sahoo, Buddhadeva, Sangram Keshari Routray, and Pravat Kumar Rout. "Complex dual-tree wavelet transform and unified power management-based control architecture for hybrid wind power system." *Sustainable Energy Technol. Assess.* 47 (2021): 101560.
14. Sahoo, Buddhadeva, Sangram Keshari Routray, and Pravat Kumar Rout. "Robust control and inverter approach for power quality improvement." Green technology for smart city and society. Springer, Singapore, 2021. 143-156.
15. Nguyen, Tu A., and M. L. Crow. "Stochastic optimization of renewable-based microgrid operation incorporating battery operating cost." *IEEE Trans. Power Syst.* 31.3 (2015): 2289-2296.
16. Sahoo, Buddhadeva, *et al.* "Neural Network and Fuzzy Control Based 11-Level Cascaded Inverter Operation." *CMC-Comput. Mater. Con.* 70.2 (2022): 2319-2346.
17. Mah, Angel Xin Yee, *et al.* "Optimization of a standalone photovoltaic-based microgrid with electrical and hydrogen loads." *Energy* 235 (2021): 121218.
18. Sahoo, Buddhadeva, Sangram Keshari Routray, and Pravat Kumar Rout. "Advanced Speed-and-current control approach for dynamic electric car modelling." *IET Electr. Syst. Transp.* 11.3 (2021): 200-217.
19. Chishti, Farheen, Shadab Murshid, and Bhim Singh. "Natural genetics adapted control for an autonomous wind-battery based microgrid." *IEEE Trans. Ind. Appl.* 56.6 (2020): 7156-7165.
20. Sahoo, Buddhadeva, *et al.* "Mathematical Morphology-Based Artificial Technique for Renewable Power Application." *CMC-Comput. Mater. Con.* 69.2 (2021): 1851-1875.
21. Chishti, Farheen, Shadab Murshid, and Bhim Singh. "Natural genetics adapted control for an autonomous wind-battery based microgrid." *IEEE Trans. Ind. Appl.* 56.6 (2020): 7156-7165.
22. Sahoo, Buddhadeva, Sangram Keshari Routray, and Pravat Kumar Rout. "Advanced control technique based neutral clamped inverter operation." 2021 1st Odisha International Conference on Electrical Power Engineering, Communication and Computing Technology (ODICON). *IEEE*, 2021.

3

A Novel Concept of Hybrid Storage Integrated Smart Grid System with Integrated SoC Management Scheme

Pritam Bhowmik[1]*, Priya Ranjan Satpathy[2], Soubhik Bagchi[3] and Buddhadeva Sahoo[4]

[1]*School of Electrical Engineering, Vellore Institute of Technology, Chennai, India*
[2]*Department of Electrical and Electronics Engineering, CBIT, Hyderabad, India*
[3]*Department of Electrical Engineering, Budge Budge Institute of Technology, Kolkata, India*
[4]*School of Electrical Sciences, IIT Bhubaneswar, Bhubaneswar, India*

Abstract
The uncertainties and system stability is decreasing day by day due to the fast penetration of renewable energies into the power distribution network. To ride through the issue of instabilities in the distribution network, the chapter proposes a novel concept of hybrid energy storage integration with the robust battery charge equalization scheme. The performance evaluation was done for the stated novel charge equalization scheme having the two kinds of high energy dense storage element in the topology. As per the topology, battery and the compressed air energy storage system were actively integrated at dc-side to ensure the high energy density of the smart grid. Furthermore, to minimize the bus voltage and frequency oscillation, a flywheel storage system integrated with the state of power management capability was introduced. The comparative performance of the proposed system is satisfactory while evaluated under critical and real-world scenarios.

Keyword: Hybrid storage, smart grid, state of charge, standalone system, cyber physical system

**Corresponding author*: pritam.bhowmik@vit.ac.in

O.V. Gnana Swathika, K. Karthikeyan and Sanjeevikumar Padmanaban (eds.) Smart Grids as Cyber Physical Systems Volume 1: Artificial Intelligence, Cybersecurity, and Clean Energy for Next Generation Smart Grids, (29–40) © 2024 Scrivener Publishing LLC

3.1 Introduction

Petroleum and natural gas are major sources of energy across the globe [1]. Due to rise in demand, these non-renewable sources will be depleted in the near future [2]. For a sustainable world, major leading countries are trying to shift to renewable sources like solar, wind, and tide to harness electricity. Renewable energy is a non-polluting source and would not get extinguished. This major advantage is attracting people to use this energy more on a daily basis and become less dependent on traditional fuels like oil and gas. The government is playing a leading role in promoting and encouraging common people to use renewable energy [3]. The Government of India is already building a number of projects and subsidizing the common people, one of which is the scheme of subsidizing up to 40% of total expenditure for the installation of solar panels on the rooftop of houses for domestic usage [4]. Although the government promotes this renewable energy, the main problem is that, technically, it does not stabilize when multiple renewable sources are integrated together because there is no direct coupling in any of these systems. These are coupled by converter, as most of the converters have a rating of 10 KVA, due to which the high inertia of wind farm or the sources which has high inertia cannot cross the converter and reach the utility. As a solution, battery, compressed air, and flywheel energy storage system (*i.e.*, BESS, CAES, and FESS) can be incorporated in the highly renewable penetrated network [5]. These help to store the energy of the transient part. Due to the interfacing converters, the inertial stability of the system is challenged. Eventually, to cope with the issue, storage elements are often introduced.

Integrated storage elements are often categorized into two groups: power dense storage elements (PDSE) and energy dense storage elements (EDSE). Power dense storage elements (PDSEs) are majorly responsible for compensating transient power demand. On the other hand, EDSEs are used as steady state power compensators. Energy dense storage elements (EDSE) units are comprised of many small storage systems like BESS and CAES. The power sharing amount the separate systems of a single energy dense unit is required to be critically regulated and monitored to ensure the equal utilization of available store energy. The process of regulating the power sharing for ensuring equal exposure of storage elements is called state of charge (SoC) management. A simple proportional controller-based charge equalization scheme is introduced in Reference [6]. Enhancing the efficiency in Reference [7], the concept of fuzzy logic-based SoC management scheme was demonstrated. The technique proposed in Reference [7] does not ensure the equalized SoC while subjected to a non-linear load. As the SoC is a slow dependent function of active power, an adoptive gain-based control scheme was proposed in Reference [8]. In all the reviewed literature [1–8], it was found that the test system was exposed to compensate only the steady state power or a variable power with a very small slew rate. The actual power demand at the utility end is complex in nature, consists of steady state as well as transient demand in it. The EDSE should not be exposed to the transient demand and PDSE should not unnecessarily be used for compensating steady state demand. Therefore, in this study, the concept of demand allocation was proposed and realized through a transfer bus. The complex demand from the utility end is reflected to the transfer bus through the proposed demand allocation system. The demand allocation system decomposes the complex demand into high frequency (*i.e.*, transient demand) and low frequency demand (*i.e.*, steady state demand). After computing the steady state demand, the proposed SoC balancing screen is responsible to provide the current references for the converters which interface BESS and CAES.

To keep the voltage regulation at the IEEE 1547 prescribed limit during the transient instances, a FESS was introduced. The performance evaluation for the proposed scheme was carried out and compared the efficiency of the same with some conventional charge management scheme.

3.2 Proposed Droop SoC- and State of Power (SOP)-Based Management Method

3.2.1 Basic Operation Mode of DESS

The primary objective of the distributed energy storage system (DESS) is to satisfy the local load demands connected in the small network and to ensure a proper use of surplus power, if available. During the charging period, the energy storage unit system (ESUS) demands for the maximum power from the network. The charging current may remain constant or decrease inversely with the present SoC. When the renewable energy source is not sufficient to compensate the total load demand, the DESU turns into the discharging mode. In discharging period, the SoC of each ESUS is balanced, and as a result, the output power is equalized. With this concept of control method, the demand is shared proportionally to ensure the equalized SoC of the ESUS.

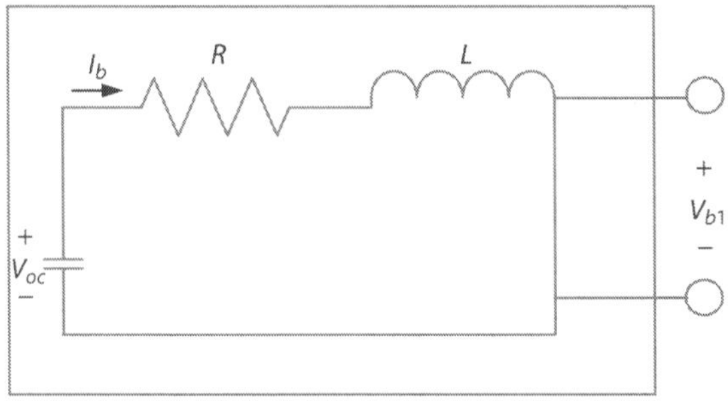

Figure 3.1 Model of cell.

3.2.2 ESUS Model

The functional model of the ESUS system is represented in Figure 3.1. The simplified schematic representation of the ESUS model was undertaken to showcase the detailing of the cell parameter, taking into consideration the internal resistance, and some offered reactance during the current flickering mode.

The mathematical expression for the SoC can be formulated based on the charge count method as depicted below [9].

$$SoC_{pri} = SoC_{pri.t=0} - C\frac{d}{dt}\left(i_{pri}\right) \tag{3.1}$$

Where SoC_{pri} is the initial condition at $t = 0$, C is the capacity of the ESUS, and i_{pri} is the primary input current of converter with respect to time.

By the applicable function of SoC and T (Ambient Temperature), both v_{oc}, z_{eq} can be calculated with a specific value. With varying T, open circuit voltage, v_{oc}, which is dependent on SoC, is not significantly changed. As a result, v_{oc}, z_{eq} can be expressed as follows:

$$v_{oc} = f_{01}(SoC) \tag{3.2}$$

$$z_{eq} = f_{02}(T) \tag{3.3}$$

f_{01} and f_{02} can be expressed as the relationship of the dependency of data, as shown below:

$$v_{oc} = ja_{v=1,3}\left(e^{av2.soc} - e^{-av4.soc}\right) \quad (3.4)$$

$$z_{eq} = a_{z1}.T^2 - a_{z2}.T + a_{z3} \quad (3.5)$$

Coefficients are shown in Tables 3.1 and 3.2 for voltage and impedance, respectively.

3.2.3 Basic Model of SoC Management Control System

A pulse width modulation-based converter can be mathematically represented as [10]:

$$v_{out.dc} = v_{dc}^{ref} - n_d x\ p_{out.lpf} \quad (3.6)$$

where $v_{out.dc}$ and v_{dc}^{ref} are the output voltages of dc and reference values, respectively, $p_{out.lpf}$ is a low pass filter output power which is filtered, and n_d is a drooping coefficient.

The current regulation among the parallelly integrated converters is specified by the input dc side voltage which is further a dependent function of the SoC. Topologically, the sharing coefficient is regulated inversely proportional with the charge content of storage element, the output power of each converter will be proportional to the SoC^b. Two parallel converter modules can be expressed as [11]:

Table 3.1 Values of coefficients of voltage.

a_{v1}/V	a_{v2}/V	a_{v3}/V	a_{v4}/V
3.4	4 x 10^{+4}	7.9 x 10^{-1}	2.5 x 10^{-1}

Table 3.2 Values of coefficients of impedance.

$a_{z1}/\Omega C^{+2}$	$a_{z2}/\Omega C^{+1}$	a_{z3}/Ω
4.6 x 10^{+7}	4.23 x 10^{+5}	1.75 x 10^{+3}

$$v_{dci} = v_{dc}^{*} - \left(n_{d0} \big/ SoC_i^b \right) \cdot p_{out.lpf} \tag{3.7}$$

Where n_{d0} is a drooping coefficient at its initial condition and b is an exponent of SoC.

Here, b is the exponent, which helps in regulating the current output and thus regulates the acceleration of the SoC equalization time.

As per the s-domain approach, the following model can be expressed as:

$$n_d \cdot \hat{p}_{in.lpf} = bSoC_i^{b-1} \left(v_{dc}^{*} - v_{dc.in} \right) . \hat{SoC_i} - SoC_i^b . \hat{v}_{dc.i} \tag{3.8}$$

where ^ denoted as perturbed values, equilibrium point values are SoC_i, v_{dci}, v_{dc}^{*}.

Neglecting the interfacing converter characteristics in the system, the power output can be expressed as simply as:

$$p_{out} = p_{in.i} = v_{in}.i_{in}.i \tag{3.9}$$

3.2.4 Proposed SoC Management Scheme and the Undertaken System

To get potentially high energy and bulk power supply in the microgrid, BESS, CAES, and FESS systems were designed and integrated to form a hybrid DESS.

Accounting for the behavioral characteristics of the interfacing switching converter, the relative inertia decreases. In the low inertia network, the probability of the sub-synchronous resonance and low frequency oscillation becomes prominent. To minimize the low frequency oscillation in the network, the FESS with high density of power is incorporated in the architecture.

In the test system, there are three energy storage devices (BESS, CAES, FESS) introduced to optimize the SoC in microgrid which are incorporated with transfer bus system through individual dc/dc converter with separate ripple filter. One part of the demand allocation system, which is interconnected with transfer bus and the main bus, is responsible for the demand allocation. The demand allocation system, bifurcates the complex demand into two subsequent quantities like transient demand and the steady state demand.

After the availability of the steady state and the transient demand, the SoC management scheme can be activated. The SoC management is solely responsible for ensuring the charge equalization between the BESS and

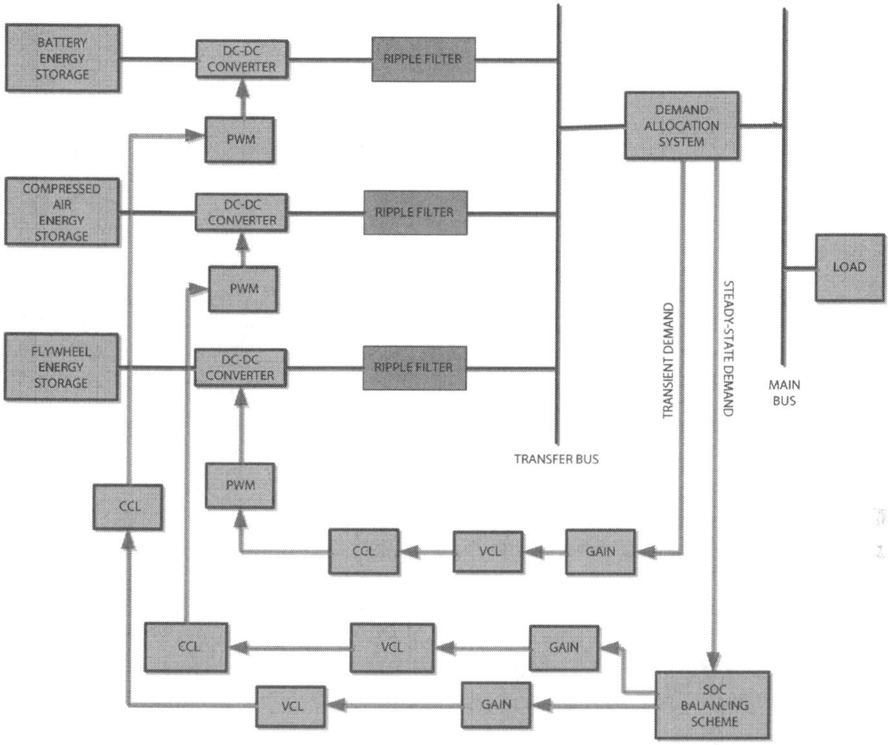

Figure 3.2 Schematic diagram of proposed SoC management scheme and the undertaken system.

CAESS. As the BESS and CAESS are low power density-based storage elements, they only compensate for the steady state part of the demand. Eventually, from the steady state demand, the reference voltage signal for the voltage control loop (VCL) can be computed. Based on the given response by the VCL, the current control behaves and the duty cycle for the dc/dc converter is subtracted. The schematic representation of the proposed technique is illustrated in Figure 3.2.

3.3 Result Analysis

3.3.1 Charging Case

The performance of the proposed technique of charge equalization was compared with respect to the conventional constant voltage-based charging method. Figure 3.3 illustrates the SoC of the two energy storage elements,

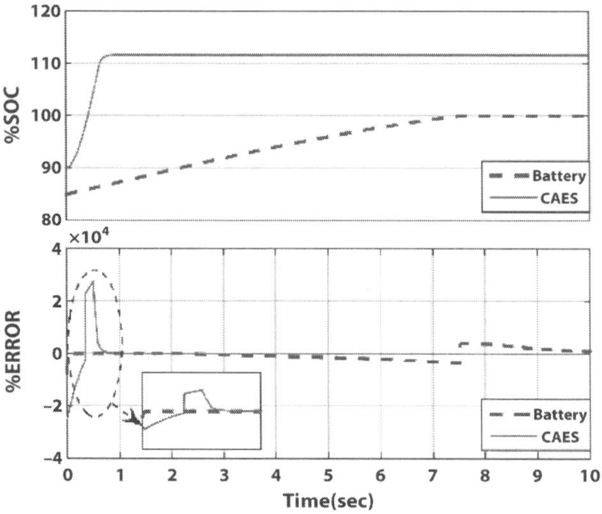

Figure 3.3 SOC and exeuted charging voltage error in conventional method.

i.e., BESS and CAES. The system was initialized with 90% and 85% of SoC for the battery and CAES, respectively. It can be observed that within 1.27 s, CAES is charged to 112.78%, which signifies the rate of charge of CAES is 10.39 times higher than the rate of charge of BESS. Due to the huge dissimilarity in charging rate, the SOC never becomes equalized for the conventional system. Moreover, an issue of overcharged compressed air-charged system can be observed. The response of the SoC and the behavior characteristic of the conventional system can be well explained through the response which illustrate the executed error in charging voltage (see Figure 3.3).

On the other hand, the undertaken system was evaluated against the same scenario to have an account of the performance. The initialization of the system was kept unchanged. It is shown in Figure 3.4 that developed topological scheme does not suffer from the overcharging issues. Furthermore, the charging rate is self-adjusted based on the dynamic SoC. Statistically, it can be observed from Figure 3.4 that the charging rate for the CAES is 1.38 while for the BESS it is 2.08. It signifies that the rate of charge of the battery energy storage system is 50.72% higher than the CAES. A huge difference in charging percentage rate is introduced due to the difference with the dynamically measured SoC. It can be also observed from Figure 3.4 that, at 7.2 s, while the equalized SoC is ensured, the charging voltage of BESS (*i.e.*, %error) becomes close to zero.

The responsibility of the FESS is to smoothen the demand of the charging current in the network. From the acquired flywheel response as shown in

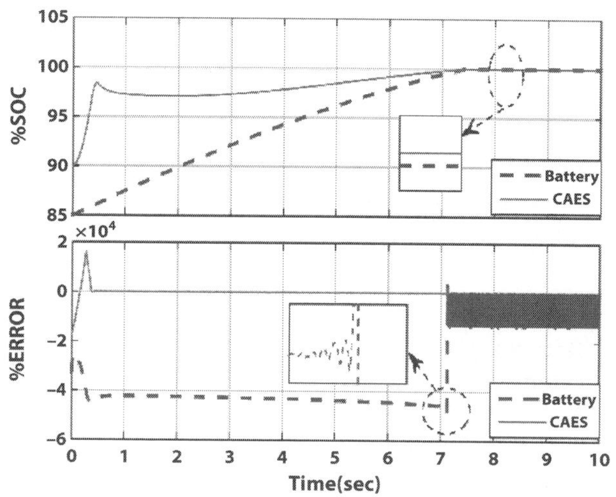

Figure 3.4 SoC and executed charging voltage error in proposed method.

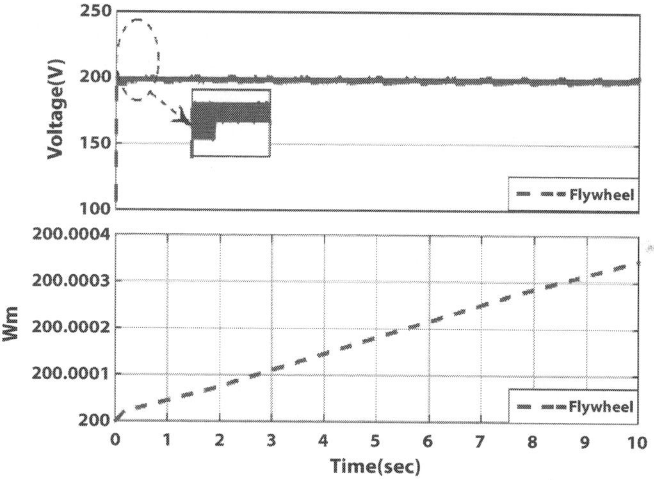

Figure 3.5 Voltage and torque response of flywheel.

Figure 3.5, during a small period of transience (*i.e.*, initialization period of 40 ms), FESS compensates for the flickered charging power.

3.3.2 Discharging Case

The performance characteristics of the proposed technique was computed in a conditional discharging scenario. The performance of the

SoC equalization for the conventional approach is shown in Figure 3.6. It should be perceived that from the mentioned illustration that the predictable control algorithm-based system drastically suffers to normalize the SoC. At 1.36 s, SoC was equalized, although the conventional technique failed to maintain the equalized charge state. Finally, there is a 51.28% of inequality observed at 10 s. Figure 3.6 also illustrates the voltage error for the self-explanation of mismatched SoC. The response acquired from the proposed system is illustrated in Figure 3.7. It was

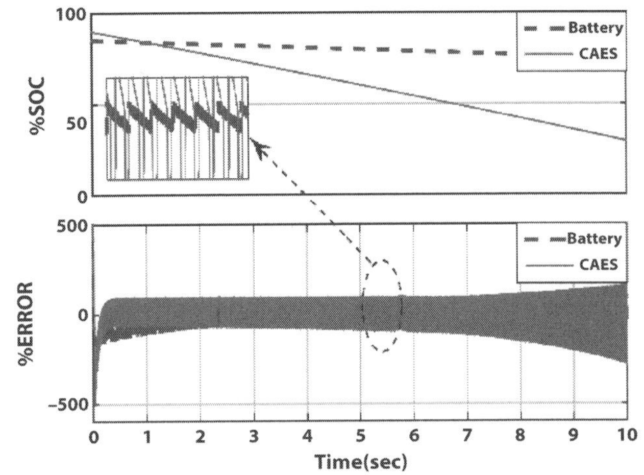

Figure 3.6 SoC and executed discharging voltage error in conventional method.

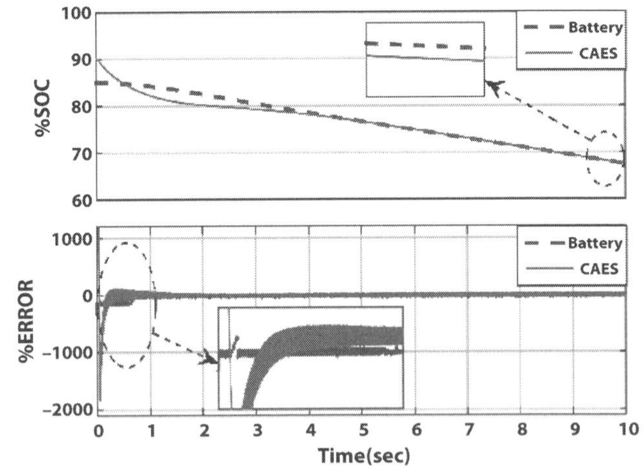

Figure 3.7 SOC and executed discharging voltage error in proposed method.

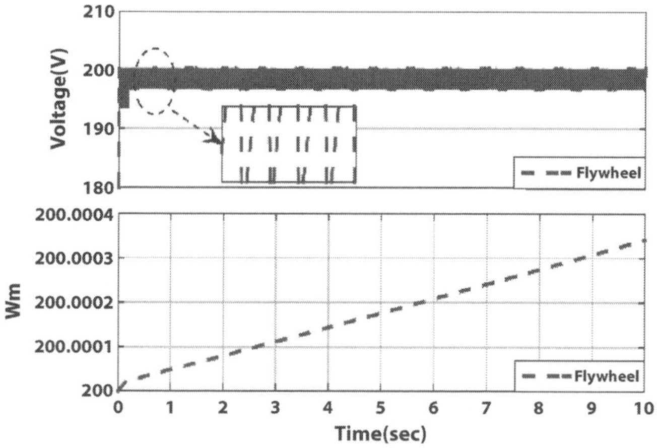

Figure 3.8 Voltage and torque response of flywheel.

observed that the equalized SoC was ensured within 3.86 s, rendering a small error of 0.92%. It can also be observed that after ensuring the equalized charge state, the proposed technique was capable in maintaining the equalization state. Therefore, from 3.86 s onwards, the battery and the CAES discharge in a critically computed proportion to maintain the equal residual charge. The equational error in the charging voltage is also negligible (see Figure 3.7).

The response of the flywheel can be observed in Figure 3.8. It is observed that, in the proposed system, the FES satisfactorily smoothened out the ripples present in the charging current. The main objective to maintain and regularize the SoC was ensured by the proposed FES integrated system.

3.4 Conclusion

The chapter deals with the issue of instabilities in the distribution network. To minimize the sensitivity of the system, the concept of hybrid energy storage integrated with state of charge (SoC) management scheme was introduced in the study. Moreover, to keep the bus voltage and frequency oscillation within the standard prescribed in IEEE 1547, a high power dense flywheel-based storage mechanism was considered. The improvement in the system performance in respect to the energy equalization time, frequency deflection, and voltage instability was verified. It was observed in the study that the proposed system can ensure the normalized SoC within a very short time with execution of 0.92% of mismatch error.

From the analysis, it can be stated that the proposed hybrid energy storage system integrated with the SoC balancing scheme is suitable for real-world applications.

References

1. Murshed, M., Alam, R., & Ansarin, A. (2021). The environmental Kuznets curve hypothesis for Bangladesh: the importance of natural gas, liquefied petroleum gas, and hydropower consumption. *Environ. Sci. Pollut. Res.*, *28*(14), 17208-17227.
2. Azam, A., Rafiq, M., Shafique, M., Zhang, H., Ateeq, M., & Yuan, J. (2021). Analyzing the relationship between economic growth and electricity consumption from renewable and non-renewable sources: Fresh evidence from newly industrialized countries. *Sustainable Energy Technol. Assess.*, *44*, 100991.
3. Sawhney, A. (2021). Striving towards a circular economy: climate policy and renewable energy in India. *Clean Technol. Environ. Policy*, *23*(2), 491-499.
4. Alhammami, H., & An, H. (2021). Techno-economic analysis and policy implications for promoting residential rooftop solar photovoltaics in Abu Dhabi, UAE. *Renewable Energy*, *167*, 359-368.
5. Choudhury, S. (2021). Flywheel energy storage systems: A critical review on technologies, applications, and future prospects. *Int. Trans. Electr. Energy Syst.*, *31*(9), e13024.
6. Olabi, A. G., Onumaegbu, C., Wilberforce, T., Ramadan, M., Abdelkareem, M. A., & Al–Alami, A. H. (2021). Critical review of energy storage systems. *Energy*, *214*, 118987.
7. Garud, K. S., Jayaraj, S., & Lee, M. Y. (2021). A review on modeling of solar photovoltaic systems using artificial neural networks, fuzzy logic, genetic algorithm and hybrid models. *Int. J. Energy Res.*, *45*(1), 6-35.
8. Zuo, H., Zhang, B., Huang, Z., Wei, K., Zhu, H., & Tan, J. (2022). Effect analysis on SOC values of the power lithium manganate battery during discharging process and its intelligent estimation. *Energy*, *238*, 121854.
9. Bhowmik, P., Chandak, S., & Rout, P. K. (2018). State of charge and state of power management among the energy storage systems by the fuzzy tuned dynamic exponent and the dynamic PI controller. *J. Energy Storage*, 19, 348-363.
10. Bhowmik, P., Chandak, S., & Rout, P. K. (2019). State of charge and state of power management in a hybrid energy storage system by the self-tuned dynamic exponent and the fuzzy-based dynamic PI controller. *Int. Trans. Electr. Energy Syst.*, 29(5), e2848.
11. Kim, J. W., Choi, H. S., & Cho, B. H. (2002). A novel droop method for converter parallel operation. *IEEE Trans. Power Electron.*, 17(1), 25-32.

4

Parameters Sensitivity of Solar Photovoltaic Array Architectures under Incremental Row and Column Shading

Priya Ranjan Satpathy[1], Sudhakar Babu Thanikanti[2], Belqasem Aljafari[1] and Pritam Bhowmik[3*]

[1]*Department of Electrical Engineering, College of Engineering, Najran University, Najran, Saudi Arabia*
[2]*Department of EEE, Chaitanya Bharathi Institute of Technology, Hyderabad, India*
[3]*School of Electrical Engineering, Vellore Institute of Technology, Chennai, India*

Abstract

The susceptibility of the solar photovoltaic (PV) array towards the unavoidable partial shading has raised serious concerns as power generation gets affected the most along with physical damage to the system. This issue can be solved through numerous mitigation methodologies, among which, array architectures gained a wide audience due to ease of implementation. The performance of these architectures largely depends on the patterns and nature of shading whose electrical parameters can be utilized to design effective partial shading algorithms. In this study, the sensitivity of the electrical parameters of the array architectures was studied under field-occurring incremental row and column-shading environment to determine the effective parameters for shading prediction. The PV array architectures are differentiated based on the junction wires and the ability of shading losses minimization. The entire investigation is conducted in the MATLAB/Simulink environment for a PV array of 7x7 size operating under incremental column and row shading. The performance and parameter sensitivity of the array architectures were studied and found that the current and voltage parameters show inconsistent behavior under shading. Also, the array encountered higher losses of 86.72% during row-level shading as compared to column-level shading at 74.05%.

**Corresponding author*: pritam.bhowmik@vit.ac.in

O.V. Gnana Swathika, K. Karthikeyan and Sanjeevikumar Padmanaban (eds.) Smart Grids as Cyber Physical Systems Volume 1: Artificial Intelligence, Cybersecurity, and Clean Energy for Next Generation Smart Grids, (41–54) © 2024 Scrivener Publishing LLC

Keywords: Photovoltaic, partial shading, parameters, array, modules, power generation

4.1 Introduction

The performance of the solar photovoltaic (PV) system directly depends on the behavior of the connected PV arrays that are formed from the connection of modules in multiple architectures depending on the components' rating [1]. The PV arrays are designed considering the size of the system and are expected to deliver the maximum power based on the field irradiance and temperature conditions. But, in reality, the arrays fail to deliver the desired power output even under operating at sufficient field irradiance and temperature conditions [2]. The primary cause of the lower power output is partial shading, in which some areas of the array experience lower irradiance levels because of the shadow cast by neighboring objects such as buildings and trees, as well as due to dust coverage and clouds [3]. However, the most common causes of partial shading are the shadows of nearby buildings and clouds that covers the array either from the column or row side (in roof top-based PV systems). The partial shading reduces the power generation, forces the lower irradiance receiving modules to turn on the bypass diodes forming non-convex curves, and physically damages the modules [4].

In general, series is the most preferred architecture adopted by multiple power plants to generate high voltage with lower current output to reduce losses in the system [5]. However, the series architecture encounters severe power losses and has higher chances of failure during partial shading [6]. Hence, as an alternative, series-parallel (SP) architectures are used to reduce the losses and increase the productivity of the system during partial shading. But studies show that the SP architecture also encounters losses during shading and hence, various other architectures were proposed [7]. Another alternative to the SP architecture that generates equivalent voltage and current to SP is bridge-linked (BL). The difference, however, is in the wiring; in the case of BL, two extra wires are linked in a ladder fashion to the connections of the modules [8].

The additional wires provide an extra low resistive path for the currents to flow without flowing through the lower irradiance receiving modules. Later on, another new architecture was developed by connecting extra wires to the junctions of all the modules connected to an array and named total-cross-tied (TCT) [9]. The schematic representation of the three architectures implemented to a 7x7 PV array is depicted in Figure 4.1 from which the additional wires in the junctions can be visualized. Various investigations were conducted on the performance of these architectures

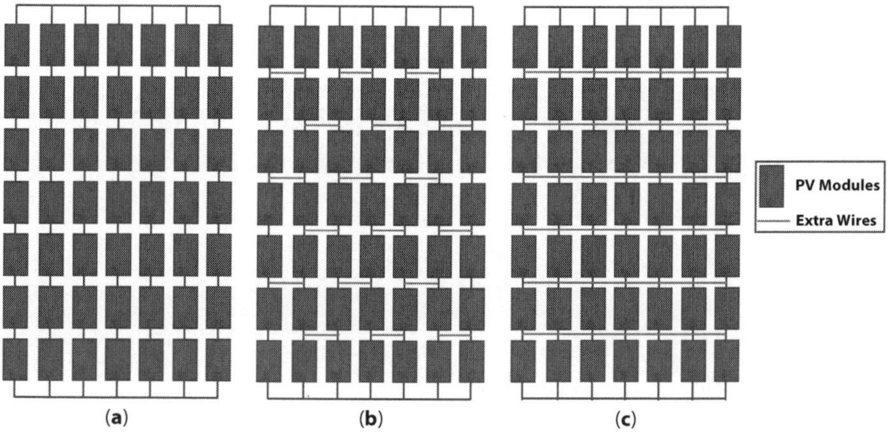

Figure 4.1 Schematic representation of a 7x7 PV array with (a) SP, (b) BL, and (c) TCT architectures.

under partial shading and most of them concluded that the TCT has a higher power generation than others [10–12]. However, some studies showed that the TCT has lower power generation than the BL and SP mainly under partial shading, caused by nearby buildings and cloud passage [13, 14].

In order to accurately detect partial shade in the system, several techniques were suggested that are based on the electrical parameters of the PV arrays [15, 16]. The parameters of the array depend on the irradiance and temperature based on which a partial shading detection algorithm was proposed [17]. Another current and voltage indicators analysis was carried out for the partial shading detection in PV systems [18]. A shading detection algorithm based on the evaluation of three indicators namely the current indicator, voltage indicator, and power indicator was proposed [19]. These algorithms use the current and voltage parameters as indicators for partial shading detection in the arrays but, all these methodologies were tested for the SP architectures.

Hence, in this study, the array architectures were investigated under incremental shading over column and row for sensitivity analysis of various electrical parameters. The variation of these electrical parameters can be utilized to determine the appropriate and effective parameter for the shading prediction in the array architectures. The PV array architectures which include BL, SP, and TCT, are differentiated based on the junction wires and the ability of shading losses minimization. The entire investigation was conducted in the MATLAB/Simulink for 7x7 size array operating under row and column incremental shading scenarios.

4.2 System Modelling and Description

A PV module is mathematically modelled using Equation 4.1, derived from the electrical circuit of the cell given in Figure 4.2 that consists of a photocurrent source (I_{Ph}), diode, series resistance (R_S), and shunt resistance (R_{Sh}).

$$I_{PV} = N_p * I_{ph} - N_p * I_0 \left[\exp\left\{ \frac{q * (V_{PV} + I_{PV} R_s)}{N_s A k T} \right\} - 1 \right] - I_{Sh} \quad (4.1)$$

$$\text{where } I_{Sh} = \frac{V_{PV} \times \frac{N_P}{N_S} + I_{PV} \times R_S}{R_{Sh}} \quad (4.2)$$

The output cell current, parallel cells count, current of diode, charge of electron, output cell voltage, series cells count, ideality factor, Boltzmann constant, temperature, reference temperature (25°C), temperature coefficient, reverse saturation current, irradiance, reference irradiance (1000 W/m^2), open-circuit voltage, and short-circuit current are denoted by I_{PV}, N_P, I_O, q, V_{PV}, N_S, A/B, k, T, T_r, K_i, I_{rs}, G, G_{ref}, V_{OC}, and I_{SC}, respectively.

The PV module rating used in the study is tabulated in Table 4.1 and such modules are used to form a 7x7 PV array with a maximum power output of 15.925 kW at standard testing conditions (STC). However, considering the field scenario, the PV arrays were investigated at 800 W/m^2 and 50°C with maximum power output under normal scenarios as 13.03 kW.

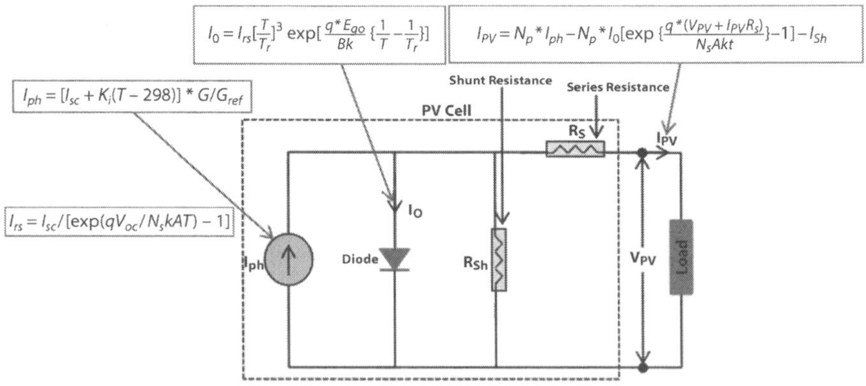

Figure 4.2 Equivalent circuit representation of a PV module.

Table 4.1 PV module specification at STC (1000 W/m^2 and 25°C).

Parameters	Ratings
Power output	325 W
Voltage output	37.80 V
Current output	8.60 A
Open-circuit voltage	46.60 V
Short-circuit current	9.20 A
Series cells count	36
Parallel cells count	0

The schematic representation of the 7x7 PV array architectures, *i.e.*, TCT, BL, and SP was previously depicted in Figure 4.1.

4.3 Electrical Parameters Estimation

The array electrical parameters that are utilized to investigate the presence of partial shading include power output (P_O), voltage output (V_M), current output (I_M), voltage at open circuit (V_{OC}), current at short circuit (I_{SC}), fill factor, efficiency, and thermal voltage (V_T). The power output of the array can be calculated from the product of the maximum voltage and maximum current, denoted as:

$$P_O = V_M \times I_M \tag{4.3}$$

Equation 4.4 can be used to determine the PV array fill factor, which has a value between 0 and 1, with 1 signifying superior performance.

$$FF = \frac{V_M \times I_M}{V_{OC} \times I_{SC}} \tag{4.4}$$

Equation 4.5, in which the irradiance received by the array's modules and the area of the modules are denoted by G and A, and can be used to compute the PV array's efficiency.

$$Efficiency = \frac{V_M \times I_M}{G \times A} \tag{4.5}$$

The thermal voltage (V_T) of the array can be calculated using Equation 4.6

$$V_T = \frac{(2V_M - V_{OC})(I_{SC} - I_M)}{I_M - (I_{SC} - I_M)\ln\left(\frac{I_{SC} - I_M}{I_{SC}}\right)} \tag{4.6}$$

4.4 Sensitivity Analysis of Electrical Parameters of PV Array Under Incremental Partial Shading

The array with TCT, BL, and SP architectures was investigated under incremental column and row shading and the behavior of the electrical parameters was studied.

The incremental row shading scenario for the 7x7 PV array is pictorially shown in Figure 4.3 (a) in which the building shade initially covers the first row of the array (PSN 2) and increases until it covers the entire array (PSN 8). Similarly, the incremental column shading is represented in Figure 4.3 (b) in which the shadow incrementally covers the first column to the last one. The modules under no partial shading were subjected to receive 800 W/m^2 whereas the modules under shading received 100 W/m^2.

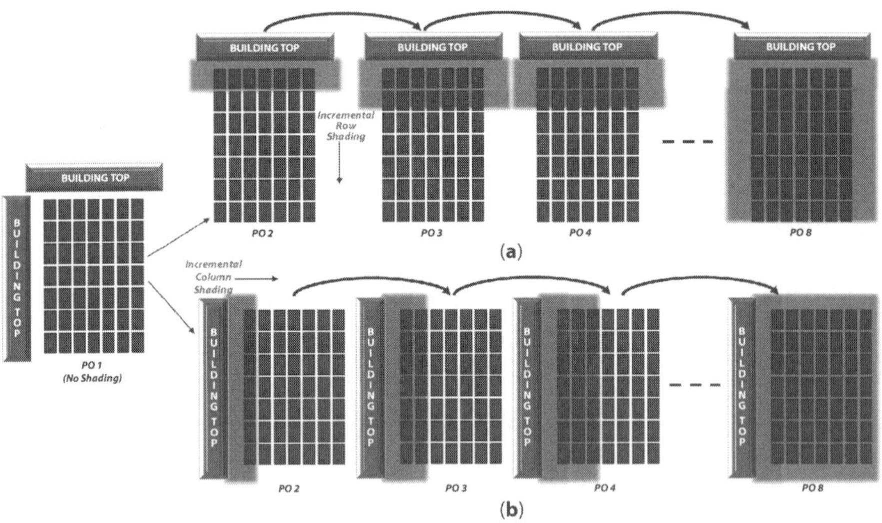

Figure 4.3 Incremental partial shading scenario for 7x7 array architectures. (a) Row and (b) column.

4.4.1 Analysis under Incremental Row Shading Scenario

The incremental shading over row (PSN 1 to PSN 8), as shown in Figure 4.3 (a), were applied to the arrays with TCT, BL, and SP architectures, and from the analysis, it is seen that all the configurations have equal power output for all the positions. The arrays I-V and P-V graphs are depicted in Figure 4.4 (a) and (b), respectively. It was observed that the short-circuit currents of the array architectures remain nearly constant at 51.6 A from PSN 1 to PSN 7 and for PSN 8, the value reduces to 6.45 A as the entire array operates under lower irradiance of 100 W/m^2. The open-circuit voltage of the array architectures reduces from 324.7 V for PSN 1 to 304.2 V for PSN 8. The power outputs of the PV array architectures were noted as 13.03 kW, 11.02 kW, 9 kW, 6.98 kW, 4.97 kW, 2.96 kW, 1.73 kW, and 1.63 kW for PSN 1, PSN 2, PSN 3, PSN 4, PSN 4, PSN 5, PSN 6, PSN 7, and PSN 8, respectively. The fill factors of the array are higher for PSN 1 and PSN 8, *i.e.*, 0.777 and 0.856, respectively, due to convexity in the characteristics

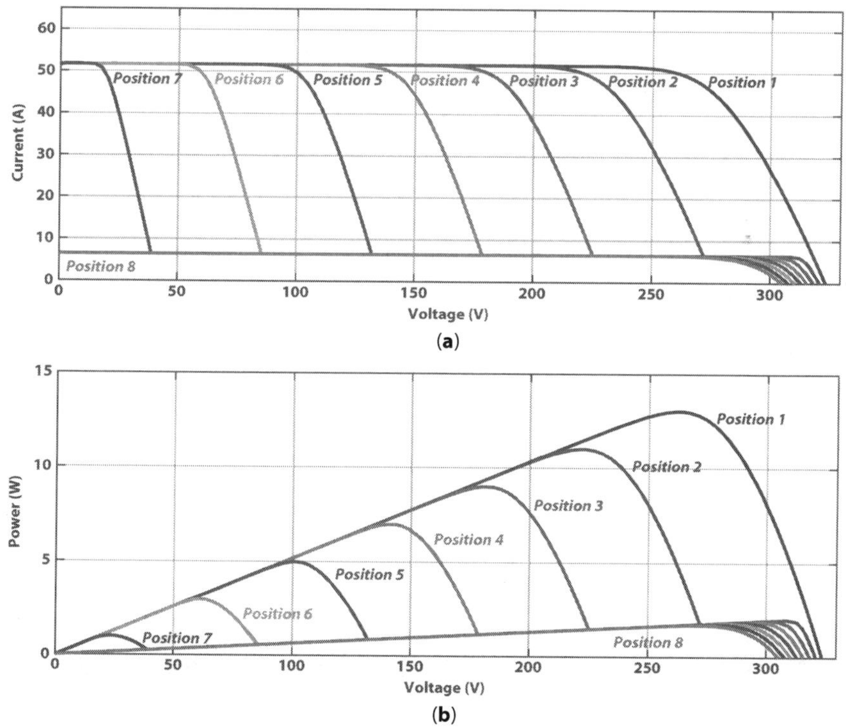

Figure 4.4 Curves representing (a) I-V and (b) P-V behavior of array for numerous positions of the incremental row shading.

Table 4.2 Data showing the parameters of array architectures for incremental shading over row.

Shading position	P_O (kW)	V_M (V)	I_M (A)	V_{OC} (V)	I_{SC} (A)	Fill factor	Efficiency (%)	V_T (V)
PSN 1	13.03	261.26	49.87	324.7	51.63	0.777	19.63	6.225
PSN 2	11.02	221	49.86	321.2	51.62	0.665	16.60	3.801
PSN 3	9	180.4	49.89	318.3	51.61	0.548	13.56	1.312
PSN 4	6.98	140.4	49.72	316.4	51.6	0.428	10.52	−1.19
PSN 5	4.97	100.32	49.54	310.4	51.57	0.310	7.49	−3.96
PSN 6	2.96	60.76	48.72	310.5	51.52	0.185	4.46	−9.31
PSN 7	1.73	276.01	6.27	307.5	51.36	0.110	2.61	9.475
PSN 8	1.68	269.47	6.23	304.2	6.45	0.856	2.53	7.262

curves as the arrays operate under one irradiance value. The detailed data of the different parameters of the array architectures for incremental row shading are tabulated in Table 4.2.

4.4.2 Analysis under Incremental Column Shading Scenario

Similarly, the incremental column shading scenarios (PSN 1 to PSN 8), as shown in Figure 4.3 (b),were applied to the array architectures with all with equal power outputs during each shading position. The short-circuit current of the array significantly decreases from 51.63 A to 6.45 A with a shading increment area over the column, as shown by the I-V and P-V curves of the PV array layouts in Figures 4.5 (a) and (b), respectively. The power output of the PV array architectures for PSN 1, PSN 2, PSN 3, PSN 4, PSN 5, PSN 6, PSN 7, and PSN 8 were noted as 13.03 kW, 11.49 kW, 9.91 kW, 8.32 kW, 6.69 kW, 5.05 kW, 3.38 kW, and 1.68 kW, respectively. The V_{OC} of the array is noted to have been reduced from 324.7 V to 304.2 V under PSN 1 to PSN 8, respectively. The fill factor of the array during all the shading positions remained at a higher value due to convexity in the characteristic curves. The data of the parameters of the PV array architectures for different positions of the column shading are tabulated in Table 4.3.

The comparison analysis of different electrical parameters of the array architectures for incremental shading over row and column is shown in Figure 4.6. The power comparison graph in Figure 4.6 (a) demonstrates that the array's power generation is higher during column shading than row. It is discovered that the array experiences losses of about 86.72% and 74.05% due to partial shading caused by row and column coverage, respectively. Additionally, it can be shown that the power curves of the

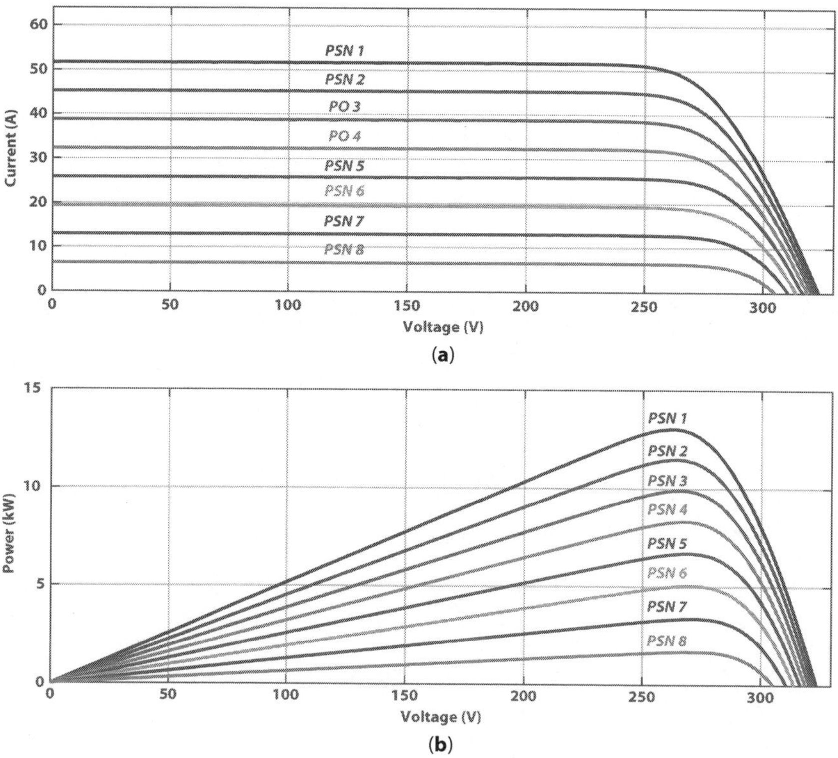

Figure 4.5 Curves representing (a) I-V and (b) P-V behavior of array for numerous positions of the incremental column shading.

Table 4.3 Data showing the parameters of PV array architectures during incremental column shading.

Shading position	P_O (kW)	V_M (V)	I_M (A)	V_{OC} (V)	I_{SC} (A)	Fill factor	Efficiency (%)	V_T (V)
PSN 1	13.03	261.26	49.87	324.7	51.63	0.777	19.63	6.225
PSN 2	11.49	264.5	43.44	322.6	45.18	0.788	17.31	7.312
PSN 3	9.91	265.38	37.34	320.4	38.72	0.799	14.93	6.909
PSN 4	8.32	266.31	31.24	318.2	32.27	0.810	12.54	6.338
PSN 5	6.69	270.61	24.72	315.8	25.82	0.820	10.08	8.781
PSN 6	5.05	271.53	18.60	313.8	19.36	0.831	7.61	8.291
PSN 7	3.38	270.85	12.48	311.9	12.91	0.839	5.09	7.099
PSN 8	1.68	269.47	6.23	304.2	6.45	0.856	2.53	7.262

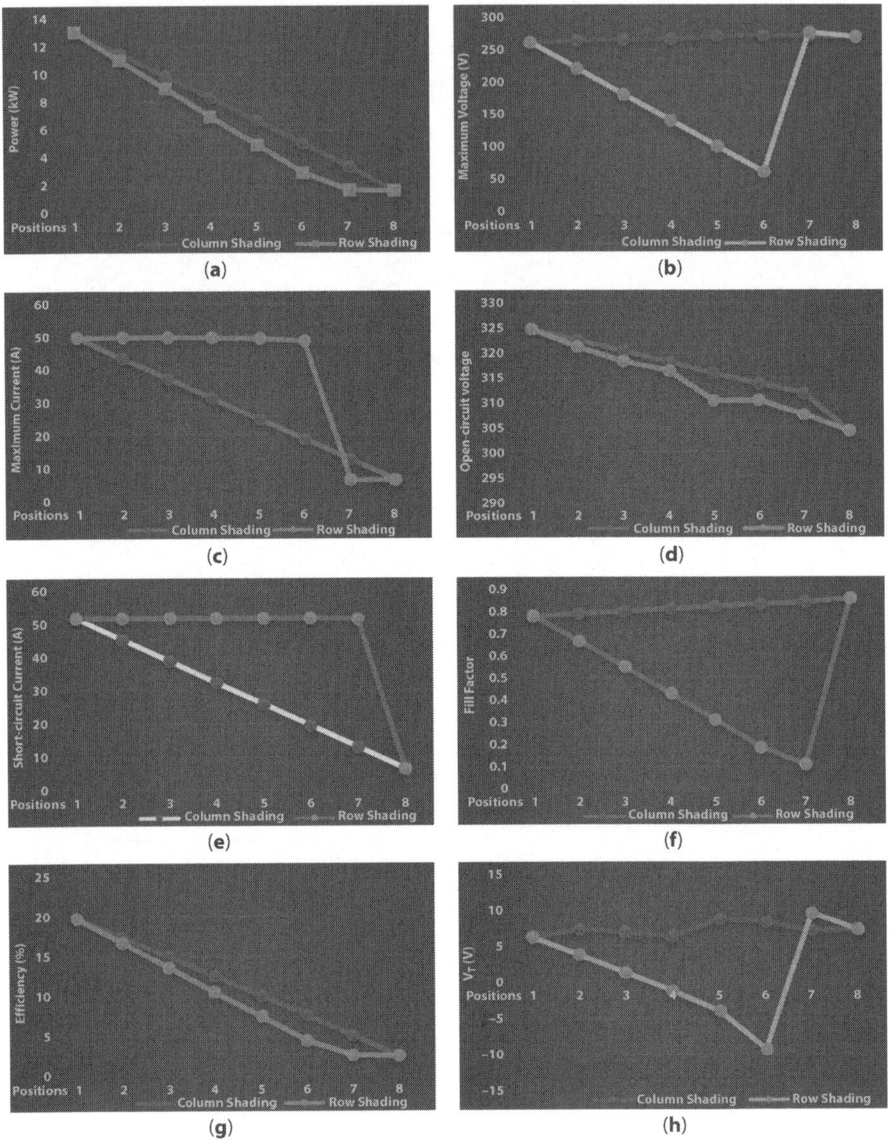

Figure 4.6 Parameters comparison of array architectures. (a) P_O, (b) V_M, (c) I_M, (d) V_{OC}, (e) I_{SC}, (f) fill factor, (g) efficiency, and (h) V_T.

PV array topologies fall with shifting placements or growing shading circumstances. The values of the voltage output during incremental column shading are graphically given in Figure 4.6 (b) that remain nearly same whereas for incremental row shading the voltage value reduces until

Table 4.4 Parameters behavior during incremental shading.

Shading	P_O (kW)	V_M (V)	I_M (A)	V_{OC} (V)	I_{SC} (A)	Fill factor	Efficiency (%)	V_T (V)
Row	↓	↓↑	-↓	↓	-↓	-	↓	↓↑
Column	↓	-	↓	↓	↓	↓↑	↓	↕

PSN 6 and then suddenly increases from PSN 7. The maximum currents comparison graph of the PV array architectures are depicted in Figure 4.6 (c). It can be observed that the maximum current of the array gradually reduces during incremental column shading whereas, for incremental row shading, it shows inconsistent behavior. It is seen from the open-circuit voltage comparison curves in Figure 4.6 (d) that the value reduces with increasing partial shading in columns and rows of the PV array. The short-circuit current comparison graph in Figure 4.6 (e) shows that the value reduces during column shading whereas, for row shading, it shows inconsistent behavior. The fill factor of the array during column shading remains nearly equal whereas the value is reduced until PSN 7 and then increases over complete row shading (Figure 4.6 (f)). The efficiency comparison of the arrays for incremental row and column shadings is shown in Figure 4.6 (g) and found to be reducing with incrementing shading. The thermal voltage has shown inconsistent behavior for both the incremental shading scenarios that can be seen from the curves in Figure 4.6 (h). A detailed description of the behavior of the parameters of array architectures with incremental shading scenarios over rows and columns is given in Table 4.4.

Therefore, based on the study above, the following parameters were evaluated as the most suitable ones that may be utilized to identify the shading in the PV array.

Incremental row shading: power output, open-circuit voltage, and efficiency.

Incremental column shading: power output, current output, open-circuit voltage, short-circuit current, and efficiency.

4.5 Conclusion

The PV array architectures were investigated under incremental shadings in this chapter, in MATLAB for 7x7 array with 15.92 kW capacity. Also, the behavior of various electrical parameters was investigated under incremental rows and columns shading. From the analysis, the following conclusions were found:

- For incremental rows and columns shading situations, all array architecture produced the same amount of power.
- Compared to column shading, row shading scenarios cause a greater reduction in the PV array's power output.
- Under scenarios of row and column shading, the PV array experiences a maximum power loss of 86.72% and 74.05%, respectively.
- Maximum power, open-circuit voltage, and efficiency are the most suitable parameters that can be utilized to estimate row-level partial shading.
- Maximum power, maximum current, open-circuit voltage, short-circuit current, and efficiency are the most suitable parameters that can be employed for the prediction of column-level partial shading.

It was found that the widely used parameters that are preferred mostly for the design of partial shading detection algorithms, *i.e.*, voltage and current, show inconsistent behavior during the incremental row and column shadings. Hence, designing proper algorithms by considering other parameters can be an efficient solution for partial shading detection in PV arrays and which can be the future scope of this investigation. Also, the reduction of power losses in the array by reducing the partial shading effect in the rows of the array can be another future scope of this investigation.

References

1. Abdulmawjood, K., Alsadi, S., Refaat, S. S., & Morsi, W. G. (2022). Characteristic study of solar photovoltaic array under different partial shading conditions. *IEEE Access*, 10, 6856-6866.
2. Şevik, S., & Aktaş, A. (2022). Performance enhancing and improvement studies in a 600 kW solar photovoltaic (PV) power plant; manual and natural cleaning, rainwater harvesting and the snow load removal on the PV arrays. *Renewable Energy*, 181, 490-503.
3. Aljafari, B., Satpathy, P. R., & Thanikanti, S. B. (2022). Partial shading mitigation in PV arrays through dragonfly algorithm based dynamic reconfiguration. *Energy*, 257, 124795.
4. Satpathy, P. R., Bhowmik, P., Babu, T. S., Sharma, R., & Sain, C. (2022). Bypass Diodes Configurations for Mismatch Losses Mitigation in Solar PV Modules. In *Innovation in Electrical Power Engineering, Communication, and Computing Technology* (pp. 197-208). Springer, Singapore.

5. Hong, S., Kang, H., Kim, G., Lee, S., Kim, S., Lee, J. H., ... & Lee, K. (2016). A series connection architecture for large-area organic photovoltaic modules with a 7.5% module efficiency. *Nat. Commun.*, 7(1), 1-6.
6. Mäki, A., & Valkealahti, S. (2011). Power losses in long string and parallel-connected short strings of series-connected silicon-based photovoltaic modules due to partial shading conditions. *IEEE Trans. Energy Convers.*, 27(1), 173-183.
7. Mansur, A. A., Amin, M. R., & Islam, K. K. (2019). Performance comparison of mismatch power loss minimization techniques in series-parallel PV array configurations. *Energies*, 12(5), 874.
8. Pendem, S. R., & Mikkili, S. (2018). Modeling, simulation, and performance analysis of PV array configurations (Series, Series-Parallel, Bridge-Linked, and Honey-Comb) to harvest maximum power under various Partial Shading Conditions. *Int. J. Green Energy*, 15(13), 795-812.
9. Pendem, S. R., Mikkili, S., & Bonthagorla, P. K. (2019). PV distributed-MPP tracking: Total-cross-tied configuration of string-integrated-converters to extract the maximum power under various PSCs. *IEEE Syst. J.*, 14(1), 1046-1057.
10. Satpathy, P. R., Jena, S., Jena, B., & Sharma, R. (2017, April). Comparative study of interconnection schemes of modules in solar PV array network. In *2017 International Conference on Circuit, Power and Computing Technologies (ICCPCT)* (pp. 1-6). IEEE.
11. Satpathy, P. R., Sarangi, A., Jena, S., Jena, B., & Sharma, R. (2018, March). Topology alteration for output power maximization in PV arrays under partial shading. In *2018 Technologies for Smart-City Energy Security and Power (ICSESP)* (pp. 1-6). IEEE.
12. Kumar, A., Pachauri, R. K., & Chauhan, Y. K. (2016, July). Experimental analysis of SP/TCT PV array configurations under partial shading conditions. In *2016 IEEE 1st International Conference on Power Electronics, Intelligent Control and Energy Systems (ICPEICES)* (pp. 1-6). IEEE.
13. Lappalainen, K., & Valkealahti, S. (2017). Output power variation of different PV array configurations during irradiance transitions caused by moving clouds. *Appl. Energy*, 190, 902-910.
14. Satpathy, P. R., Babu, T. S., Shanmugam, S. K., Popavath, L. N., & Alhelou, H. H. (2021). Impact of uneven shading by neighboring buildings and clouds on the conventional and hybrid configurations of roof-top PV arrays. *IEEE Access*, 9, 139059-139073.
15. Dhimish, M., Holmes, V., Mehrdadi, B., Dales, M., Chong, B., & Zhang, L. (2017). Seven indicators variations for multiple PV array configurations under partial shading and faulty PV conditions. *Renewable Energy*, 113, 438-460.
16. Satpathy, P. R., & Sharma, R. (2020). Parametric indicators for partial shading and fault prediction in photovoltaic arrays with various interconnection topologies. *Energy Convers. Manage.*, 219, 113018.

17. Gosumbonggot, J., & Fujita, G. (2019). Partial shading detection and global maximum power point tracking algorithm for photovoltaic with the variation of irradiation and temperature. *Energies*, 12(2), 202.
18. Silvestre, S., Kichou, S., Chouder, A., Nofuentes, G., & Karatepe, E. (2015). Analysis of current and voltage indicators in grid connected PV (photovoltaic) systems working in faulty and partial shading conditions. *Energy*, 86, 42-50.
19. Chaibi, Y., Malvoni, M., Chouder, A., Boussetta, M., & Salhi, M. (2019). Simple and efficient approach to detect and diagnose electrical faults and partial shading in photovoltaic systems. *Energy Convers. Manage.*, 196, 330-343.

5

Controlled Smart Robotic Arm for Optimized Movement in Pharma Application

Deepa Thangavelusamy*, Kripalakshmi Thiagarajan, S. Angalaeswari, D. Subbulekshmi, A. R. Kalaiarasi, Sambit Pattnaik and Preetam Singh Chauhan

School of Electrical Engineering, Vellore Institute of Technology, Chennai Campus, Vandalur-Kelambakkam Road, Chennai, India

Abstract

Rising casualties resulting from medical errors have posed a significant threat to people across the globe due to the shortage of an exemplary automated system for ensuring optimal drug composition synthesis. This project proposes a smart robotic arm solution to monitor and detect the endpoint color at the titration inflection point and categorize the samples using an integrated mobile app. This robot has a color sensor, a power unit, actuators (DC servo motors), an end effector (impactive gripper), and an Arduino microcontroller (to control DC servo motors and color sensor) with embedded software. The embedded color sensor performs color detection at the inflexion point during stoichiometric titrations for medicines and then segregates it based on a delineated logic. A mobile platform is also designed to facilitate the robotic arm movement, providing a smoother operating range. For optimized and smooth operation, the interfaced mobile app offers users various control angles at mobility junctions such as gripper, wrist, and shoulder. Additional sensors are implanted to trace the object in 3D space and manual intervention for security.

Keywords: Innovative robotic arm, inflection point, color sensor, impactive gripper, stoichiometric titration, mobile platform, interfaced mobile app, automated system

**Corresponding author*: deepa.t@vit.ac.in

O.V. Gnana Swathika, K. Karthikeyan and Sanjeevikumar Padmanaban (eds.) Smart Grids as Cyber Physical Systems Volume 1: Artificial Intelligence, Cybersecurity, and Clean Energy for Next Generation Smart Grids, (55–68) © 2024 Scrivener Publishing LLC

5.1 Introduction

Global industrial automation has risen significantly over the last few decades. Global industrial behemoths prefer it because it aids in reducing human labor and, as a result, the cost of production. Furthermore, automation enhances the delivery of high-quality products with greater consistency. The robotic arm became one of the most valuable pieces of technology unveiled in the twentieth century, swiftly becoming a cornerstone in many manufacturing areas. It can be used for various tasks and processes that would be too time-consuming, difficult, or dangerous for humans to accomplish [1]. When users think of robotic arms, they might suppose of the automotive industry. Still, they can also be used for a diverse array of other transformative technologies other than just welding and painting auto parts. As a result of these factors, robots have become an essential element of the automation industry.

A robot is an electromechanical device operated by software programs and employs motors, actuators, and sensors [2]. They perform a wide range of operational activities such as welding, picking up objects, coloring, and serving several needs in various working environments. While operating in a comparable pattern to the human arm, robot arms can still have a far more comprehensive range of motion because their design is entirely up to the creator's imagination [3]. For instance, the joint that connects the segment. The edge of the robotic arm that is engineered to do the work for which it was designed is the end effector. It can be tailored for virtually any task, for instance, gripping like a hand, painting, tightening screws, and so on. Overall, it is like collaborating with a human-like arm that can lend the user more freedom while also automating and simplifying your tasks at varying phases of a trial [4]. It would be highly beneficial to understand the human-machine link when carrying out a task. These robots can be constructed at a reasonable cost while retaining high accuracy and control [5, 6]. The robotic arms can rotate and move like a hinge.

Four-degree-of-freedom robotic arms can be used in the industrial sector to transport light materials, such as medical tubes and flasks [6–8]. Furthermore, to make the robot mobile, users can choose a bot at its base. In this invention, a unique monitoring action of the proper composition of medicine for color identification at the inflexion point during the stoichiometric titration of chemical components in medicines is used to organize the accumulated samples at different angles in the sample holder for more uncomplicated segregation and time savings.

The vision was to build a prototype of a joint arm configuration with four degrees of freedom while interconnecting modern I.T. trends to make mass batch production convenient by incorporating integrated mobile app technology. Pharmaceutical companies can use robots to migrate compounds or biological samples to combine novel substance elements or test the pharmaceutical benefit of existing concoction matter. The robots can pick and position fluids and strong augmentations and warm/cool, blend, shake, and evaluate them [9–11]. Innumerable research group robots were referred to as auto-samplers because their primary function is to provide consistent trials to scientific devices.

5.2 Description of the Prototype

The proposed smart robotic arm, as shown in Figure 5.1 is trained to perform similar functions to that of a human arm. The said architecture is almost comparable to the human hand, which accomplishes necessary tasks to interact with the outside world. The robotic arm with four degrees of freedom was configured to be sufficiently accurate across all movement patterns in the requisite axis while still being cost-effective, which is also a major determinant [12–16]. The multitude of actuators increases as the number of degrees of freedom multiplies, rendering it even more cost-effective.

This robotic arm's primary function is to monitor the inflexion point during the titration and organize the samples based on their color. Our primary importance has been in the pharmaceutical industry, where this robotic arm interprets and closes the nozzle of the burette when a specific

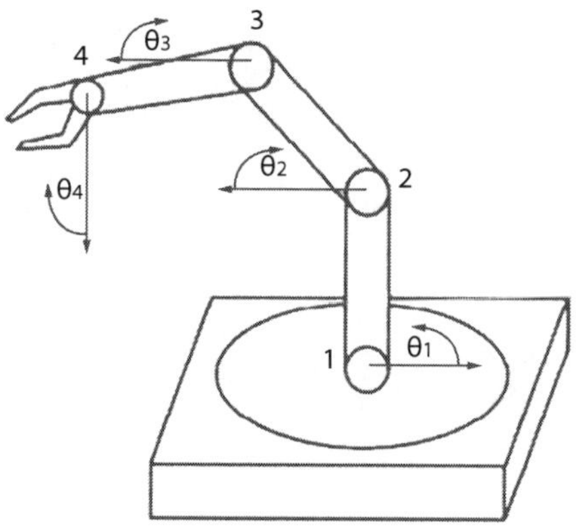

Figure 5.1 Mechanical projection of the prototype robotic arm model.

characteristic color in the titration process in the production process of medicine is attained [17]. This is accomplished through the use of a color sensor, which continuously monitors the change in the color during the titration, and when the precise characteristic color is attained, it directs the robotic arm to close the nozzle and pick the beaker/test tube containing the colored solution by rotating at a precise angle [18–20].

5.3 Segments of the Prototype

The modules of the robotic arm mainly consists of linking the connections of the control circuit of the mechanism and designing the app for the user interface to provide a range of control positions.

5.3.1 Designing the Circuit of the Prototype

This project's circuit architecture is primarily comprised of servo motors, actuators, and a microcontroller, as shown in Figure 5.2. The connectivity path with the smartphone-based application is established by interfacing an Arduino microcontroller with a Bluetooth module (HC-05). Six digital pins on the Arduino board are linked to the control pins of the six servo motors, as shown in Figure 5.3. Because the Arduino could not withstand the amount of electricity that all servos can draw, a 5 V-2 A external power source was used to power them. So once everything is connected, the next step is to program the Arduino with the linked codes and make the app.

Figure 5.2 Circuit connections with the color sensor, servo motors and Arduino microcontroller.

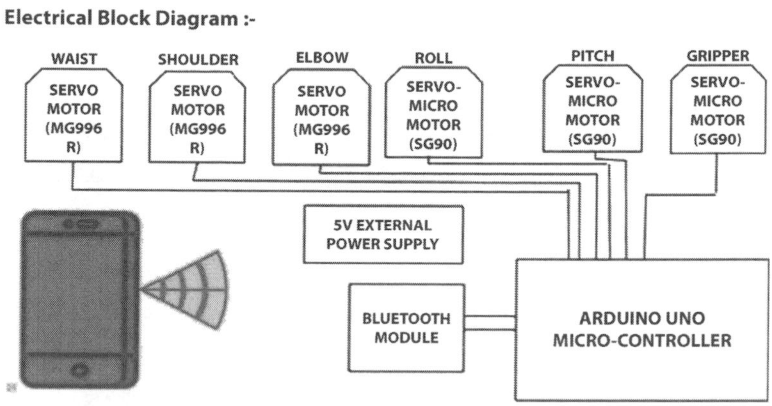

Figure 5.3 Control schema of the robotic arm.

5.3.2 Designing the Mobile App for User Interface

The user application was designed using the MIT App Inventor online open source platform, as shown in Figure 5.4 and Figure 5.5. Two buttons at the top make it possible to connect your smartphone to the Bluetooth module. The graphic of the robot arm is on the left side, while six sliders for directing the servos and one slider for speed control are on the right side. Each slider has a distinctive starting value, minimum value, and maximum

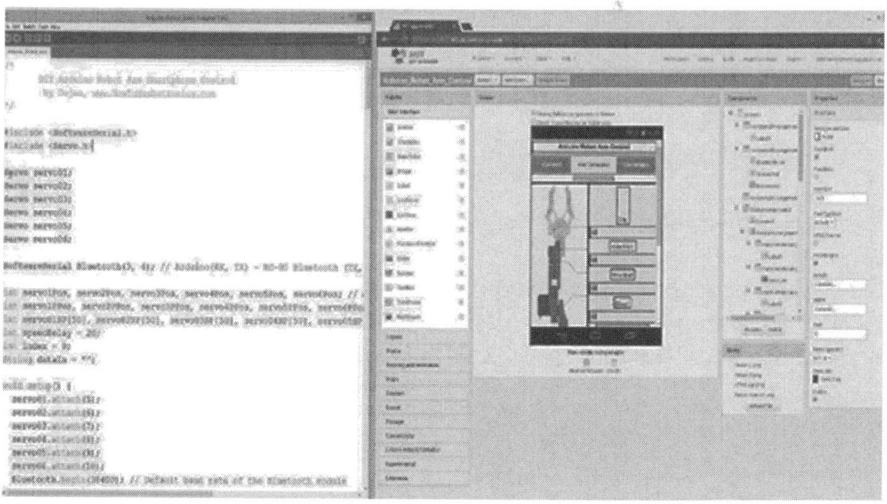

Figure 5.4 Linking the mobile app with Arduino microcontroller for controlling the robotic arm.

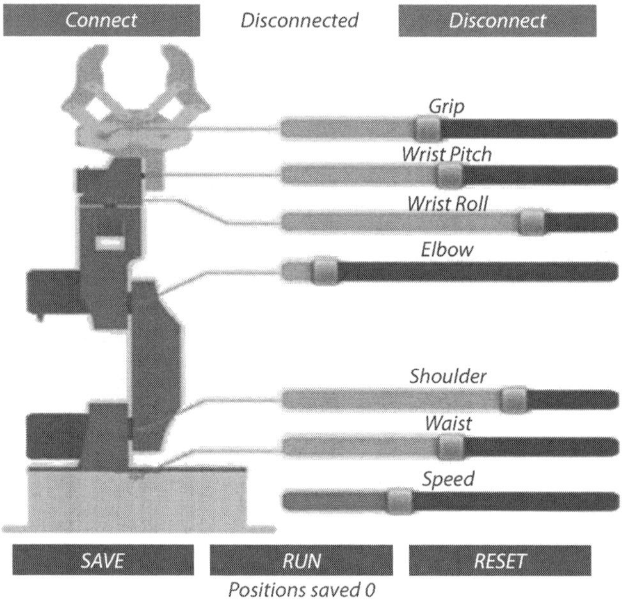

Figure 5.5 The mobile app interface design.

value to support the robot arm joints. The robot arm is configured to run periodically using three buttons at the bottom of the app. Then there are the servo position control slider blocks, and the robot arm programmed buttons blocks. So, if the slider's position alters, the application uses the Bluetooth function Send Text to send a text to the Arduino, which consequently regulates the movement range. The requisite operations were then initiated by associating the Arduino code with the app leveraging interface libraries.

5.4 Design Specifications

The robotic arm has four servo motors, each with one degree of independence (DOF), and they all operate on 5 V. A robotic arm system is typically constituted of linkages, joints, actuators, sensors, and controllers. Ultrasonic sensors with a range of 21 cm and a Bluetooth module were used to control the bot. The bot is composed primarily of a DC motor, an Arduino Mega, and an L293D Motor Driver, as shown in Table 5.1.

There is also a color sensing module for the robot being used to pick and position by the prototype arm. The Color Sensor, predicated on the TCS3200, is a full-fledged color detector capable of detecting static color.

Table 5.1 Key components of the design.

SR. no.	Components	Specifications
1.	Arduino ATmega2560	Operating voltage-5 V Digital I/O Pins 54 (of which 14 provide PWM output)
2.	Servo Motor	Voltage-4.8 V-6 V
3.	Ultrasonic sensors	Supply voltage-5 V Global current consumption 15 mA Range-21 cm
4.	DC motor	Operating voltage-3 V RPM-2400
5.	Color sensor	TCS 3200 Single-Supply Operation (2.7 V to 5.5 V)

The sensor's yield is a square wave with a frequency approximately equal to the incident intensity of the light. It also has onboard light emitting diode (LED) lights for fill illumination. It comes with a TAOS TCS3200 RGB sensor chip and four white LEDs. The TCS3200 can detect and quantify a nearly enormous variety of visible colors [21–23].

This design is practically identical to the human hand, which perform functions to interact with the physical world. There are two types of joints: prismatic and rotary joints, and they connect adjoining links—the degrees of freedom of the robotic arm. A robotic arm with four degrees of space was designed because it is acceptable for all movement patterns in the requisite axis and is also economically viable, which was a crucial factor in the prototype's architecture. The number of transducers increases as the number of degrees of freedom multiplies, rendering it more economically feasible [23–25]. The amount of independent joint variables defines the number of degrees of freedom in a robotic system, as shown in Table 5.2. Each part of the arm provides a wide range of movement patterns through its assembly and supports a wide range of operation.

The area that the gripper can cover is called the workspace of the robotic arm. It depends on several specifications like the length of the links, the DOF, the angle at which the object will be picked up, and the main configuration of the robot [21, 22]. Support links the arm to the base. Popsicle sticks are used as the mainframe of the arm. A servo motor is connected at each joint so that the arm can move itself to pick up the object. The claw is designed to be extended maximum so that the thing can be picked up

Table 5.2 Angle of rotation of the arm for different degrees of freedom.

Degrees of freedom (DOF)	Degrees
Base	(0'–120')
Shoulder	(100'–125')
Elbow	(0'–15')
Gripper	(70'–175')

easily [26–28]. A thread is connected with the claw to control it properly. The other end of the thread is associated with a servo motor. By rotating the servo, the claw is operated.

The selection of a suitable motor is of prime importance and uses the above calculations with presumed values. The model is based on the motor's torque and speed calculation. Proper joint force calculation ensures that the engine selected can support the weight of the robotic arm and the load that must be picked [29]. When the user demands significant torque and concise speed, servo motors are obligated. Torque is the force's propensity to rotate an object about an axis. It is given mathematically as the cross product of lever arm distance and force, which produces rotation. Based on the calculated moment, a suitable servo motor with a torque of 2.5 kg/cm, *i.e.*, 0.245 Nm, was selected to match the torque demand of the robotic arm. They can turn 0'–180' with a speed of 0.1 s/60' and a weight of 9 g each.

5.5 Simulation Analysis

The simulation was carried out using SolidWorks by first creating the design for each segment of the robotic arm corresponding to each degree of freedom region and assembling it as a unit to get the desired design position, as shown in Figure 5.6. Each part resembles the range of degrees of rotation with each node as a junction to provide optimized movement. The object to be picked is maintained at a specified point from the bot anywhere on the periphery of the region of the arms' operational range. The base servo motor is then programmed to gently rotate from its initial position of 0° to 180°, permitting the color sensor to detect the presence of any colored object on its trajectory. Two mounted servo motors on the arm will slowly flex the bot down to the defined level, and the fourth will be used to open or close the arm's grip on the test tube containing the colored drug solution.

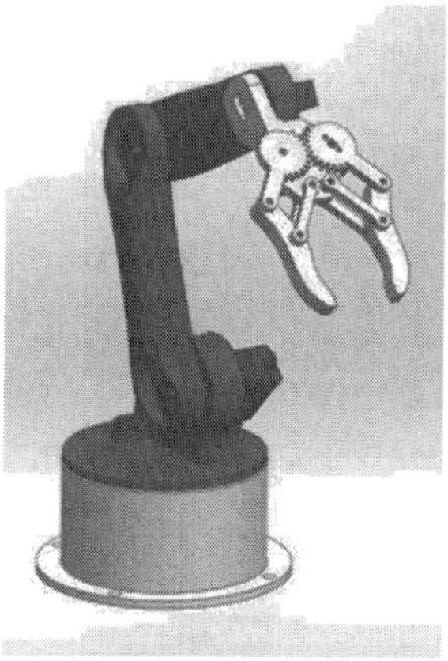

Figure 5.6 Assembled simulated robotic arm using SolidWorks.

5.6 Hardware Analysis

The prototype model, as shown in Figure 5.7 presents the notable hardware results:

a) The color sensor can detect any colored object within the range of 10 cm.
b) The user can rest the position of the arm at any time with app control.

The concept arm, as shown in Figure 5.8 and Figure 5.9 can provide the user with the optimal range of movement at each degree of freedom

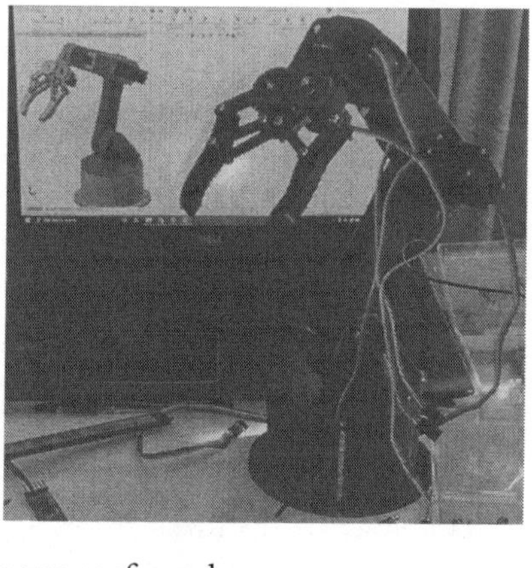

Figure 5.7 Smart robotic arm configured.

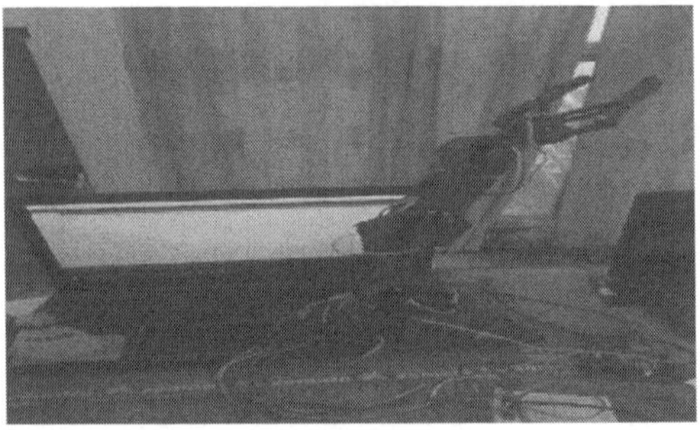

Figure 5.8 Robotic arm with range of movements.

Figure 5.9 Robotic arm with gripper lift movement.

junction of the arm, whether the gripper, shoulder, or wrist. The arm could indeed pick and position test tubes and small to medium-sized flasks used during titrations, which makes up a sizeable substantial part of the intended application.

5.7 Conclusion

The smart robotic arm is committed to delivering a cost-effective and optimal approach to the drug manufacturing industry's challenges to facilitate mankind during any outbreak. The arm is capable of executing titrations with specificity by clearly identifying the inflexion point of stoichiometric titrations.

By incorporating a transportation functionality into the configuration, the arm is apt to provide a more time-efficient approach to today's drug delivery process. With much more extensive applicability into the arm, such as the integration of machine vision and augmented reality, it advocates solutions to a broad range of needs of today's drug delivery subject and assists in the enhancements of drug processing infrastructure.

After validating and carrying out the prototype tests, it satisfies the aforementioned strategic objectives.

References

1. I. Mihara, Y. Yamauchi, and M. Doi, "A real-time vision-based interface using motion processor and applications to robotics," in *Systems and Computers in Japan*, vol. 34, pp. 10-19, 2003
2. Neo Ee Sian; Yokoi, K.; Kajita, S.; Kanehiro, F.; Tanie, K., "Whole body teleoperation of a humanoid robot - development 2017 International Conference on Advanced Computing and Communication Systems (ICACCS -2017), Jan. 06 – 07, 2017, Coimbatore, INDIA of a simple master device using joysticks,", *IEEE/RSJ International Conference on Intelligent Robots and Systems*, vol.3, no., pp.2569,2574 vol.3, 2002 doi: 10.1109/IRDS.2002.1041657
3. R. J. Wang, J. W. Zhang, *et al.*," The Multiple-Function Intelligent Robotic Arms", *FUZZ-IEEE Journal*, Korea, 20-24 August 2009, pp. 1995-2000.
4. S. Yu, X. Yu, B. Shirinzadeh, and Z. Man, "Continuous finite-time control for robotic manipulators with terminal sliding mode," *Automatica*, vol. 41, pp. 1957-1964, 2005.
5. P. Singh, A. Kumar, and M. Vashisth, "Design of a Robotic Arm with Gripper & End Effector for Spot Welding,", *Univers. J. Mech. Eng.*, vol. 1, pp. 9297, 2013
6. K. C. Gupta and K. Kazerounian, "Improved numerical solutions of inverse kinematics of robots," *IEEE Int. Conf. Robot. Autom.*, vol. 2, March 1985, pp. 743–748
7. A. Elfasakhany, E. Yanez, K. Baylon, and R. Salgado, "Design and development of a competitive low-cost robot arm with four degrees of freedom," *Mod. Mech. Eng.*, vol. 1, p. 47, 2011.
8. M L. B. Duc, M. Syaifuddin, *et al.*," Designing 8 Degrees of Freedom Humanoid Robotic Arm", *International Conference on Intelligent and Advanced Systems*, Kuala Lumpur 25-28 November 2007, pp. 1069-1074.
9. De Schutter J. and Van Brussel, H., (1988) "Compliant robot motion - II – a control approach based on external control loop", *Int. J. Rob. Res.*, Vol. 7(4), pp. 18-13.
10. Dario, P., Guglielmelli, E., Allota, B. and Carrozza, M.C. (1996) "Robotics for Medical applications", *IEEE Rob. Autom. Mag.*, pp. 44-56
11. Abidhusain Syed, Zamrrud Taj H. Agasbal, Thimmannagouday Melligeri, Bheemesh Gudur, "Flex Sensor Based Robotic Arm Controller Using Micro Controller", *J. Softw. Eng. Appl.*, 2012, pp. 364-366
12. Kurbanhusen Mustafa, S., Yang, G., Huat Yeo, S., Lin, W. and Chen, M., 2008. "Self-calibration of a biologically inspired 7 DOF cable-driven robotic arm.", *IEEE/ASME Trans. Mechatron.*, 13(1), pp.66- 75.

13. Cunningham, A., Keddy-Hector, W., Sinha, U. Jamster: "A Mobile Dual-Arm Assistive Robot with Jam box Control.", *IEEE International Conference on Automation Science and Engineering*, CASE 2014, 509–514. EEE Computer Society, Taipei, Taiwan.
14. S. Waldherr, R. Romero, and S. Thrun, "A gesture-based interface for human-robot interaction," in *Auton. Rob.*, vol. 9, no.2, pp. 151-173, *Springer*, 2000.
15. Mohammed, A.A. and Sunar, M., 2015, May. Kinematics modeling of a 4-DOF robotic arm. In, *2015 International Conference Automation and Robotics (ICCAR), IEEE*, pp.87-91.
16. Chu, X., Fleischer, H., Klos, M." Application of Dual-Arm Robot in Biomedical Analysis: Sample Preparation and Transport.", *IEEE Instrumentation and Measurement Technology Conference I2MTC* 2015, pp.500–504.
17. Chu, X., Fleischer, H., Roddelkopf, T. "An LC-MS Integration Approach in Life Science Automation: Hardware Integration and Software Integration.", *IEEE International Conference on Automation Science and Engineering* CASE 2015, 979–984. IEEE Computer Society, Gothenburg, Sweden.
18. R. Dillmann, "Teaching and learning of robot tasks via observation of human performance," in *Rob. Auton. Syst.*, vol. 47, no. 2-3, pp. 109-116, 2004.
19. S. Calinon, and A. Billard, "Active teaching in robot programming by demonstration," in *16th IEEE International Symposium on Robot and Human Interactive Communication*, RO-MAN 2007, pp. 702-707, Jeju Island, Korea, 2008
20. F. Boyer, S. Ali, and M. Perez, "Macrocontinuous dynamics for hyper redundant robots: application to kinematic locomotion bioinspired by elongated body animals," *IEEE Trans. Robot.*, vol. 28, no. 2, pp. 303–317, Apr. 2012.
21. Kuma R., Kalra P and Prakash N., "A virtual RV-M1 robot system", *Rob. Comput. Integr. Manuf.g*, Vol. 26 No. 6, pp. 994-1000, 2011.
22. Raza ul Islam, J. Iqbal, S. Manzoor, A. Khalid and S. Khan, "An autonomous image-guided robotic system simulating industrial applications", *7th IEEE International Conference on System of Systems Engineering (SoSE)*, Genova, Italy, pp. 314-319, 2012.
23. Popovic N., Williams S, Schmitz-Rode T., Rau G., Disselhorst-Klug C., "Robot based methodology for a kinematic and kinetic analysis of unconstrained, but reproducible upper extremity movement", *J. Biomech.*, Vol. 42 No. 10, 2009.
24. Sahu S., Biswal B. and Subudhi B., "A novel method for representing robot kinematics using quaternion theory", *IEEE Conference on Computational Intelligence, Control and Computer Vision in Robotics & Automation*, pp. 6-82, 2008.
25. Tsai L.W., "Robot Analysis: The mechanics of serial and parallel manipulators", *John Wiley & Sons*, Feb. 1999.

26. Mark S., Seth H. and Vidyasagar M., "Robot modelling and control", *John Wiley & Sons*, 2006.
27. J. J. Craig, "Introduction to robotics: Mechanics and control", *Prentice Hall*, 2nd edition, pp.84.
28. Man C., Fan X., Li C. and Zhao Z., "Kinematics analysis based onscrew theory of a humanoid robot", *J. China Univ. Min. Technol.*, Vol. 17 No.1, pp. 49–52, 2007.
29. Clothier K.E. and Shang Y, "A geometric approach for robotic arm kinematics with hardware design, electrical design and implementation", *J. Rob.*, Article ID 984823, Vol. 2010.

6

An Exploration of Internet of Everything in Smart Universe

Karmel Arockiasamy[1]*, Kanimozhi G.[2] and Umamaheswari E.[3]

[1]*Vellore Institute of Technology, Chennai, India*
[2]*Centre for Smart Grid Technologies, Vellore Institute of Technology, Chennai, India*
[3]*Centre for Cyber Physical System, Vellore Institute of Technology, Chennai, India*

Abstract

The Internet of Everything (IoE) is a concept that extends the focus of the Internet of Things (IoT) on machine-to-machine (M2M) exchanges to reflect a more dynamic system that also involves people and processes. Hospitals, agriculture, cities, transit networks, retail, home automation, and so on are all examples of IoT applications. This article looks at the latest IoT innovations in healthcare, city management, agriculture, security, and anonymity. This study also discusses several pertinent issues and roadblocks that may be classified as possible work in the IoE sector.

Keywords: Authentication, data integrity, elliptic curve cryptograph, health care, Internet of thing, radio-frequency identification, security, wearable sensors

6.1 Introduction

Different studies in the emerging field of nanotechnology have made ultra-compact formats and low-priced sensors [1] possible. In the next decade, it is projected that a trillion new sensors will be installed that are "energy neutral." Data storage is competitive and the study of big data is at the lowest point in history. The lowest cost per megabyte is for wireless communication [2, 4]. Few underdeveloped nations ignored the "infrastructure of the wired economy" and departed directly to radiocommunication. Artistic movement research in cognitive computing, artificial intelligence (AI), and

**Corresponding author*: karmel.a@vit.ac.in

O.V. Gnana Swathika, K. Karthikeyan and Sanjeevikumar Padmanaban (eds.) Smart Grids as Cyber Physical Systems Volume 1: Artificial Intelligence, Cybersecurity, and Clean Energy for Next Generation Smart Grids, (69–110) © 2024 Scrivener Publishing LLC

machine learning can turn the Internet of things (IoT) into applications for the Internet of everything (IoE). The IoE will create a new understanding of our intricate society's interreliance and propose ways to regain life's excellence. The withdrawal of the words we use today would be a decent pointer. Similarly, the terms "smart shirts" and "wearable" will be abandoned and never mentioning the World Wide Web again. We are not going to talk about independent vehicles but just get in the car instead.

Cisco defines the Internet of Objects as "the intelligent link between people, processes, data, and things. The three levels namely, the perception layer, the transport layer, and the application layer forms the IoT structure [27]. The perception layer, which collects physical data, is made up of a large number of smart devices with sensors, and this data is transmitted to the application layer through the appropriate route through the transport layer. The application layer supply the needed marketing and business opportunities. Wearable health sensors are deployed in urban centers to catch victims in emergency situations such as natural disasters. Healthcare data will be linked to transportation, and which then will be linked to security systems; safety systems will be linked to environmental monitors. The environmental data will be linked to education, and so on. Climate data, satellite images, and routine natural testing of plant seedlings appear to be combining with current records in the coming years to shed light on the building of seeds for the future. In the field, the plant-by-plant mind will improve respect at no other level seen at any other time [16].

The IoE is built on three pillars: everyday smart and connected objects; an information-centric network that presents all of the current internet's bandwidth and security issues; and a machine-controlled real-time insights generated by ubiquitous distributed sensors and smart objects. For both general-purpose and specialized chip designs, these pillars present unique problems and possibilities. Despite the fact that the world's population is the tenth largest economy, India has the world's second-largest economy. Regardless, India should place a greater emphasis on dependable growth of components such as infrastructure, industries, and hospital services for the finest financial development [12].

6.2 Related Work

6.2.1 Smart Infrastructure

Backbone components with Information and communication technology (ICT) as a backend, such as physical, electrical, and digital components, make up a smart city's infrastructure. The ICT infrastructure includes all communication infrastructure components, such as wireless hot spots, fiber optics, and so on. Middleware plays a vital role in smart infrastructure:

a. Supervisory and operational workers have easy access to data from anywhere for better decision-making, resulting in a positive influence on smart cities.
b. Smart buildings, as distinct to green buildings, are also a part of smart infrastructure.
c. Intelligent buildings rely on smart grids to distribute energy efficiently and reliably from a variety of sources.

Intelligent transportation systems (ITS), such as automobile navigation and computerized toll collection, are the greatest illustration. Vehicular ad hoc network (VANET) have evolved to give a broad range of wellness and data applications to travelers and drivers. Furthermore, VANETs can capture a large amount of data from moving cars for a variety of purposes, including safety aid, traffic control, and monitoring the environment [5]. Virtual to virtual (V2V) communication was tried in numerous industries, including Toyota [6], Honda [7], Volvo [8], BMW [9], and academics [10], including privacy and security problems [11]. The platform for designing smart building sensors is shown in Figure 6.1, which may be utilized to increase building sustainability and energy efficiency, decrease maintenance costs through predictive maintenance, maximize space use, and improve comfort, tenant security, and safety.

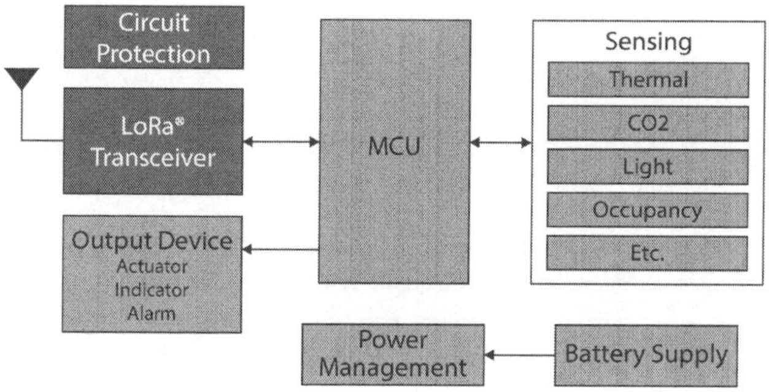

Figure 6.1 Springboard for designing smart building sensors [53].

6.2.2 Smart Building

Building owners, operators, and managers have long benefited from the operational and financial efficiencies of automating infrastructures like security and access control, lighting control, heating, fire detection, water monitoring, ventilation, and air conditioning (HVAC).

6.2.3 Smart Healthcare

Facility owners, operators, and managers have long profited from the operational and economic savings of automating infrastructures such as security and access control, lighting control, heating, fire detection, water monitoring, ventilation, and HVAC. Tasks like monitoring building temperature, security, and predictive maintenance are made easier to manage via mobile devices and computers thanks to smart buildings, which use the IoT to connect building operations. Only your creativity limits the sorts of IoT sensors that may be utilized in building automation, facilities management, and smart building applications [30].

The challenges that IoT medical services confront, such as protection, security, low-control activities, and wearability are addressed, and recommendations for possible research themes are offered. The approach for monitoring blood glucose levels in diabetic patients requires patients to manually obtain blood glucose readings at regular times [17]. Two types of blood glucose irregularity are evaluated, namely missed blood glucose readings and aberrant blood glucose levels and the severity of the irregularity is assessed. This structure has been shown workable, but it might be enhanced even further by computerizing blood glucose predictions. A device for detecting heart ambushes is gathered using moment elements and a bespoke radio cable [18].

To measure heart growth, which is dealt with by a microcontroller, an electrocardiogram (ECG) sensor is used. The customer's mobile phone receives this data via Bluetooth, where the ECG data is managed and displayed in a client application. The makers recognize that allowing heart attack seek programming will enhance the gadget. Measurement of respiratory function, which is best-known to aid in the arousal for a heart ambuscade [19], can be used to improve assistance. SPHERE [20] might be a gadget that uses wearable, environmental, and vision-based sensors to track development and prosperity in a changing environment. The intention behind is to enable better prepared patients—and more importantly, patients who are unceasingly exhausted—to stay in the comfort of their own homes while their wellbeing is being monitored [3]. This takes into consideration administrative and professional initiatives that exacerbate any issues. Machine learning might be good to learn, according to investigators taking a shot at the problem and to settle on choices around the social protections of the patient.

6.2.4 IoE in Healthcare Networks

The IoT healthcare network handles the topology, architecture, and platform of the IoT healthcare network. Few architectures that serve as a kick-off point for the creation of potential IoT networks have been proposed [13, 14]. The Internet-of-Things network for healthcare (IoThNet) topology regulates the plan of action of various parts of the social insurance network of the IoT and reveals allocated circumstances of steady conditions for medicinal services. The design of IoThNet implies a structure for the guarantee of the physical components of IoThNet, its utilitarian association, and its instructions and techniques for functioning. The phase of IoThNet is the combination of the device phase and the model of the figuring phase. For the purposes of automated monitoring, patient position monitoring, etc., a new smart healthcare system (SHS) architecture [21] has been suggested.

The SHS architecture comprises harmonized system of nomenclature (HSN), IoT smart gateway, and data visualization and management user interfaces. In order to allow interoperation between smart mobile technologies, a complex network architecture based on certain paradigms such as CoAP, 6LoWPAN, REST was introduced. In addition, an ultra-low-power HSN was introduced for the collection of real-time physiological parameters of vital patients. The results demonstrate the suggested system's ability to not only conduct patient identification and follow-up within healing institutions and nursing foundations, but also to provide effective remote patient tracking and timely crisis care management.

6.2.5 IoE Healthcare Services

In addition to basic IoT protocols, telemetry transport message, restricted application protocol, constrained application protocol [22]-based messaging architecture with global IoT healthcare standards were developed and put into use. In terms of packet transfer rates and metrics for packet loss ratio, the recently developed protocol outperforms established protocols like the message queuing telemetry transport protocol. By merging two IoT systems, namely UniversAAL [25] and BodyCloud based on SaaS Architecture [24], a new IoT-based healthcare platform [23] was established to provide active and assisted living healthcare services. BodyCloud offers services for cross-disciplinary deployment, large-scale data exchange, and decision support. Body cloud and UniversAAL IoT platforms have been compared on a layer-oriented basis.

6.2.6 IoE Healthcare Security

Privacy, decision, authentication, accessibility, fresh knowledge, non-repudiation, permission, resilience, blame resistance, self-healing is the critical security prerequisites for IoT healthcare administration as shown in Figure 6.2. Memory limitations, computational impediments, healthiness impediments, flexibility, skillfulness, media communications, gadget collection, topology organization, multi-protocol arrangement, energetic security overheads, and secure bundles shift are some of the security issues in IoT healthcare administrations [44]. Threats based on host properties such as hardware compromise, user5 compromise, device compromise, and so on; For defense, detection, and reaction services, an intelligent collaborative security model [28] was developed. A survey of RFID-based body-centric systems [25] and user environment knowledge set is presented. Innovation in RF recognizable evidence (RFID) is currently being established to provide some portion of the physical IoT layer in smart environmental conditions for individual healthcare services through the minimal expense, autonomous resources, and dispensable sensors. Figure 6.3 represents the springboard for designing smart healthcare sensors [55] that can be used for connecting hearing aids, food safety refrigeration, fall detection, Patient monitoring of chronic health conditions, medical refrigeration, and safety monitoring.

Sensors are used to monitor the wellbeing of users and to cause remote assistance. Chemical interactive materials (CIM) are united into the

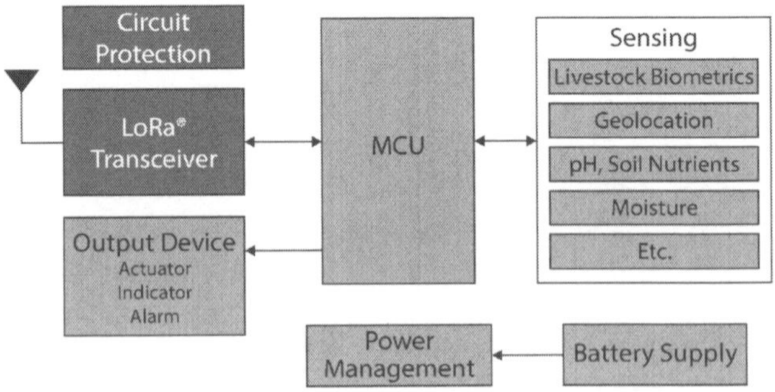

Figure 6.2 Springboard for designing smart healthcare sensors [55].

radio-wire of the mark to differentiate volatile mixtures. Essential execution was explicitly obtained by stacking a "PEDOT: PSS" collapsed repair tag. This CIM ended up being adequate for a solitary drop to achieve 0.5 dB/RH2 moisture affectability. Innovation in RFID was shown to be conceivably useful by calling body prothesis, sutures, stents, or orthopedic settlement to batch with the human well-being state from within. Tissue re-growth and prosthesis displacement, for example, may all be checked continually or on demand by the IoT system in order to achieve the optimistic objective of monitoring biophysical processes in advance. Tags would be implanted inside the prosthesis in this situation, transforming them into professional multi-functional devices that would produce data in addition to the original therapeutic purpose.

The development of a useful linking correspondence using user control that complies with regional emission regulations is the key test in the design of implantable labels. Ambient tags, wearable labels, and a mix of both types of tags are used to identify and order in-house human activity in the examples below. Authentication and privacy are important considerations when designing an RFID protection system. A secure elliptic curve cryptography-based RFID authentication system [27] was developed to be coupled with the ID verified transmission protocol in order to establish mutual authentication. The suggested scheme [26], which is resistant to many forms of attacks, is divided into two parts: the organization phase and the validation phase. The relationship between the user and the back-end worker is solid in the suggested plot, but the relationship between each tag and the user is unpredictable.

6.2.7 IoE in Smart Countries

The challenges [3] associated with the design of smart cities include design cost, operating cost, higher operating efficiency, reduction of carbon dioxide emissions, city growth, disaster resilience, dependable communication, information security, operating cost, reduction of city waste, the city's long-term sustainability, system failure, data volume, and public safety. Big data is critical in the development of intelligent cities. Big data includes things like enterprise data, database, location, email, public data, HTML, social media, clickstream, transactions, photos, and sensor data. India developed an e-wellbeing plan between 2000 and 2002 in order to enhance the use of data and communication innovation [28, 61, 62] in the well-being sector.

In mid-2008, Australia established a critical system to handle national coordination and cooperative effort in e-wellbeing [28, 33, 61, 62]. The Ministry of Internal Affairs and Communication (MIC) of Japan issued the u-Japan Policy [28, 29, 64] in mid-2004 to begin the recognition of system pervasiveness. In mid-2008, Australia created a critical framework for the management of national cooperation and joint e-wellness efforts [28, 33]. The u-Japan Policy [28, 29, 64] was submitted by the Japanese Ministry of Internal Enterprises and Communications (MIC) in mid-2004 to begin recognizing the omnipresent scheme. In 2003, Germany sponsored the enactment of eHealth [28, 33, 34] exercises regulating human services administrations. By presenting an open biological IoT structure in 2014, the Korean government [35] developed IoT administrations and products. In 2010, China's industry and information technology service [28] proposed a combined national important IoT strategy. The US Federal Trade Commission [28, 36, 37] announced the IoT strategy and its regulatory implications in 2014. In 2011 [38], the World Health Organization encouraged the use of mHealth in industrialized nations to combat tobacco use.

6.2.8 Smart Agriculture

Cultivation is the vital and perhaps most significant occupation of every nation since it jointly equalizes the basic crude materials of the industry with the need for food. The introduction of shrewd creativity in farming hones must be changed to concentrate on a successful way to achieve production. The fertilizer and excrement should be correctly related to the use of computerized technologies in the countryside in the midst of agribusiness honey. This allows the analyst to create a sensitive web application that can reliably predict the form of trim that is ideal for production by using soil testing. This prediction model may be built to collect the necessary data from the sensors that will be used to identify the crop that is most likely to grow at that time. Agricultural fields might be worked, cultivated, managed, and installed totally by task forces of collaborating autonomous machines through land and air in the not-too-distant future. They'll also be working at all hours of the day and night. Driverless farm trucks that can follow pre-altered courses are currently being sent wherever crops are growing. Over fields that determine soil quality and crop success, robots murmur. Land sensors monitor the amount of soil and water changes, starting water frameworks and fertilizer applications. The development of intelligent agricultural sensors [54] for agricultural weather stations, autonomous irrigation, greenhouse automation, field equipment monitoring, animal monitoring, soil monitoring, precision farming, pest control, and water management is based on Figure 6.3.

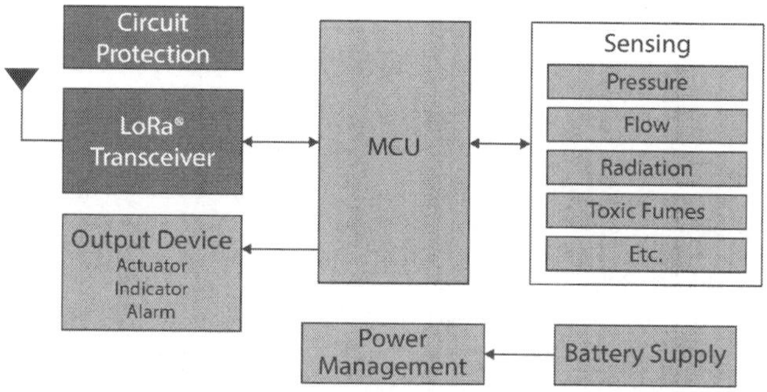

Figure 6.3 Springboard for designing smart agriculture sensors [54].

The World Bank reports that if the general population continues to climb at its current pace, they need to deliver half more food by 2050. In any event, changes in yields could decrease by more than a quarter due to the effects of ecological changes. As a result, self-driving work trucks, flying robots, ground-based sensors, and encased aquaculture assets might all assist farmers produce more nutrition at a lower cost, and even more sensitively. Meteorological and climate data analysis provides for near-constant data on crop quality and quantity, as well as the possibility of early warning on caution/ready conditions, allowing for simple mediation planning and attempts. A few records are inferred from meteorological and agronomic data that are regarded to be significant factors in determining product yield, such as editing water satisfaction, surplus, and excess dampness, and so on. After the inquiry was completed, numerous gadget executions tended to learn calculations employing IoT in order to restore correct trim forecasts [39–41] based on a three exhortation frameworks that mostly relied on prior knowledge. It uses random woodland computation to predict suitable changes based purely on modern-day Nitrogen, Phosphorus, and Potassium (NPK) took a soil toll since it is significantly superior to trusting Bayes and ID3 at executing an irregular woodland collection of rules. Random forest collection of rules, on the other hand, does not negotiate with unique variables with a large number of classes.

Other edit cycle guidance machine that uses the frequent pattern (FP) tree was shown. It also includes the application of adequate methodologies to suggest appropriate fertilizers. Anyhow, this process will degrade the state of the soil by stretching it to match its richness with the conditions needed for that purpose. Another framework presents frameworks for predicting viable crops based mostly on soil NPK levels. The framework incorporates a notice of SMS carrier use inside the enabled trim. Gullible Bayes measurements are used to predict the best appropriate adjustment. In any event, Gullible Bayes has a lesser effectiveness than the KNN set of rules.

6.2.9 Smart Grid

For a secure and efficient world, the smart grid (refer to Figure 6.4) is critical. Smart grid as shown in Figure 6.5, is an IoT-based application that allows suppliers and customers to communicate electronically and gain a variety of benefits as a consequence. Early identification of electrical current as a result of accidents (lighting, traffic difficulties, earthquakes, etc.) is possible thanks to IoT for smart grid. Events may be handled and solutions sought using this detection. Transmission lines, software, substations, smart counters, and other components make up this network. Many governments have invested in smart networks as technology has advanced.

Smart networks are especially essential in Europe because of their poor geographical organization. In Europe, smart meters should account for 80% of all meters by 2020. Due to India's enormous population, networks are poor. There are significant transmission and distribution losses. As a result, conducting smart network research in India is quite vital. However, because of its dense population, it is a country that places a high value on smart grid investments. The rise of sophisticated technological solutions such as smart grids, as well as the growth of IoT technology, will handle energy consumption and production concerns in the most efficient way possible.

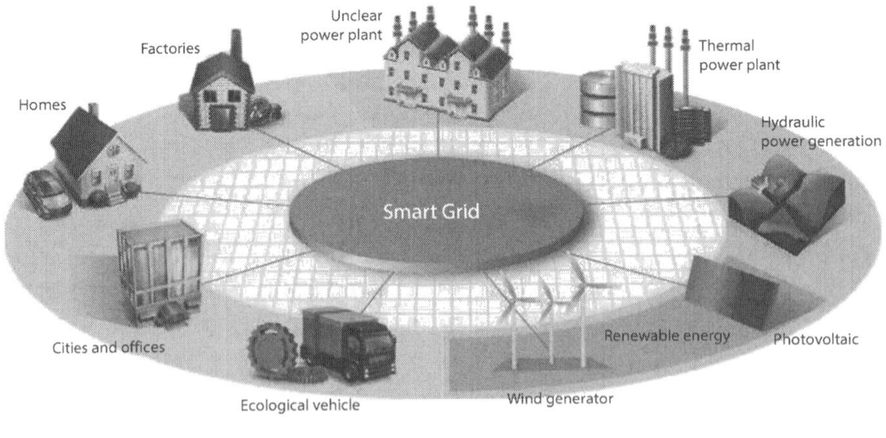

Figure 6.4 Smart grid [29–31].

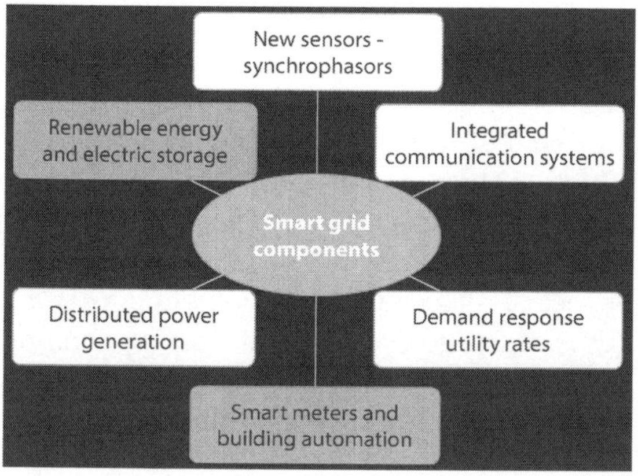

Figure 6.5 Smart grid components [29–31].

To identify power outages, restore them quickly, decrease transmission and distribution losses, boost efficiency, and produce new energy sources, present grid systems must be transformed into smart grids. The following are some of the reasons why smart grid IoT apps are required:

a. To ensure efficient energy management: energy usage and expenditures are considerably decreased because of smart networks. Smart city applications and smart lighting, for example, can be employed. This helps to conserve electricity. Lighting can be given depending on the time of day or the weather. In a short amount of time, lighting disruptions can be removed. When individuals are not at home, energy usage may be reduced in smart homes. Smart energy utilization saves energy and provides a slew of other advantages.
b. Makes use of clean energy: batteries are becoming less and less vital in our lives as a result of smart grid technology. In terms of carbon efficiency, it is a good system. Green technology should be employed in IoT systems for more sustainable solutions. Everyone wishes for a clean planet. Environmental cleanliness is essential for today's and future generations' health. Energy savings entail astronomical expenditures.
c. Low cost: smart grids allows us to aware of a power outage as soon as it occurs and identify where the electricity is being shut off, lowering expenditures. As a result, IoT in the sector of energy will result in significant cost reductions.

d. Increased energy consumption: with the advancement of technology and the rise of new habits, as well as the growth in population, there is a growing demand for energy. Smart grids may be the answer to this growing energy demand.
e. Rapid solutions to regional issues: smart cities make it easier to obtain more information on regional issues in a shorter amount of time. Smart networks may be used to construct systems that adapt rapidly to droughts and forest fires. These are only a handful of the scenarios that demonstrate the value of smart grids. There are several other advantages, which will continue to grow as technology progresses.

6.2.10 Industrial IoT

Manufacturers may acquire real-time information about their equipment by using powerful IoT sensors, control, and software applications. As a result, IoT allows you to remotely monitor and instruct devices to perform autonomous actions and identify problems. This is especially beneficial in environments where human life is at risks, such as mines, chemical facilities, nuclear power plants, and the disposal of hazardous waste and radioactive material. While IoT has a bright future, widespread adoption will necessitate a robust ecosystem of technologists, academics, OEMs, and government, regulatory organizations, industrial consortiums, and venture capitalists working together to establish sustainable business models and a suitable environment for development [32, 33].

Revolutions have happened on a regular basis in the industrial business. Machines reduced the quantity of labor that individuals were required to complete. The demand for physical labor fell, but machine efficiency increased. Previously, humans had to operate this machinery, but that is no longer the case, and even human operators are being replaced. This is far from finished. Almost everything will be intelligent in the future. IoT, big data analytics, and AI are all regarded as technical achievements. IoT has shown itself to be very useful in industries that need quick development and high-quality items. Similar needs exist in the industrial sector, and IoT may be able to help.

Benefits of IoT in the manufacturing industry:

1. Cost –savings - optimizing asset and inventory management (leading to lower carrying inventory and search durations), limiting machine downtime, more flexible operations, and greater and more effective energy adoption help companies

minimize operating expenses and establish new income streams. These organizations may focus on marketing product usage and post-sale product services rather than just selling things since they have efficient and smart products.

2. Massive customization - enabling mass customization needs a significant rise in the types of generated stock-keeping units, which leads to an increase in inventory and diversification. Another stumbling block arises from the complexities of manufacturing activities since the creation of a particular number of things by one stock-keeping unit might be followed by the manufacture of more items by another stock-keeping unit.
3. Improved security – the IoT makes the workplace safer. The IoT allows for the assessment of employees' health as well as the monitoring of dangerous activities that might result in injuries through the use of wearable devices. Along with protecting the safety of the employees, IoT also focuses on the safety concerns that arise in potentially risky locations. The oil and gas business is one such example. Because gas moves through a network of pipes, IoT is used to analyze gas leaks in this business.

6.2.11 IoT in Education

The internet's growth in recent times can be called exponential as there are a wide variety of applications on the basis of IP (internet protocol). For example, smartphones, tablets, laptops, and desktop computers. In the era of online learning, these devices have proved to be the most effective means of conveying knowledge to anyone and everyone who wants to learn. In the last two decades, the methods of teaching and learning have evolved drastically explaining the stronghold the internet has on this sector [34]. Considering education as an independent sector, student-teacher interaction is one of the major aspects IoT has influenced. Increased engagement from students and the convenience of teachers have been spotted after the use of this concept in the educational sector. Instructors at the post-secondary education level, reveal that there are motivations as well as challenges when IoT is included as a part of the curriculum for students choosing the field of computer science [35].

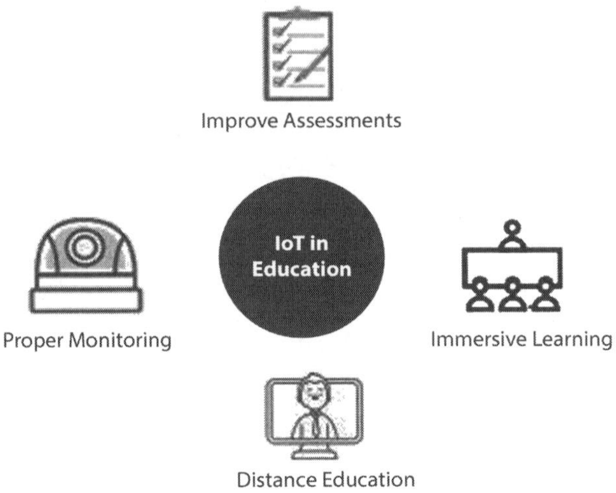

Figure 6.6 Effect of IoT in education.

6.2.12 Use Cases

6.2.12.1 Smart Classrooms

Smart classrooms and smart books are impressive inventions that stem from IoT. They have been a prime research area, especially in the past decade. Multiple devices like smart projectors, attendance systems, interactive whiteboards, laptops, etc. are a part of this system. In Reference [36], Saini *et al.* introduce a survey to bridge the recurring gaps in smart classroom analysis and implement a comprehensive smart classroom to provide a complete solution to all the minor problems in the existing systems. A smart classroom is intended to connect faculties and students, assist in the teaching process and help collect feedback from students and faculty. Apart from monitoring the teaching methods, a smart classroom also has a provision to monitor and condition the environment in which students are being taught. This additional support guarantees a dependable outcome. Likewise, digital examinations present a way for effective monitoring and marking for both the student and faculty. Figure 6.6. describes a few of the positive areas of impact with regards to IoT in the educational sector.

6.2.12.2 Smart Books

From reading books with wear and tear to reading their digitized versions, the convenience of students has gradually risen. In Reference [37], a smart book system is proposed that is also context-aware. The intended system allows e-books to respond to the user's context based on their situation or react based on the user's interests in multiple domains. This was further enhanced by combining it with a few active elements and constraints, like rules to control the content being read. Moreover, the enhancement facilitates the user to go beyond just reading and finishing a chapter. This context-sensitive behavior demonstrates how IoT can provide an immersive experience to the user.

6.2.12.3 Augmented and Virtual Reality in Education

Implementation of Augmented Reality (AR) in education facilitates the use of IoT. Students are exposed to an enveloping method of learning. Augmented reality devices are classified into three types, namely, wearable, handheld and fixed. An AR environment can be simulated in a simple device such as a mobile phone making it more accessible. In Reference [38], Slavova and Mu mention that the use of virtual reality (VR) is found to increase the students' attention span. These devices can be used by students to experience or visualize hazardous places like construction sites, look into historical literature and distant places [39]. For new teachers, training can be given through this medium to get them used to new technology before their classes. This human-centric technology with the help of IoT can be used to bring about changes in the style of teaching. Moreover, it will improve the learning skills of students, catering to each student's unique method of grasping information.

6.2.12.4 Smart Campus

In Reference [40], Curtin University proposed their vision for the year 2030 to make their entire campus a smart campus. The benefits of this proposal would be presenting a seamless, reactive, and ingenious digital surrounding. This idea also projected a spike in employability rates. In this plan, the buildings are to be remade, and features like high-speed Wi-Fi and rich AV systems are planned to be installed. Besides, a living laboratory system where students can complete hands-on and immersive learning experiences is proposed as a part of the design. Also, a living laboratory can help students gain familiarity with the theoretical concepts that are taught as a part of their curriculum. The campus environment is constantly monitored using multiple sensors and controlled according to the comfort of the users. The data collected from these sources can be used for further analysis to make the plan more engaging and user-centric.

6.2.12.5 Assisted Learning for the Disabled

Internet of things (IoT) contexts can be employed anywhere including learning materials or schemes. For the students with disabilities, the inclusion of IoT systems in their learning environment can help ease their difficulties and allow them to engage and learn in full capacity. Visual assistants, voice assistants, captivating interfaces, etc., are a few of the generic ways of introducing IoT to the students. An aspect of personalization is delivered along with promising results. It opens a wide array of opportunities for students to take on their individual learning challenges. Students with learning-related disabilities (*e.g.*, dyscalculia, dysgraphia, dyslexia, ADHD, etc.) can be given more attention in areas where they need assistance and hence bridge the gap between their current learning style and learning necessities [41].

6.2.12.6 Distance Learning

In today's world, there are many online educational tools that give additional information and assistance for handling a lecture at institutions. Multiple sources provide informative content on the web. These are collected, categorized, and labeled in such a way that the user who needs them will be able to identify related content. Online meeting environments are being used as a substitute to classrooms where instructors teach using virtual whiteboards or slide decks. Nevertheless, these tools cannot be a perfect replacement for a real-life experience. Students are able to learn, submit their work, and take up assessments through online mode. In Reference [42], Tew *et al.* recommend a platform by using high-level IoT devices to enhance the standard of education provided. The proposed system that senses the student's environment and their status gives a comprehensive analysis of the data received and is also stored in the cloud.

6.2.12.7 Advantages of IoT in Education

a) The use of IoT in education increases students' attention span and encourages them to perceive information in a creative manner.
b) It prepares students to adjust and evolve along with technology in the coming years.
c) The ability of the technology or devices to be customized according to student's personal needs.

d) Time constraints due to students' emergencies or commitments can be handled effectively. Instructor/Student data can be monitored, stored, and analyzed efficiently.

6.2.12.8 *Disadvantages of IoT in Education*

a) Employing IoT in this sector increases users' dependence on these devices that ultimately if they crash the repercussions may be radical [43].
b) When these methods are constantly changing, overstimulation of the mind can be anticipated making conditions worse even though the results assured might be different

6.2.13 IoT in Waste Management

Waste management has become a global issue in recent years. According to a World Bank study, the world generates 2.01 billion tons of municipal solid waste each year, with at least 33% of it not being treated responsibly. By 2050, it is predicted to reach 3.40 billion tons. The IoT is one technology that cities may employ to enhance waste management. Furthermore, IoT management solutions have shown to be beneficial in this industry for optimizing and automating processes. However, adoption is significantly greater throughout the supply chain. However, the same logistic technology may be used to return the waste from such things. There are several IoT-based waste management options available, but they are hampered by perceived barriers.

6.2.14 Route Optimization

Day, week, and season all affect how many people use disposal units. It might be advantageous for municipalities or waste management companies to add a sensor to a dumpster to measure how full it is. Therefore, these "smart" dumpsters could give waste collectors real-time fill level information. In order to prioritize areas that need cleaning while avoiding disposal units that are still available, the IoT business solution may use this data to determine the best routes for garbage collectors to travel. As a result, empty trash cans are not used in the waste collection process, which is more effective and saves both labor and gasoline.

Evreka, situated in Turkey, is one real-life example of an IoT management system. This firm works with towns all throughout North America and Asia to provide unique garbage collecting solutions. They attach heat-resistant, shock-resistant, and water-resistant sensors to garbage cans. These sensors gather data on location, temperature, and fill ratio before sending it to the cloud. There, the data aids in the optimization of vehicle routes as well as the estimation of visit frequency and fill levels.

6.2.15 No Deliveries were Missed

The "smart" dumpsters' collected data also helps to cut down on missed collections. The authorities will be informed instantly if the sensors determine that the trash can is full. The next pickup at this location can then be planned by the IoT waste management system. By doing this, garbage can overflow is prevented and the waste management process is streamlined. Ecube Labs is a good example. CleanCUBE, the company's solar-powered trash compactor, can store up to eight times more rubbish than non-compacting bins, resulting in an 80% reduction in collection frequency. CleanFLEX, an ultrasonic fill-level sensor, monitors a container's fill level and can assist waste management firms to improve collection efficiency by up to 50%.

6.2.16 Recycling in an Effective and Efficient Way

A major source of concern is the rising number of consumer electronics devices that end up in landfills. This is because they frequently contain both dangerous substances and valuable components. For instance, the battery of an old smartphone could leak lithium into the groundwater. Contrarily, using this device makes it less difficult to recover precious materials like gold than it is to mine them in the first place. Thanks to the IoT, businesses can recycle e-waste for valuable and scarce resources by using digital twins of sanitation systems.

6.2.17 IoT Management Systems that are Automated

Waste categorization is another area where IoT might be beneficial. Sorting and classifying garbage takes a lot of time and effort. The automated categorizing of this trash is possible with next-generation "digital bins." Bin-e, a Polish corporation, is an excellent example. It makes waste processing more efficient by combining AI-based item detection, fill level management and data processing. Its "Intelligent Garbage Bins" can distinguish between four types of waste: glass, paper, plastic, and metal. When each waste category is filled, the containers compress the trash and inform sanitation workers.

6.2.18 Analyzing Data Quickly

Throughout the day, the linked gadgets keep track of how quickly the bins fill up. These gadgets also keep track of how often they empty the dumpsters in different places. The significance of IoT management systems is demonstrated by how they interact with other smart city technologies. More correlation and optimization can be shown according to the combined statistical power. The possibilities are endless: improved garbage bin distribution, elimination of improper disposal techniques, and even waste reduction at the landfill.

6.2.19 IoT in Water Management

In a smart city environment, the management of all the natural resources (*e.g.*, air, water, fuel, etc.) constructively plays an important role. According to the reported statistics, about 71% of our planet's surface is covered by water. Out of this large volume, nearly 96.5% is considered ocean water and about 3% is the source of freshwater. But not all sources of freshwater are accessible to us. It was reported that only 0.08% of that is consumed by human beings [15]. This sparing amount of water has to be managed productively to ensure that it is sufficient and remains sufficient for the entire population. Employing the concept of IoT here, using sensors to manage the resource ensures the just use of the available water.

Multiple discussions concerning the necessity for a better water management system to preserve the planet have been put forth. The recent events of climate change stand as additional proof so as to why technology should be employed in saving the plant. Figure 6.7 highlights the most important reasons for having a proper water management system.

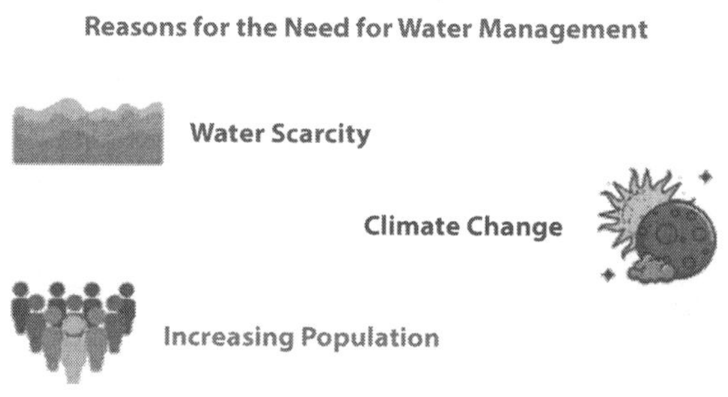

Figure 6.7 Common motives for effective water management.

6.2.20 Use Cases

6.2.20.1 Water Management in Group Residential Areas

In Reference [41], a water management system using IoT is constructed. The system is used in an apartment setting where multiple blocks are a part of a single community. It uses an Arduino to measure the flow of water directly from the main pipe. Then the pipe splits the water flow into each part of the block. This is sensed either by a flow meter (YF S201) or by utilizing a solenoid. The data generated, *i.e.*, the flow rate (number of liters of water per hour), is sent for storage to the cloud by employing IoT space. The stored data can be fetched using a mobile application by the respective apartment authorities and the water utilization rate of the entire apartment can be monitored. This way, the residents using more water can be notified immediately. In case the water utilization exceeds the limit, the solenoid can cut off the supply.

As a slight variation in Reference [45], Wadekar *et al.* mention their system where the water level is identified and can be visualized through a smartphone. In addition to that, the motor can be turned on or off depending on the context obtained from the water level.

6.2.20.2 Water Management in Campuses

An IoT-based water distribution system is introduced in Reference [46]. It was implemented for a large university campus. This system is highly efficient due to being low-cost. The authors have chosen to highlight two particular components, a sub-GHZ based campus-wide wireless network which connects an ultrasonic-based water level sensor. The data collected using these sensors are uploaded to the cloud via the network which is then used for further analysis and visualization. This system can measure depths accurately to approximately 10 m which was tested in multiple overhead tanks (OHTs) on the university campus. Aside from the water level, it can also measure the volume of water present in these OHTs, thereby letting the person in charge know the status of water levels all around the campus. As a result, the final product had optimized circuitry and returned reliable results according to the algorithm.

6.2.20.3 Water Management in Industries

Water is an essential component in multiple industries. With the booming of industries, the necessity for the preservation of our water resources is succeeding in acquiring higher interest. Several water management systems have been suggested in multiple publications over the years. These systems used technologies to approach the problem. At the industry level, not only water level and water distribution are important but also the quality of water is also an influential factor. The study in Reference [47] first presents the basic architecture and the numerous elements of an IoT-based water management system. The water quality is measured by several parameters like salinity, turbidity, and pH level. An IoT and machine learning-based system was taken as the best possible solution for an effective water management system.

6.2.20.4 *Water Management in Irrigation*

Given the fact that the agricultural sector is the biggest user of the available freshwater, the functional use of the right amount of water is the key to attaining complete stability. In precision irrigation that is used in agriculture, it is necessary to increase the crop yield while creating a decrease in expenditure. The strong use of IoT technologies guarantees a method for producing the specific measure of water required for the crops. IoT is a logical option for a smart water management system, in spite of not reaching its full potential in practice. The SWAMP design in Reference [48] explains an Internet of things based smart water management system especially for precision irrigation in the agricultural sector. It also includes a practical approach. In SWAMP, the water management capability is divided into three phases: the water reserve phase, water distribution phase, and water consumption phase. The results from this project reveal that this design will be capable of proving to be adequate in performance. Some specifications can be reengineered for optimization and scalability.

6.2.20.5 *Water Management for Underground Water Source*

In a smart city, one of the most common sources of water being used by citizens is underground water. The presence of groundwater ensures that the citizens remain content. Additionally, it also promises sustainability, and it is the only source that causes the slightest damage to the environment. Big data and IoT with sensors and other tangible devices gather data that provides practical technology explications. In proper water management systems, smart water meters report the quality of water, the amount of use, presence of any leaks, or possible water contamination. In Reference [49], a supervisory controller and data acquisition (SCADA) approach is considered for water management using IoT. The data collected from these IoT devices is checked to see if there are any damages to the physical system that misreports values. The analysis of this data requires the application of big data analysis. The test outcomes confirm that the implementation intends to monitor and control the use of water by both industries and individuals to obtain a firm control on conserving water.

6.2.20.6 *Advantages of IoT in Water Management*

a) It can be used to cross-verify certain facts and events due to the availability of historical data [50].
b) The involvement of IoT makes the systems transparent and accessible to any authorized individual.
c) Suitable use of freshwater (a depleting resource) is observed and achieved due to this technology.
d) The possibility of analysis paves the way to predict the usage for forthcoming months and be prepared for unprecedented events.
e) The concept is innovative, sustainable and a requirement for the current scenario.

6.2.20.7 *Disadvantages of IoT in Water Management*

a) The instance of the overall system being too demanding or expensive can be seen.
b) In some systems, one failed node can create a catastrophic impact on the other nodes or the entire system.
c) Since the most intricate concepts are in the process of development, the deployment of a fully-fledged system will be time-consuming.

6.2.21 IoT in the Food Industry

The food and beverage sectors are getting smarter as a result of IoT applications such as the connection of IoT smart linked goods that collect data during the supply chain of the products and smart algorithms [51–53].

6.2.21.1 *Accessibility to Customers*

Customers in the digital era expect to be able to obtain all information from their cellphones. Their cellphones will provide them with information that will make them more aware of the world, as well as their lifestyle choices and surroundings. Smartphones can manage a variety of digital tags, including barcodes, quick response (QR) codes, and RFID tags. Users will be able to access QR code-enabled goods on the site more rapidly, delivering precise information about them. Customers need complete visibility of the purchased items in order to be confident in the delivery of their orders. It is also a critical responsibility for businesses to gain client loyalty by allowing customers to track the progress of delivery of products at each step.

6.2.21.2 Quality Food Assurance

Customers put a high priority on the quality of food, and businesses do not want to take any chances. It is also crucial for any company's reputation; therefore, it performs everything that should be accomplished. Increasingly sophisticated product storage aids businesses in the safe movement of goods, allowing them to constantly maintain product quality. Internet of things (IoT)-connected devices may also detect any changes in storage temperature and assist businesses in troubleshooting in the shortest time feasible using simple approaches.

6.2.21.3 Improving Food Safety

Food safety is one of the most important concerns for businesses, and it serves as a link between customers and the organization. One of the main reasons for enhanced food safety is the usage of the IoT to monitor their facilities. Food firms may benefit from smart sensors and cloud-based analytics to provide new results. While manufacturing and transporting, these sensors can provide information on biological and chemical processes. Smart sensors also help organizations save money on staff performing the same job.

6.2.22 Transparent Supply Chain Management

For businesses, a digitally integrated supply chain guarantees exceptional traceability and clarity. It establishes a strong link between firms and their customers, making them more devoted to them. Increasing customer support for supply chain transparency will also give businesses more practical and strong tools. Clients can simply examine the comprehensive food and beverage distribution network with the enterprise thanks to the deployment of IoT technology in the industry. Consumers may simply obtain complete information about a product by scanning the provided QR code. For example, a code on the candy wrapper will provide details on the ingredients, manufacture date, packaging techniques, and more! As a result, clients can make an informed purchase choice regarding the goods swiftly and easily.

6.2.22.1 Recall of Goods

According to research, one out of every ten people dies as a result of tainted food, therefore these figures should make producers nervous. As a result, food and beverage firms need cutting-edge technologies in order to maintain food quality. The Internet of Things substantially aids brands and manufacturers in recognizing faulty items in the shortest amount of time possible, allowing them to recall the same product at any point. Companies may also obtain comprehensive information on the ingredients in food and beverage items and use the internet to communicate that information with their customers.

6.2.22.2 Energy Conservation

One of the most important methods for lowering a company's high electricity expense is energy conservation. Tracking power use using traditional means is a difficult challenge for food and beverage firms. Many businesses continue to rely on outdated equipment that places additional strain on the budget and has a significant impact on sales. The use of IoT devices in the company's electrical infrastructure can dramatically reduce costs. The IoT-enabled gadgets will turn off the power when they are not in use.

6.2.22.3 Effective Inventory Control

The backbone of the internet of things is smart linked goods. Smart packaging combined with IoT will greatly improve inventory management efficiency. As a result, the IoT will bring more cost-effective, intelligent, and timely solutions. Furthermore, these solutions will be cost-effective in terms of avoiding overstocking and reducing waste.

6.2.22.4 Forged Product Identification

Forged items have become one of the most important causes for businesses to sever ties with their consumers. As a result, they are the source of consumer and corporate mistrust. Companies may also use the IoT to track all of their products and stay up to date on them at the same time. As a result, it assists businesses in distinguishing between genuine and counterfeit items by scanning digital tags on food packaging, which confirms the products' legitimacy. As a result, both consumers and distributors will gain greatly.

6.2.22.5 *Logistics that are More Efficient*

Many food and beverage firms are using the internet of things in their operations for logistics. Internet of things (IoT)-based logistics also aids businesses in automating shipments while also addressing replenishment queries. Radio-frequency identifier transmitters help to better control the distribution chain. As a result, enterprises do not get data on product waste, and the food supply chain is well monitored, with remarkable advantages beginning with delivery and ending with the shipment.

6.2.22.6 *Operational Efficiency*

Because this is a buzzword among businesses, most manufacturers are attempting to obtain it, which is not an easy undertaking. The organization can better manage its operations by reducing manual procedures and bringing together business units and operations at the same time. The IoT has also reduced operating costs for businesses, and it has made it easier to spot problems with food and beverage vending machines. It keeps track of the user's response and sends out a warning to replace the vending machines as soon as possible.

6.2.23 IoT in the Banking Sector

Since IoT has taken root in multiple sectors, both in emerging and existing businesses. This opens up new limits of innovation which allows businesses to reshape and restructure customer experiences, including identifying the strengths and weaknesses of business strategies. This next stage of digital metamorphosis will mold the lives of businesses and their consumers [54]. The use of the internet connects objects with people and places. IoT can be used in the financial sector to investigate the influence of the emerging digital trends on the procedural system of a conventional bank. Figure 6.8 summarizes the bank authorities profits for using IoT in banking.

In the past two decades, the conventional banking systems have to reevaluate their process of operation and services to respond to those changes. Here, the internet of things serves as one of the essentials that support banks to achieve their complete digital change.

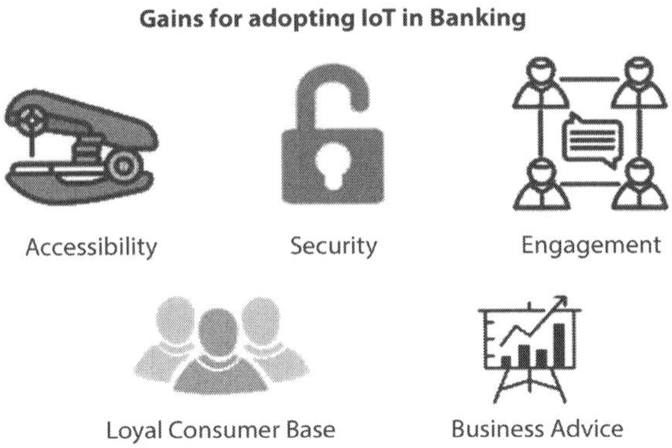

Figure 6.8 Gains for adopting IoT in the banking sector.

The shift in common working models available today demands the use of advanced technical solutions. The increased use of IoT devices/gadgets and the web by customers has inspired an expansion in IoT knowledge. It transforms lives and improves the way businesses are ventured. Banks have to utilize the existing IoT knowledge and convert it into practical data.

6.2.24 Use Cases

6.2.24.1 Debt Collection

Handling debt collection accounts from individuals and business customers requires substantial effort and additional expenses for the financial institutions (mostly banks) that are lending the money [55]. Observing the work and supply chain movement of the person/company that owes the institution is simplified by the use of IoT by installing sensors and providing them with networking capability. This can help the banks in determining the status of the debtor, their current financial situation, and other influencing factors. In addition to that, the additional overhead charges and bouncing of cheques can be prevented. Furthermore, a chain of automated teller machines (ATMs) and related banking devices can be used to monitor the debtor's expenses. The capability of the debtor to pay back the loaned money before the tenure given to them expires.

6.2.24.2 Heist Prevention

In the era of digitization, and automated transactions everyone requires money without having to go to the bank every time. Thus the invention of ATMs was a huge success, as they were very much accessible. Many of these machines were installed all around cities by banks, which made their customers content. On the flip side, these ATMs were a target of thievery again owing to their number and accessibility. As time passed by, due to the prospect of thievery, the security of the customer became the superior goal. Existing protection frameworks remain superficial. In Reference [56], a venture plan that uses a security framework for an ATM that works using NodeMCU. The aim of this project ensures the safety of the ATMs using a vibration sensor, hence employing the concept of IoT. If a robber enters and attempts to damage the ATM, the installed vibration sensor vibrates and gives a warning to the NodeMCU. This sets off the lock system of the ATM, which bolts the entrance, hence, letting the robber trapped. Apart from this, the design also has a provision to sprinkle chloroform to subdue the person and ring the alarm. This information is then sent to security officials who can open the door using a passcode set by the bank. Figure 6.9 illustrates this system's design in the form of a block diagram.

6.2.24.3 *Fraud Detection*

As we know, the banking sector is one of the most complicated and fine-tuned sectors that can affect the performance of multiple other sectors in case of any error. Similar to several other industries, big data in banking is uncertain and questionable. Hence, managing the data and real-time

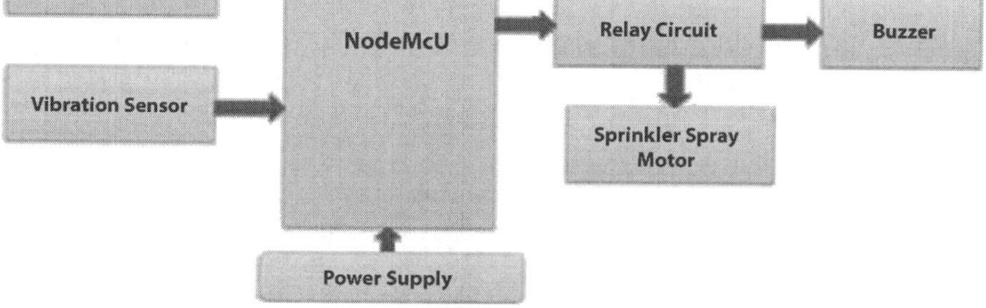

Figure 6.9 Block diagram used to implement system design [56].

observations is necessary to eliminate fraud. The chances for financial fraud are high due to the volume of data coming in every day through multiple sources. This data can be unformatted, unstructured, incredible, irrelevant, etc. The data should be cleaned, structured, and categorized properly since this information will be used for further analysis. The IoT proves to be the best solution to collect, store and share data on behalf of banks. All the necessary data can be connected using the internet through various devices/things. In Reference [57], a Hadoop-based distributed file system is discussed along with MapReduce. This is used as architecture for collecting, storing, and recovering data from multiple huge datasets that are accumulated through the internet employing the concept of IoT. Consequently, analyzing this data for anomalies or patterns can aid in fraud detection.

For example, larger financial organizations, like HSBC have successfully realized AI-based anti-fraud methods. Since fraud prevention is a high-priority task, the concept of IoT is crucial [55].

6.2.24.4 *Emergence of FinTech*

In modern times, most financial companies have moved towards introducing new technology in the financial sector and termed it as financial technology or FinTech. It is encompassed with the concept of IoT. The fundamentals of both IoT and FinTech have to be blended to create new business models thereby leading to better business environments. Numerous corporations are feigned due to their financial-level investments. Now, an expanding number of companies use IoT and generate new enumerated merits. The direct supervisors of current money-related organizations (banks) are appalled by the means of budgetary reform. To create a compelling business idea and action, FinTech and IoT are merged based on every business's unique needs [58].

6.2.24.5 *Employee Training*

In banking, high-quality service is given the most preference to deliver complete client satisfaction. By many technical schemes, banking firms regulate their agents/employees to adhere to their vital objectives. The difficulty here is that there are no perfect metrics or automatic and intelligent methods to assess the particular employee's performance and behavior. After the training period, the company can keep track of the employee's performance by keeping scores. In Reference [59], an effective tool is developed to assess a particular employee's performance in a particular bank. The tool's design is based on IoT by implementing the appropriate evaluation metrics. Additionally, the aim was to determine the important information from an employee's profile and working style so that the supervisor can set achievable and clear goals for the employee. An evaluation model is developed that takes employee performance data and performs data analysis of all the employers and employees to determine their performance for effective decision making. At the end of testing, the experimental outcomes revealed that the IoT-based tool produces exceptionally better performance than the currently existing tools applied in the employee training process.

6.2.24.6 *Advantages of IoT in Banking*

a) A banking firm's inventory can be tracked and managed remotely when IoT-related methods are applied.
b) The use of IoT ensures that certain work in the banking sector can be completed remotely and hence avoids unnecessary commute.
c) IoT-based banking methods give customers satisfaction and also protection for their funds.
d) Data collection is made more straightforward and manageable.
e) Since different branches of a particular bank can be connected, a real-time data transferring system is achieved.

6.2.24.7 *Disadvantages of IoT in Banking*

a) The use of different versions of devices/things might raise compatibility issues that can make the entire implementation ineffective.
b) When a device failure occurs, the data received is compromised which results in incorrect analysis and decisions.
c) In some rare circumstances, the security of the IoT system can be jeopardized putting sensitive data at risk.
d) One specific disadvantage is the loss of employment in the banking sector when certain operations become automated.

6.2.25 IoT in Government Sectors

The IoT-enabled powerful abilities in real-time environment sensing and reacting may encourage digital metamorphosis thereby helping us to unlock the true capacity of digital/smart government. In a data-driven society, the government has to deliver policies and assistance of public concern and public benefit. Nevertheless, we can see that there exists an indispensable requirement for IoT analysis on government policies and use. New frameworks need to be developed to assist the governing party to help to make decisions that favor the citizens. The IoT is used in many major domains in multiple world governments. The outcomes of its use show some proof that it helps the leaders make better decisions and make better plans. Many governments have started their research in the field of IoT by funding their scientists to accommodate the growing need for evolution. National IoT policies are still urgently needed, though, to encourage systemic IoT use across all application domains [60].

6.2.26 Use Cases

6.2.26.1 Public Healthcare

Governments treat health as one of the sustainable developing areas in their countries all around the world. The IoT has a lot of uses in this particular sector which is not being studied up to its potential. The purpose of governments now shifting towards investing in technology is to prioritize the use of IoT to its fullest in the healthcare sector to elevate the life status of their citizens. In Reference [61], a study is implemented to carry out descriptive research according to statistics compilation. According to the examination methodology (FAHP), a single cross-sectional review is conducted. Right after the process of data collection, various methods of analysis were carried out to determine the priority of IoT usage. From these results, the two principles: "economic prosperity" and "quality of life" gained the most priority for using IoT in the public healthcare sector. Besides, the other top preferences for IoT in healthcare were UV radiation, fall detection, and dental health.

6.2.26.2 Public Transportation

Governments that want to incorporate smart cities as a part of their mission would use IoT. With the inception of IoT within smart cities, the ultimate intention of smart cities is simple to accomplish. A key feature in the concept of smart cities is public transportation. Any designed smart city must have an Intelligent government transportation system. With the shortage of relevant information systems in the transport system, commuters are encountering immense difficulties in overpopulated cities. Commuters are made to wait for an extended period of time because they are not aware of the status of an upcoming bus or train to reach their respective destination. Here, with

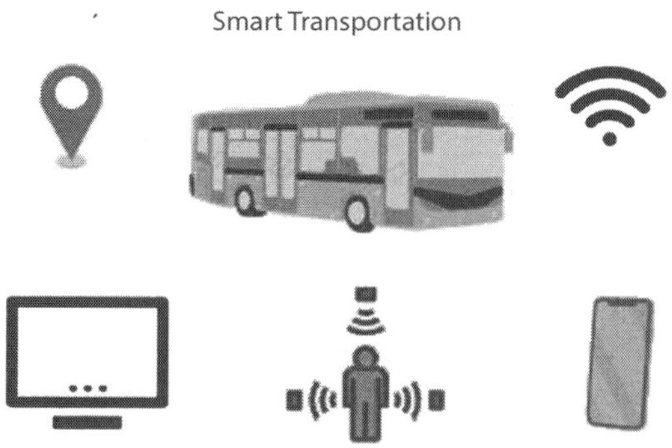

Figure 6.10 Features in a smart transport system.

the use of technology in the transportation system a commuter may travel with ease in their city of residence. A system is proposed in Reference [62], where the recent location of the buses is displayed along with the arrival and departure points. This can be notified to the riders using monitors or display boards in the bus terminus. It can also be extended as a mobile application if needed. The designed system has a Global Positioning System (GPS) system that traces the position of the vehicle and a sensor installed in the bus lets the riders know passenger count. The collected data is processed using a Raspberry Pi computer and stored in a server. Figure 6.10 lists a few features in a smart transportation system.

6.2.26.3 Disaster Management

Governments also use technologies like IoT to keep them prepared in case of any emergency situations. Disaster management focuses on reducing any significant damage from disasters (natural/manmade), ensuring prompt and fitting support to the civilians, and attaining efficient and agile restoration.

These goals need a prepared and capable rescue service post disasters. Various kinds of data about the influence of the disaster are needed for preparing an effective and instant aid operation. The IoT technology present now is considerably sophisticated and possesses the capacity to be beneficial during disasters. In Reference [63], the necessities for planning a rescue operation for natural disasters are discussed. An IoT-based solution is proposed after considering the necessities. The solution is verified utilizing the task-technology fit (TTF) method for examining the importance of the selection of IoT as a part of the plan for disaster management. Outcomes from the study confirmed the focus dimensions of TTF constructs. The paper presents meaningful benefits in the development of suitable constructs for modeling TTF with IoT for the purpose of disaster management.

6.2.26.4 *Public Safety*

Internet of things (IoT) applications have proven to instantly aggregate and interpret data about any issue, supporting the officials to identify hazardous incidents in a better way, determine how to react, and report resolutions to those affected. Sensors can be installed to record and communicate early signs of an emergency or violation. For example, a device called ShotSpotter detects gunshot noises up to 10 ft [64]. It automatically alerts the police who dispatch officers. The device can promote response time and also decrease the dependence on any bystanders to register a crime.

6.2.26.5 *Advantages of IoT in Government Sectors*

a) Any collected data can be accumulated and displayed from the database as per the citizens' request.
b) Better planning systems and decision-making systems for governments can be modeled and installed to use.
c) All the available resources are utilized to their optimal level and also energy conservation is ensured.
d) The convenience of the citizens is guaranteed.
e) The infrastructure of cities is improved along with efficient services.

6.2.26.6 *Disadvantages of IoT in Government Sectors*

a) Security of human lives is a part of IoT implementation but the reliability of this fact is questionable.
b) The data collected from individuals are susceptible to attacks hence rising privacy issues.
c) Since the coverage of these connected technologies is huge the difficulty in modeling efficient systems increases.

6.2.27 IoT in Underwater Vehicle

A typical IoT-based marine environment monitoring system employs a variety of sensors to detect and track chemical and physical parameters. An advanced IoT-based marine environment monitoring and protection system would also be able to control some items, gadgets, or machinery inside the monitored marine environment to change physical and chemical parameters to improve the marine environment. Ocean sensing and monitoring, water quality monitoring, coral reef monitoring, marine (offshore or deep-sea) fish farm monitoring, and wave and current monitoring are all examples of IoT-based marine environment monitoring applications. An ocean sensing and monitoring system is a generic marine environment monitoring system that has been utilized by oceanographic and hydrographic research vessels for a long time. Temperature, pH, turbidity, conductivity, and dissolved oxygen (DO) are among the parameters monitored by water quality monitoring systems in ocean bays, lakes, rivers, and other bodies of water. Typically, a coral reef monitoring system examines both coral reef ecosystems and the surrounding environment [65].

There are two types of underwater vehicles: manned and unmanned systems. Unmanned systems towed by a ship, for example, can be classified into numerous sub-classes. A submerged system with its own power supply and operated by an onboard computer is known as an autonomous underwater vehicle (AUV). Autonomous underwater vehicles, which are also known as remotely operated vehicles (ROVs), submersible gadgets, unmanned underwater vehicles (UUVs) or remote-controlled submarines, may follow a predefined path. The embedded systems of AUVs are divided into three categories based on their functionality: propulsion, drive, and navigation and positioning systems and systems for mapping and sampling [66].

6.2.28 IoT in Criminology and Emergency Management

The Internet has changed from something new to something needed over the last three decades. It can be difficult to predict if a hyper-connected environment will be secure for the next 30 years. Smart cities and deepfake, however, are expected to be two of the largest dangers to information security in the future years, according to researchers. Smart meters and lights, for example, already have voice assistants and applications. The IoT will be integrated with technology and the urban environment to create a smart city. For the elderly who are bewildered, there are traffic lights that alter intensity based on human activity (detected by cellphones) and automated maps [67]. Ms. Taddeo warns that cybercrime and cybersecurity in 2050 should not come at the expense of privacy. She cites cutting-edge systems that can "monitor all of your movements while you're connected—any keystroke, movement on the trackpad, or where your eyes are moving." According to her, the danger of ubiquitous data collecting and continual verification is that we'll end up with a panopticon in which monitoring networks are expanded to guard threats of all sizes [68].

6.2.28.1 Cyber Crime Attacks

Malware: a program that performs a destructive function on a specific computer or network, such as captivating over a program or operating data.

Phishing: by misleading the receiver of an email into giving sensitive information or downloading malware by clicking on a notification hyperlink, an email-born attack is launched. An intruder identifies and gets access to a site between the sender and the receiver in a "man in the middle" (MitM) attack.

Integrity: receiver of electronic messages, with the ability to alter them while they are in route. Both the sender and the recipient claim to have had face-to-face conversations. In the military, a MitM attack can be employed to overwhelm an opponent.

Trojan Horses: the Trojan Horse, named after the Trojan horse of ancient Greek history, is a sort of malware that seems to be one object on a target network, such as a standard piece of code that permits malicious programs to function while on the host network.

Denial of service (DdoS) attack: an intruder takes control of a large number of machines (potentially thousands) and allows them to activate features on a target system, such as a website, causing it to fail due to requirement overload.

Cyberterrorism: the term "cyberterrorism" refers to criminal activity that takes place through the internet. Individuals and organizations have utilized privacy to frighten others of various races, sects, nationalities, or values since the creation of the Internet.

Cyberstalking: the use of electronic means, such as the Internet, to threaten or abuse an individual or group of persons is known as cyberstalking. Cyberstalking includes attacks, sexual harassment, false charges, fraud, slander, libel, vandalism, and data theft.

6.2.28.2 Crime Harvests and the IoT [69]

Physical theft prospects may dwindle when the market becomes saturated, because few individuals will desire to buy stolen goods at this stage. Many things, notably electronic and (say) automotive goods, are, nevertheless, always being developed in order to improve usefulness and sales. Furthermore, commodities might be exported to places where the market is yet untapped. As a result, the risk of theft does not necessarily diminish as a result of market penetration. There are already indicators that IoT devices are being used for criminal purposes.

In 2016, attackers built networks of hacked internet-connected cameras and video recorders by exploiting the inadequate security of hundreds of thousands of internet-connected cameras and video recorders. These botnets were then used to perform distributed denial of service (DDoS) assaults, which entail flooding internet services with requests until they become unresponsive owing to a lack of resources. The 2016 assaults, which targeted DYN servers in the United States using Mirai malware, affected web services including Twitter and Netflix, and were notable because they were the first documented attack to exploit hacked IoT devices (as opposed to infected computers). The quantity and type of crimes involving IoT devices are still restricted.

6.2.28.3 *Digital Device Forensics [70]*

Digital forensics is worried about advanced proof. Computerized proof, otherwise called electronic proof, is handily changed. Every admittance to a record on an advanced gadget (PC, cell phone, IoT gadget) changes the document's last access time and accordingly adjusts the document in a manner that could establish proof. This is an illustration of a change that is not malignant or critical, yet could be viewed as proof altering. A significantly more major issue is the chance of malignant modifications. Methodology for electronic proof and dealing with are being used to guarantee the uprightness and ease of use of advanced proof in court. The way that computerized proof can be found on a wide range of gadgets in an assortment of configurations adds to the trouble. Personal computers (PCs), PDAs, computerized cameras, GPS gadgets, and some other gadget (thing) that stores information in advanced structure are instances of these gadgets (and it appears to be that it will ultimately incorporate everything). Every one of these gadgets may store information in an alternate manner on an alternate medium. Luckily, there is some accepted normalization, which makes things somewhat simpler. Distinctive programming utilizes various information configurations to make computerized records on gadgets.

6.2.28.4 *The Need for IoT Forensics*

There is an absence of a technique, structure and tools for IoT forensics; since IoT is all over, it very well may be hard to set up which purview a case may fall under as there will frequently be more than one included.

6.2.28.5 *Evidence Identification, Collection, and Preservation*

The initial phase in IoT forensics is to recognize the accessible sources of proof. The examiner should figure out which gadgets caught appropriate information. The inquiry that should be responded to is the manner by which the IoT connects with its environmental factors. The examiner can then figure out which of the accessible sources to utilize. Besides, data on where and how information is saved should be gotten. Prior to get-together proof, information assortment requirements (physical, restrictive principles, and lawful) ought to be thought of.

6.2.28.6 *Evidence Analysis and Correlation*

Conglomeration and investigation permit bits of proof to be gathered and used to decide somebody's personality and activities. That is something to be thankful for in case the personality of the individual being examined is known early, however it is hard to know.

6.2.28.7 *Opportunities of IoT Forensics*

IoT gives more proof sources than standard advanced forensics. Associated things give an abundance of forensically important information. All gadgets that can gather, cycle, store, or trade information is potential wellsprings of proof. Indeed, even the littlest sensor that communicates a solitary worth of estimation of a solitary actual amount could be huge. Each of the information gathered by the IoT frameworks can be utilized to take a composite picture of occasions. For instance, the area of a suspect at a particular time can be dictated by corresponding information from different IoT gadgets from different areas regularly visited by the suspect.

6.3 Conclusion

The imperative creativity today is to a great extent in view of the tipping focuses revealed in the above discussion. In order for social recognition to become broad, it will require some investment, especially from the point of view of safety and security concerns. These are obstacles that open new business doors for the company that decides to unravel them.

References

1. Andrade, R. M., Carvalho, R. M., de Araújo, I. L., Oliveira, K. M., & Maia, M. E. (2017, July). What changes from ubiquitous computing to internet of things in interaction evaluation?. In *International Conference on Distributed, Ambient, and Pervasive Interactions* (pp. 3-21). Springer, Cham.
2. Zhu, Zhen, and Ren-Gen Huang. "Study on the IoT architecture and access technology." *2017 16th International Symposium on Distributed Computing and Applications to Business, Engineering and Science (DCABES)*. IEEE, 2017.
3. Choudhary, Gaurav, and A. K. Jain. "Internet of Things: A survey on architecture, technologies, protocols and challenges." *2016 International Conference on Recent Advances and Innovations in Engineering (ICRAIE)*. IEEE, 2016.
4. Fischer, J. E., Colley, J. A., Luger, E., Golembewski, M., Costanza, E., Ramchurn, S. D., ... & Froehlich, J. E. (2016, September). New horizons for the IoT in everyday life: proactive, shared, sustainable. In *Proceedings of the 2016 ACM International Joint Conference on Pervasive and Ubiquitous Computing: Adjunct* (pp. 657-660).
5. Burd, B., Barker, L., Divitini, M., Perez, F. A. F., Russell, I., Siever, B., & Tudor, L. (2018, January). Courses, content, and tools for internet of things in computer science education. In *Proceedings of the 2017 ITiCSE Conference on Working Group Reports* (pp. 125-139).
6. Al-Emran, M., Malik, S. I., & Al-Kabi, M. N. (2020). A survey of internet of things (IoT) in education: opportunities and challenges. Toward social internet of things (SIoT): Enabling technologies, architectures and applications, 197-209.
7. Saini, M. K., & Goel, N. (2019). How smart are smart classrooms? A review of smart classroom technologies. *ACM Comput. Surv., 52*(6), 1-28.
8. Beer, Wolfgang, and Alexander Wagner. "Smart books: adding context-awareness and interaction to electronic books." *Proceedings of the 9th International Conference on Advances in Mobile Computing and Multimedia*. 2011.
9. Slavova, Y., & Mu, M. (2018, March). A comparative study of the learning outcomes and experience of VR in education. In *2018 IEEE Conference on Virtual Reality and 3D User Interfaces (VR)* (pp. 685-686). IEEE.
10. Wang, M., Callaghan, V., Bernhardt, J., White, K., & Peña-Rios, A. (2018). Augmented reality in education and training: pedagogical approaches and illustrative case studies. *J. Ambient Intell. Hum. Comput., 9*(5), 1391-1402.
11. McRae, L., Ellis, K., & Kent, M. (2018). Internet of things (IoT): education and technology. *Relatsh. between Educ. Technol. Students with Disabil. Leanne, Res,* 1-37.
12. Lenz, L., Pomp, A., Meisen, T., & Jeschke, S. (2016, March). How will the Internet of Things and big data analytics impact the education of learning-disabled students? A Concept Paper. In *2016 3rd MEC International Conference on Big Data and Smart City (ICBDSC)* (pp. 1-7). IEEE.

13. Tew, Y., Tang, T. Y., & Lee, Y. K. (2017, December). A study on enhanced educational platform with adaptive sensing devices using IoT features. In *2017 Asia-Pacific Signal and Information Processing Association Annual Summit and Conference (APSIPA ASC)* (pp. 375-379). IEEE.
14. Kuyoro, S., Osisanwo, F., & Akinsowon, O. (2015, March). Internet of things (IoT): an overview. In *Proc. of the 3th International Conference on Advances in Engineering Sciences and Applied Mathematics (ICAESAM)* (pp. 23-24).
15. Rajurkar, C., Prabaharan, S. R. S., & Muthulakshmi, S. (2017, March). IoT based water management. In *2017 International Conference on Nextgen Electronic Technologies: Silicon to Software (ICNETS2)* (pp. 255-259). IEEE.
16. Wadekar, S., Vakare, V., Prajapati, R., Yadav, S., & Yadav, V. (2016, October). Smart water management using IOT. In *2016 5th International Conference on Wireless Networks and Embedded Systems (WECON)* (pp. 1-4). IEEE.
17. Verma, P., Kumar, A., Rathod, N., Jain, P., Mallikarjun, S., Subramanian, R., ... & Sundaresan, R. (2015, October). Towards an IoT based water management system for a campus. In *2015 IEEE First International Smart Cities Conference (ISC2)* (pp. 1-6). IEEE.
18. Singh, M., & Ahmed, S. (2021). IoT based smart water management systems: A systematic review. *Mater. Today: Proc.*, *46*, 5211-5218.
19. Kamienski, C., Soininen, J. P., Taumberger, M., Dantas, R., Toscano, A., Salmon Cinotti, T., ... & Torre Neto, A. (2019). Smart water management platform: IoT-based precision irrigation for agriculture. *Sensors*, *19*(2), 276.
20. Nie, X., Fan, T., Wang, B., Li, Z., Shankar, A., & Manickam, A. (2020). Big data analytics and IoT in operation safety management in under water management. *Comput. Commun.*, 154, 188-196.
21. Dogo, E. M., Salami, A. F., Nwulu, N. I., & Aigbavboa, C. O. (2019). Blockchain and internet of things-based technologies for intelligent water management system. In *Artificial Intelligence in IoT* (pp. 129-150). Springer, Cham.
22. Khanboubi, F., Boulmakoul, A., & Tabaa, M. (2019). Impact of digital trends using IoT on banking processes. *Procedia Comput. Sci.*, *151*, 77-84.
23. Vijay kumar, Sadanand. (2019). IoT Applications in Finance and Banking. 10.6084/m9.doi.one.IJRAR19K3720.
24. Sugapriya, M. (2020). Heist Tracking and Prevention in ATM Utilizing IOT and Blynk Server.
25. Boumlik, A., & Bahaj, M. (2017, April). Big data and iot: A prime opportunity for banking industry. In *International Conference on Advanced Information Technology, Services and Systems* (pp. 396-407). Springer, Cham.
26. Suseendran, G., Chandrasekaran, E., Akila, D., & Kumar, A. S. (2020). Banking and FinTech (Financial Technology) Embraced with IoT Device. In *Data Management, Analytics and Innovation* (pp. 197-211). Springer, Singapore.

27. Alti, A., & Almuhirat, A. (2021). An Advanced IoT-Based Tool for Effective Employee Performance Evaluation in the Banking Sector. *Ingénierie des Systèmes d Inf., 26*(1), 103-108.
28. Chatfield, A. T., & Reddick, C. G. (2019). A framework for Internet of Things-enabled smart government: A case of IoT cybersecurity policies and use cases in US federal government. *Gov. Inf. Q.*, 36(2), 346-357.
29. M. Kumar, A. F. Minai, A. A. Khan and S. Kumar, "IoT based Energy Management System for Smart Grid," *2020 International Conference on Advances in Computing, Communication & Materials (ICACCM)*, 2020, pp. 121-125, doi: 10.1109/ICACCM50413.2020.9213061.
30. R. Morello, C. De Capua, G. Fulco and S. C. Mukhopadhyay, "A Smart Power Meter to Monitor Energy Flow in Smart Grids: The Role of Advanced Sensing and IoT in the Electric Grid of the Future," in *IEEE Sens. J.*, vol. 17, no. 23, pp. 7828-7837, 1 Dec.1, 2017, doi: 10.1109/JSEN.2017.2760014.
31. S. M. A. A. Abir, A. Anwar, J. Choi and A. S. M. Kayes, "IoT-Enabled Smart Energy Grid: Applications and Challenges," in *IEEE Access*, vol. 9, pp. 50961-50981, 2021, doi: 10.1109/ACCESS.2021.3067331.
32. C. L. Izidoro, M. A. O. Rocha, J. D. Spacek, J. M. Neto, A. D. Spacek and O. H. Ando Junior, "Development of an Industrial IoT Based Monitoring System for Voltage Regulators," in *IEEE Lat. Am. Trans.*, vol. 19, no. 8, pp. 1410-1416, Aug. 2021, doi: 10.1109/TLA.2021.9475872.
33. W. Ali, I. U. Din, A. Almogren, M. Guizani and M. Zuair, "A Lightweight Privacy-Aware IoT-Based Metering Scheme for Smart Industrial Ecosystems," in *IEEE Trans. Ind. Inf.*, vol. 17, no. 9, pp. 6134-6143, Sept. 2021, doi: 10.1109/TII.2020.2984366
34. Burd, B., Barker, L., Divitini, M., Perez, F. A. F., Russell, I., Siever, B., & Tudor, L. (2018, January). Courses, content, and tools for internet of things in computer science education. In *Proceedings of the 2017 ITiCSE Conference on Working Group Reports* (pp. 125-139).
35. Al-Emran, M., Malik, S. I., & Al-Kabi, M. N. (2020). A survey of internet of things (IoT) in education: opportunities and challenges. Toward social internet of things (SIoT): Enabling technologies, architectures and applications, 197-209.
36. Saini, M. K., & Goel, N. (2019). How smart are smart classrooms? A review of smart classroom technologies. *ACM Comput. Surv., 52*(6), 1-28.
37. Beer, Wolfgang, and Alexander Wagner. "Smart books: adding context-awareness and interaction to electronic books." *Proceedings of the 9th International Conference on Advances in Mobile Computing and Multimedia*. 2011.
38. Slavova, Y., & Mu, M. (2018, March). A comparative study of the learning outcomes and experience of VR in education. In *2018 IEEE Conference on Virtual Reality and 3D User Interfaces (VR)* (pp. 685-686). IEEE.
39. Wang, M., Callaghan, V., Bernhardt, J., White, K., & Peña-Rios, A. (2018). Augmented reality in education and training: pedagogical approaches and illustrative case studies. *J. Ambient Intell. Hum. Comput., 9*(5), 1391-1402.

40. McRae, L., Ellis, K., & Kent, M. (2018). Internet of things (IoT): education and technology. *Relatsh. between Educ. Technol. Students with Disabil. Leanne, Res*, 1-37.
41. Lenz, L., Pomp, A., Meisen, T., & Jeschke, S. (2016, March). How will the Internet of Things and big data analytics impact the education of learning-disabled students? A Concept Paper. In *2016 3rd MEC International Conference on Big Data and Smart City (ICBDSC)* (pp. 1-7). IEEE.
42. Tew, Y., Tang, T. Y., & Lee, Y. K. (2017, December). A study on enhanced educational platform with adaptive sensing devices using IoT features. In *2017 Asia-Pacific Signal and Information Processing Association Annual Summit and Conference (APSIPA ASC)* (pp. 375-379). IEEE.
43. Kuyoro, S., Osisanwo, F., & Akinsowon, O. (2015, March). Internet of things (IoT): an overview. In *Proc. of the 3th International Conference on Advances in Engineering Sciences and Applied Mathematics (ICAESAM)* (pp. 23-24).
44. Rajurkar, C., Prabaharan, S. R. S., & Muthulakshmi, S. (2017, March). IoT based water management. In *2017 International Conference on Nextgen Electronic Technologies: Silicon to Software (ICNETS2)* (pp. 255-259). IEEE.
45. Wadekar, S., Vakare, V., Prajapati, R., Yadav, S., & Yadav, V. (2016, October). Smart water management using IOT. In *2016 5th International Conference on Wireless Networks and Embedded Systems (WECON)* (pp. 1-4). IEEE.
46. Verma, P., Kumar, A., Rathod, N., Jain, P., Mallikarjun, S., Subramanian, R., ... & Sundaresan, R. (2015, October). Towards an IoT based water management system for a campus. In *2015 IEEE First International Smart Cities Conference (ISC2)* (pp. 1-6). IEEE.
47. Singh, M., & Ahmed, S. (2021). IoT based smart water management systems: A systematic review. *Mater. Today: Proc.*, *46*, 5211-5218.
48. Kamienski, C., Soininen, J. P., Taumberger, M., Dantas, R., Toscano, A., Salmon Cinotti, T., ... & Torre Neto, A. (2019). Smart water management platform: IoT-based precision irrigation for agriculture. *Sensors*, *19*(2), 276.
49. Nie, X., Fan, T., Wang, B., Li, Z., Shankar, A., & Manickam, A. (2020). Big data analytics and IoT in operation safety management in under water management. *Comput. Commun.*, 154, 188-196.
50. Dogo, E. M., Salami, A. F., Nwulu, N. I., & Aigbavboa, C. O. (2019). Blockchain and internet of things-based technologies for intelligent water management system. In *Artificial Intelligence in IoT* (pp. 129-150). Springer, Cham.
51. Rajendran, T. H. Prasath, S. Revathi and K. Rajesh, "Basic Food Safety Monitoring And Enhancement in Coffee Industry Using IOT," *2021 International Conference on Advance Computing and Innovative Technologies in Engineering (ICACITE)*, 2021, pp. 145-148, doi: 10.1109/ICACITE51222.2021.9404609.
52. Manjunath and P. G. Shah, "IoT Based Food Wastage Management System," *2019 Third International Conference on I-SMAC (IoT in Social,*

Mobile, Analytics and Cloud) (I- SMAC), 2019, pp. 93-96, doi: 10.1109/I-SMAC47947.2019.9032553.
53. N. Kuntagod, S. Podder, S. S. S. Abbabathula, V. Subramanian, G. Mathew and S. K. Mani, "A Framework for Enabling Safe and Resilient Food Factories for Public Feeding Programs," *2021 IEEE/ACM 3rd International Workshop on Software Engineering for Healthcare (SEH)*, 2021, pp. 1-4, doi: 10.1109/SEH52539.2021.00008.
54. Khanboubi, F., Boulmakoul, A., & Tabaa, M. (2019). Impact of digital trends using IoT on banking processes. *Procedia Comput. Sci.*, *151*, 77-84.
55. Vijay kumar, Sadanand. (2019). IoT Applications in Finance and Banking. 10.6084/m9.doi.one.IJRAR19K3720.
56. Sugapriya, M. (2020). Heist Tracking and Prevention in ATM Utilizing IOT and Blynk Server.
57. Boumlik, A., & Bahaj, M. (2017, April). Big data and iot: A prime opportunity for banking industry. In *International Conference on Advanced Information Technology, Services and Systems* (pp. 396-407). Springer, Cham.
58. Suseendran, G., Chandrasekaran, E., Akila, D., & Kumar, A. S. (2020). Banking and FinTech (Financial Technology) Embraced with IoT Device. In *Data Management, Analytics and Innovation* (pp. 197-211). Springer, Singapore.
59. Alti, A., & Almuhirat, A. (2021). An Advanced IoT-Based Tool for Effective Employee Performance Evaluation in the Banking Sector. *Ingénierie des Systèmes d Inf.*, *26*(1), 103-108.
60. Chatfield, A. T., & Reddick, C. G. (2019). A framework for Internet of Things-enabled smart government: A case of IoT cybersecurity policies and use cases in US federal government. Government Information Quarterly, 36(2), 346-357.
61. Alansari, Z., Soomro, S., Belgaum, M. R., & Shamshirband, S. (2018). The rise of Internet of Things (IoT) in big healthcare data: review and open research issues. *Progress in Advanced Computing and Intelligent Engineering*, 675-685.
62. Vakula, D., & Raviteja, B. (2017, December). Smart public transport for smart cities. In *2017 International Conference on Intelligent Sustainable Systems (ICISS)* (pp. 805-810). IEEE.
63. Sinha, A., Kumar, P., Rana, N. P., Islam, R., & Dwivedi, Y. K. (2019). Impact of internet of things (IoT) in disaster management: a task-technology fit perspective. *Ann. Oper. Res.*, *283*(1), 759-794.
64. (online)https://www2.deloitte.com/us/en/insights/focus/internet-of-things/iot-in-government.html
65. Yanjun Shi, Xueyan Sun and Weiming Shen,Guobao Xu, "Internet of Things in Marine Environment Monitoring: A Review", *Sensors*, 2019. doi:10.3390/s19071711
66. Pedro José Bernalte Sánchez, Mayorkinos Papaelias, Fausto Pedro García Márquez, "Autonomous Underwater Vehicles: Instrumentations and Measurements", *IEEE Instrum. Meas. Mag.*· April 2020.

67. Afrah Almansoori, Cornelius Ncube, and Said A. Salloum, "Internet of Things Impact on the Future of Cyber Crime in 2050", *AICV* 2021, pp. 643–655.
68. Dixon, W., Samartsev, D, 3 technologies that could define the next decade of cybersecurity. World Economic Forum (2019).
69. John M. Blythe, Shane D. Johnson, "A systematic review of crime facilitated by the consumer Internet of Things", *Secur. J.* (2021) 34:97–125. https://doi.org/10.1057/s41284-019-00211-8]
70. asa Mrdovic, "Security of Ubiquitous Computing Systems", IoT Forensics, S, First Online: 15 January 2021, pp 215-229.

7

An Intelligent Smart Grid Switching System for an Efficient Load Balancing Through Machine Learning Models

Aditya Sundarajan[1], Jaideepnath Anand S.[1], Ruthul Jindal S.[1] and Maheswari R.[2]*

[1]School of Electronics Engineering, Vellore Institute of Technology, Chennai, India
[2]Centre for Smart Grid Technologies, SCOPE, Vellore Institute of Technology, Chennai, India

Abstract

In modern times, electricity has become a necessary resource for all needs and purposes ranging from mundane tasks to high-end industrial works. Due to the fluctuating electricity demand, the existing electric grid network has questionable efficiency in terms of power generation and distribution. To effectively manage and distribute power to all consumer bases, it needs to efficiently gauge power demands for different sectors and be able to cater to them. This gives the need to develop a system that can predict future power consumption which helps necessary departments to use said data and come up with an efficient distribution plan or make changes to the current distribution network to cater to these needs. The model presented here is based on long short-term memory (LSTM) and recurrent neural network (RNN) machine learning neural networks. Recurrent neural network (RNN) helps with sequence data modelling. The functioning of a feed-forward neural network can be an analogy to a human control system called the brain. The system used time-series modelling to use past consumption data to map out usage patterns and provide a reliable power consumption forecast which will be detrimental in planning for meeting future power demands. The output is finally compared with actual data from past consumption to assess its accuracy in forecasting. The cutting-edge electrical utility sector is rapidly developing into a finely tuned fusion of best-in-class digital innovations with the actual basis, referred to as a cyber physical system (CPS). These CPSs, which are completely

**Corresponding author*: maheswari.r@vit.ac.in

O.V. Gnana Swathika, K. Karthikeyan and Sanjeevikumar Padmanaban (eds.) Smart Grids as Cyber Physical Systems Volume 1: Artificial Intelligence, Cybersecurity, and Clean Energy for Next Generation Smart Grids, (111–122) © 2024 Scrivener Publishing LLC

dependent on a vast intelligent computing algorithm-based communication network, are sensitive to various technological and socioeconomic factors that could compromise their stability and are also susceptible to digital attacks. This model produces high-confidence power consumption patterns and with quality datasets as inputs, the system can produce an extremely reliable power consumption forecast which makes the planning process for improving the power supply chain more effective. The future scope for this model would be to automatically feed the output forecast from this model into an AI-based grid monitoring system that can distribute power to necessary outlets based on built-in parameters. This eliminates human intervention, thereby reducing response times in power distribution load balancing and with access to accurate and real-time data, seamless power distribution without disturbances can be achieved.

Keywords: Smart grid, microgrid, recurrent neural network, long short-term memory, time series forecasting

7.1 Introduction

Power stations are primary generators of electricity. It is generally known that these stations have several technical difficulties and that their power-producing costs are considerable. With the current trend in electricity usage, it is expected that there will be a shortfall in electricity generation. In recent times, the electric grids are supplying the necessary power required for all day-to-day needs in promising ways like domestic load, industrial load, or commercial load [1]. Due to the huge electricity demand, the existing electric grid holds a lot of trouble both in terms of generation and distribution. Microgrids are self-generating small clusters of energy systems that could supply the electricity in distinct topographical environments such as education institutions/universities, private/public health centers, shopping malls, and so on [1, 5]. In this research, the main eye of the exploration was focused upon the possibility of integrating microgrids into smart grids by predicting the values of outsourced smart grids in the future using appropriate tools and algorithms.

7.2 Backbone of Work

RNN:

Recurrent neural networks (RNNs) help with sequence data modelling. The functioning of a feed forward neural network can be an analogy to a human control system called the brain. To simply put, RNNs are widely used and trained to have an accurate prediction of a variety of outcomes compared to the other existing neural network algorithms. The RNN's primary working structure involves protecting the weights of one layer, and then driving it back as an input to predict the other layer's output.

LSTM:

Long short-term memory networks (LSTMs) are an extension of RNNs, which were introduced primarily to solve situations where RNNs fail. An RNN is a network that uses the current input while taking into account feedback from the previous output and temporarily storing it in memory (short-term memory). The areas of voice processing, non-Markov control, and music production are the most popular of many applications. Long short-term memory networks (LSTMs) do not require maintaining a limited number of prior states, as a hidden Markov model (HMM) does [2, 6]. There is access to a wide range of characteristics from LSTMs, including linear regression (LR) and Input/Output (I/O) stages. There is therefore no need for accurate adjustments. Like Backpropagation through time (BPTT), the time complexity of changing each unit is reduced to O(1) using LSTM.

7.3 Theory Behind Smart Grids and Integration in the Field

Smart grid data is used for many power consumption prediction tasks. Data is treated as sequential data. Autoregressive integrated moving average (ARIMA), support vector machine (SVM), linear regression, and artificial neural networks (ANN) are most commonly used to test and predict power consumption. A model with minimal or no feature selection requirements could greatly enhance prediction accuracy [3]. As a result, deep learning models are better at making accurate predictions using time series data. Deep learning neural networks (DNNs) that predict the future based on past inputs have proven to be capable of producing more accurate results when given sequential data. However, these learning models have performed poorly in terms of mid-term and long-term predictions [4].

The best-known deep neural networks with feedback loops from previous inputs are RNNs and LSTMs. Long short-term memory and recurrent neural networks have been shown to outperform other feedforward DNN techniques without feedback loops. Before moving on to computationally intensive and complex models, there is a complex version of deep learning models with feedback loops [5, 6]. It is useful to test simple models first before testing more complex deep learning models.

7.4 Phases of Data Through the Smart Grids

Data reformatting: after the data was retrieved and loaded into the compiler from the dataset, reformatted the date and time columns of the data, and sorting it out based on that variable to have an easier, more efficient, and less time complex-based working algorithm to train the model.

Data preprocessing: sometimes, data has a tendency to be incomplete. So that there are slim chances of arousal of problems during training of the model, data cleaning tries to eliminate noise, fill in missing or null values, and resolve inconsistencies in the data. The essential time series preprocessing steps in this section will thereby be outlined. This phase has three subphases: data cleaning, data transformation, and data reduction.

7.4.1 Data Cleaning

Finding missing data in the dataset and filling it up if needed in case of noisy data such as "####" or any form of abrupt data in the score, regression or clustering algorithms can be used to smoothen the same.

7.4.2 Data Transformation

This subphase involves using the dataset's attributes to be constructed or renewed from another set of attributes and normalizing it according to the model.

7.4.3 Data Reduction

This step involves reducing unnecessary amounts of data for training the model in a quicker and an efficient way that would thereby also lead to predicting the model accuracy degrading and lessening of the confidence intervals. Figure 7.1 shows spread *vs.* density across grids in the dataset (American Electric Power (AEP) dataset in Megawatts (MW)) after preprocessing phase to be visualized.

And Figure 7.2 shows a more detailed, scaled and magnified energy distribution of Figure 7.1 to have a better and scaled visualization of the trends.

Data Analysis and Visualization
This step involves basic data analysis for better apprehension of the user and how the model tends to work. The analyzed data is then fed into the visualization console segment for the outputs such as energy consumed per year and energy vs time as shown in the source code. Figure 7.3 shows

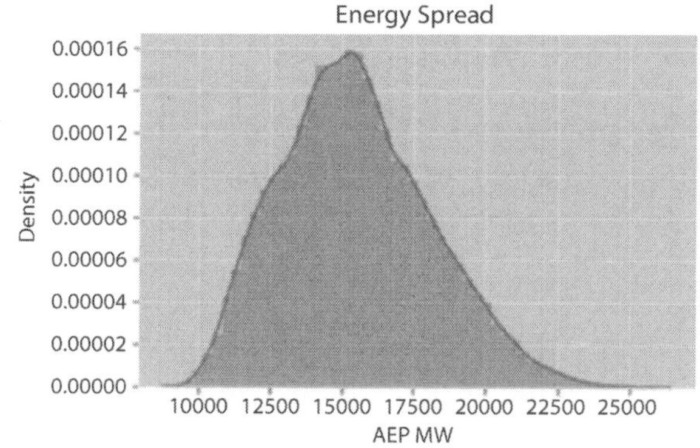

Figure 7.1 Energy distribution across grids.

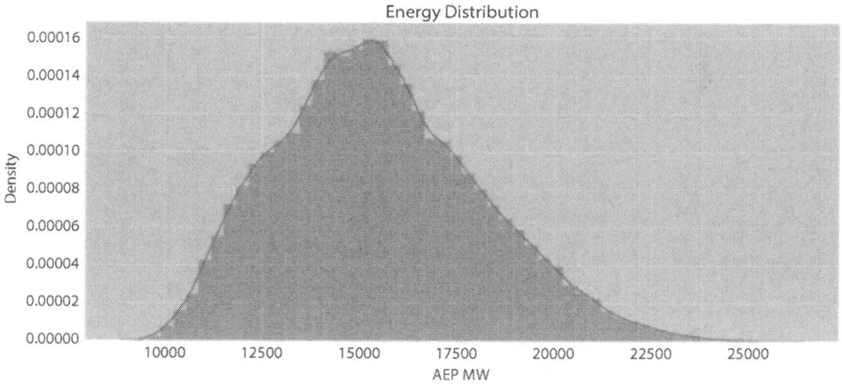

Figure 7.2 Scaled energy distribution graph.

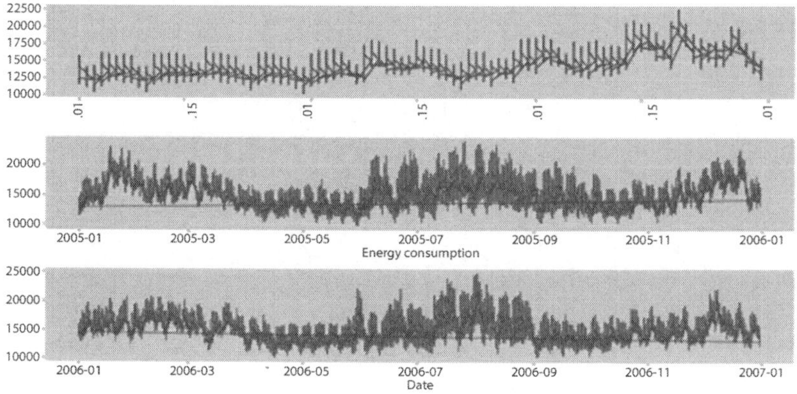

Figure 7.3 Energy consumption per hour.

the consumed energy levels that were analyzed using specific and precise data analysis methods from the dataset input.

Model Building (LSTM – RNN)
LSTM, which stands for long short-term memory model, is a special type of RNN that conserves long-term memory by using memory units that can update in the before-hidden state. The internal structure of the LSTM model includes four major units: input gate, output gate, forget gate, and cell status. The three gates are introduced to control the maintenance and the update of information in cell status. The working of LSTMs can be described with the following equations [7.1–7.7]:

$$ft=\sigma(wf[ht\text{-}1,Xt]+bf) \tag{7.1}$$

$$it=\sigma(wi[ht\text{-}1,Xt]+bi) \tag{7.2}$$

$$ot=\sigma(wo[ht\text{-}1,Xt]+bo) \tag{7.3}$$

$$at=\tanh(wa[ht\text{-}1,Xt]+ba) \tag{7.4}$$

$$ct=ft{*}ct\text{-}1+it{*}at \tag{7.5}$$

$$ht=ot{*}\tanh(ct) \tag{7.6}$$

where sigmoid is the activation function of the neural network.

Parameters of the LSTM Model Assigned
For a model to work it, has to be assigned certain trainable weights and parameters. The number of parameters and weights used here is:

$$N = 4np + 4pp + 4p + mp + m \tag{7.7}$$

where n represents number of inputs, m represents number of outputs required, and p represents the number of hidden LSTM cells.

Testing and Training the Model
The last and most impo rtant phase of the research is to train the model after testing it with dummy values. The training dataset accommodates a list of examples used during the learning curve which is thereby used to change and set the network parameters, and to accurately predict the

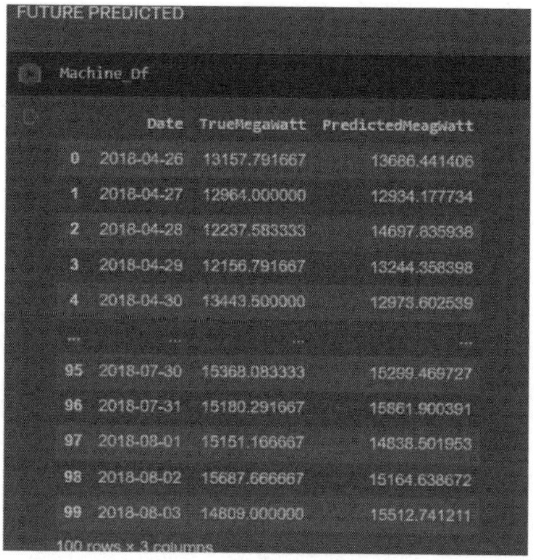
FUTURE PREDICTED

Machine_Df

	Date	TrueMegaWatt	PredictedMeagWatt
0	2018-04-26	13157.791667	13686.441406
1	2018-04-27	12964.000000	12934.177734
2	2018-04-28	12237.583333	14697.835938
3	2018-04-29	12156.791667	13244.358398
4	2018-04-30	13443.500000	12973.602539
...	...	...	...
95	2018-07-30	15368.083333	15299.469727
96	2018-07-31	15180.291667	15861.900391
97	2018-08-01	15151.166667	14838.501953
98	2018-08-02	15687.666667	15164.638672
99	2018-08-03	14809.000000	15512.741211

100 rows × 3 columns

Figure 7.4 Comparison between actual energy readings and predicted energy readings.

efficient combinations of variables which will provide a highly accurate score. Finally, the test set is deployed to evaluate competing models. There can be add-ons of confidence metrics to evaluate the trained model for certain number of trusted inputs. The first few sets of values known as 'head-values' in the predicted algorithm output is shown in Figure 7.4.

7.5 Flowchart of the Proposed Smart Grid System

Figure 7.5 shows the program progress throughout the smart grid system in flowchart form. Feature scaling involves normalizing the range of features. In the data sampling block, null values are removed to improve data accuracy, and data and time are reformatted. The sampled data is further split into testing and training datasets which is later integrated into the model [7, 8]. After the machine learning (ML) model is constructed , the model accuracy is validated and if the score is within an expected range, the program produces the final output data with actual power consumption and predicted power consumption data in the same graph.

7.6 Work Done

To predict future power consumption numbers, a dataset of past power consumption data is taken to visualize the growing or decreasing power usage

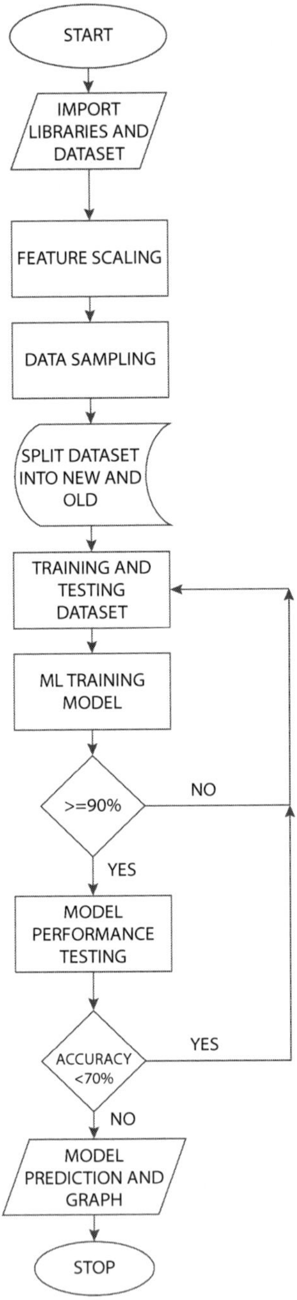

Figure 7.5 Flowchart of program.

and to train an ML model that can forecast possible power consumption data with a high confidence level [9]. First, import all required libraries and import the dataset. Then explore the dataset and visualize power usage patterns for the past 15 years. Based on this data, whether power consumption will increase or decrease is predicted. Next, visualize power usage distribution and a scaled version of the data with respect to time. After resampling the dataset into the old and new datasets, split the dataset into training and testing datasets, respectively. The training dataset goes through feature scaling. It is a method used to normalize the features or attributes that remains idle within a fixed bracket of the data. Now the ML model takes in the data and predicts future values. For this model, LSTM and RNN are used. Recurrent neural networks (RNNs) are Deep Learning (DL) models used in problems involving sequential data. Long term-short term memory (LSTM) is a model that tends to amplify the record of memory of the RNNs. These RNNs hold memory for a short time and in that they approve previously determined data to be deployed in the current neural networks (NNs). After the training phase of the model, verification of the confidence level of the model and visualization of the final data is carried out [10]. Additionally, superimposed actual and predicted power consumption data for the next two months is also visualized.

7.7 Working with Dataset—Dataset Description

The dataset consists of 100 rows and four columns for test set and 4995 rows and one column for training set. This dataset describes the AEP energy consumed across a city on an hourly basis. This dataset is classified as numerical dataset and also a multivariate dataset. The column '1' is left blank intentionally as serial count [11]. Column '2' consists of the date of each consumption recorded. Column '3' reads the actual Megawatt values of power consumed, and column '4' shows the predicted power values in Megawatt [12].

7.8 Tools Used for Implementing the Proposed Algorithm

Google Colab:

Google Colab is a completely free, cloud-based environment for coding notebooks in Python. Colab supports many well-known machine-learning libraries and can be easily loaded onto a laptop.

One can use Google Colab as a programmer to perform the following tasks.

- Build and run Python, R, Julia codes.
- Math equations and functions are easily evaluated and tracked.
- Create, share, and upload notebooks among your teammates and colleagues.
- Python code can be accessed from any device through Google drive.
- Publishing and importing of notebooks from GitHub repositories is made easier.
- Add external datasets such as those from Kaggle and add other custom datasets, too.
- Incorporate various libraries such as Keras, TensorFlow, PyTorch, and OpenCV.
- Free cloud and graphics processing unit (GPU) service which helps clients to run the code in the notebook.

7.9 Results

Figure 7.6 shows the final outcome of the research where future predicted nominals and values of the grids that is being integrated is interpreted, calculated, and visualized. It shows an overlay of actual data (green) with ML model predicted data (red). It can be seen that the predicted model produces high-confidence data with some variances compared to actual consumption data.

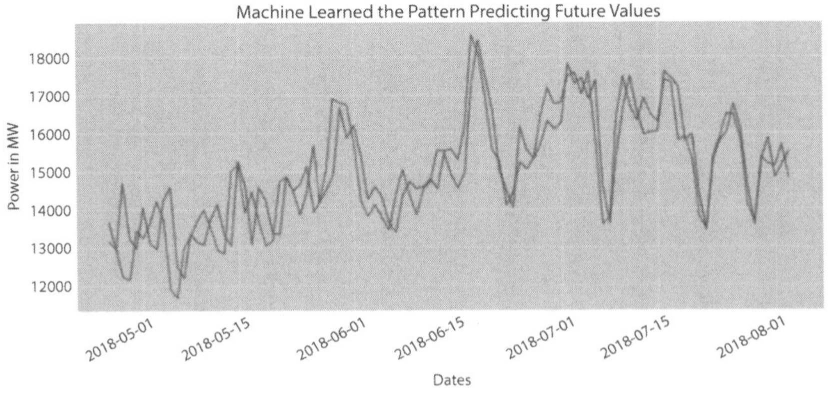

Figure 7.6 Pattern predicting future values.

7.10 Inference of the Solution

From the above research, it can be concluded and inferred that the team successfully analyzed, visualized, and predicted future energy consumptions by smart grids using LSTM accurately for a given duration using RNN.

7.11 Conclusion and Future Work

In conclusion, according to the problem statement of integrating micro-grids and smart grids for efficiency purposes, a successful and functional ML model was designed that would predict the future energy consumption by smart grid using the LSTM architecture of machine learning and RNN architecture, too. A more "in-develop" state of approach would be to have higher GPU models that process microgrid and smart grid data separately and ultimately connect them based on their efficiency.

References

1. Hartono, B. S., Yan Budiyanto, and Rudy Setiabudy. "Review of microgrid technology." 2013 international conference on QiR. *IEEE*, 2013.
2. Anestis. G. Anastasiadis, Antonis. G. Tsikalakis, dan Nikos. D.Hatziargyriou, "Environmental Benefits from DG Operation When Network Losses are Taken into Account", Distres Conference 11-12 December 2009 Nicosia Cyprus.
3. Singh, Piyush P., and Binoy Krishna Roy. "Comparative performances of synchronisation between different classes of chaotic systems using three control techniques." *Annu. Rev. Control* 45 (2018): 152-165.
4. Ph. Degobert, S. Kreuawan and X. Guillaud "Micro-grid powered by photovoltaic and microturbine", *ICREPQ*'06, 2006.
5. R. Lasseter, A. Akhil, C. Marnay, J. Stephens, J. Dagle, R. Guttromson, A. Sakis Meliopoulous, R. Yinger, and J. Eto, "Integration of Distributed Energy Resources – The MicroGrid Concept". CERTS MicroGrid Review Feb 2002.
6. Nikos Hatziargyriou, Hiroshi Asano, Reza Iravani, dan Chris Marnay. "Microgrid, An Overview of Ongoing Research, Development, and Demonstration Projects", *IEEE Power Energy Mag.*, July/august 2007.
7. A Comprehensive Review of Recent Advances in Smart Grids: A Sustainable Future with Renewable Energy ResourcesI brahim Alotaibi 1, Mohammed A. Abido 1,2, Muhammad Khalid 1,2 and Andrey V. Savkin 3, 2020.

8. A review of the development of Smart Grid technologies, Maria Lorena Tuballa a,n, Michael Lochinvar Abundo b, a College of Engineering, University of San Carlos, Talamban, Cebu City 6000, Philippines b Energy Research Institute @ ERI@N, Clean Tech One, 1 Cleantech Loop, #06-09, 63714 Singapore, 2016.
9. Smart Grid – The New and Improved Power Grid: A Survey Xi Fang, Student Member, IEEE, Satyajayant Misra, Member, IEEE, Guoliang Xue, Fellow, IEEE, and Dejun Yang, Student Member, IEEE, 2018.
10. Kumar, Abhishek, and Muhammad Zafar Ullah Khan. "Microgrids technology: A review paper." *Gyancity Journal of Electronics & Computer Science (GJECS)* 3.1 (2018): 11-20.
11. Microgrids: A review of technologies, key drivers, and outstanding issues Author links open overlay panel AdamHirsch[a] , YaelParag[a], JosepGuerrero[b], July 2018.
12. Yoldaş, Yeliz, *et al.* "Enhancing smart grid with microgrids: Challenges and opportunities." *Renewable Sustainable Energy Rev.*, 72 (2017): 205-214.

8

Hybrid Energy Storage System for Battery-Powered Electric Vehicles

G. Jegadeeswari* and D. Lakshmi

Dept. of Electrical and Electronics Engineering, AMET Deemed to be University, Chennai, Tamil Nadu, India

Abstract

Hybrid energy technologies are now widely employed for power production in many transport networks in order to minimize emissions of greenhouse gases and global climate change. Considering the distance from the shore power grid, there is a necessity for a stand-alone hybrid generation system in maritime transportation to fulfill power demand and fulfill the environmental constraints of particular ports. Due to these benefits, battery-powered electric cars are more environmentally friendly than conventional automobiles. Hybrid energy, which includes solar, wind, and hydro turbine, has been used to charge the battery. In most cases, the movement of the vehicle is not utilized. This chapter makes use of it to charge the battery by installing the hydro generator under the ship/boat. It is important to note that the used renewable energy system (RES) is made up of solar panels and a wind turbine. The single phase induction motor (SPIM) was employed as the vehicle's primary propulsion unit. The induction motor's speed is controlled by an Arduino microcontroller. The suggested system's hardware was developed, and the performances for speed control of an induction motor without and with a controller are contrasted.

Keywords: Hybrid shipboard power system, single phase induction motor, Arduino, batteries, maritime, renewable energy system, marine grid system

**Corresponding author*: jegadeeswari.dharan@gmail.com

O.V. Gnana Swathika, K. Karthikeyan and Sanjeevikumar Padmanaban (eds.) Smart Grids as Cyber Physical Systems Volume 1: Artificial Intelligence, Cybersecurity, and Clean Energy for Next Generation Smart Grids, (123–136) © 2024 Scrivener Publishing LLC

8.1 Introduction

Nowadays, induction motors are widely used in the industrial applications since it is less costly and maintenance-free than any other machines. Moreover, the performance of speed is not up to the mark, it is complex in structure, and the outcome is non-linear in nature. By varying the supply frequency, the speed may vary in this type of machines. Since there is no easy way to alter the system frequency, this drawback is overcome by the entry of modern power electronics.

On the other hand, electric traction uses either alternating current (AC) or direct current (DC) power. Due to their variable frequency and voltage, inverter is the only feasible mode of possibility. As previously stated, the three-phase converter used in the traction is driven by AC or DC motor.

Alternating current can be linked by the inverter through transformer and the rectifier section. To enhance the speed regulation of induction motor, a controller is a must, to enhance the better speed regulation in the wide speed.

In this proposed system, an neural network predictive (NNP) controller is developed with respect to regulates and the speed control of the induction motor for adjusting its reference speed in a great manner. It will be clearly demonstrated that this NNP controller is used to stabilize the controlled system and reach its tracking aim with very high accuracy.

Since the 1800s, the electric motor unpredictably preceded the internal combustion system which exists in electric cars. This will only disappear in the face of inexpensive and broadly available fuel. In the olden days, the starting system of the combustion system is done by the electric motor and battery system. However, many electric vehicle offshoots such as elevators, cable cars, submariners, and subways, have lived on to the present generation.

Aside from serving as the safest mode of transportation (none or extremely low emissions when considering the pollution of power plants, which is to be used for power generation), electric vehicles are also most cost effective and maintenance-free solution. Over 75% of the battery energy is utilized by the electric vehicles, which is more efficient. Around 20% is utilized by the combustion vehicle, which is performed by productive labor. Compared to fuel cell and gasoline, electric vehicles give the best solution and enhanced performance.

Fan Zhang *et al.* (2016) proposed a literature review on electric vehicle technologies and its applications. Nowadays, clean and renewable energy has been a major concern and challenge throughout the world. Availability of vehicles gives much benefit to humans but it also produces the major pollution to the environment. Due to this, it is hard to reduce the number of automobiles in society. So, it is evidence that, electric car is the only solution to this cause and it reduces the overall pollution. Overall, electric vehicles are friendlier, cleaner, and more cost effective and energy efficient than conventional vehicles.

Energy management of hybrid electric vehicles: a research of energy optimization of fuel cell hybrid power system based on genetic algorithm was proposed by Xueqin *et al.* (2020). The suggested work's major goal was to analyze and choose the rules and parameters of algorithm approaches in order to research and explain the influence of optimization techniques and genetic algorithms. This study intended to examine and give structured regulation for optimizing control approaches and fuel cell design. Fuel cell plays an important part in hybrid power systems because of its energy use and efficiency, which extends the efficiency of the fuel cell and delivers the benefits of future energy management systems in an optimum way.

Realistic data-powered analytical energy management for hybrid electric vehicles was presented by Tang *et al.* (2020). This work mainly focused on energy saving, especially in hybrid electric cars, an analytical energy management technique, which is based on the trip route. A short-term speed predictor was used here for extreme machine learning. Cost is the major optimization factor which is incorporated in the battery-powered vehicles. After valid evidence, it clearly states that training datasets, when compared with real time data, is more accurate than using typical datasets. The prediction of speed is more educated on data and route information is clear in this system and it is more accurate. The life of the battery was also investigated. The balance between the battery ageing is more economical, and that can be achieved by altering the battery temperature and its weighting cofactor.

Kanumilli *et al.* (2016) presented a plug-in electric solar car. This can be extended with an electric car with solar panel. It is driven by battery connected automobiles. Nowadays, an electric car plays a vital role in the market especially in the field of research and development (R&D) for development of solar-powered electric cars. These cars are emission free and are also the best solution in the future. The current scenario of this study is to develop a design which includes the plug-in electric car, which is driven by solar-powered electric car, and this is purely available in the market.

Chetan Mahadik *et al.* (2014) proposed an improved method and efficient control of electric bicycle system with real-time information sharing. This work discussed the operation of an electric bicycle system using an optimized technique. The aim of this proposed system was to establish a conventional bicycle into electric bicycle through various techniques like regenerative braking and brushless Direct current (BLDC) motor is used for a control technique. Moreover, a real-time sensing and crowd is used to improve the performance of the bicycle. Some of the issues were related with electric bike structure, including bike and battery checking, electric-assisted range, and recharging technique procedures. This work deliberated the method necessities for organizing an electric bike and emerging on battery management, operational ideas, and system design. The best solution to this problem was offline battery management, which helped in sun charging, and dynamic usage was possible.

Ke Bao *et al.* (2012) proposed an integration of battery charge and its discharge control for energy management in electric vehicle application. He proposed an analysis of energy control at a charging station. An AC converter was used for grid interface and a DC converter was used for electric drive vehicle (EDV) battery management which comprises the charging station. For grid side converter, a DC control technique was used for DC-link voltage management, AC system bus voltage, and reactive power. For electric vehicle converter side, constant voltage and constant current control technique were used for charging and discharging. This work mainly focused on the energy management which required for charging and discharging of electric vehicles as well as grid to vehicle (G2V) and vehicle to grid (V2G) necessities.

Farooq M *et al.* (2014) proposed a less cost effective solar Photovoltaic (PV) automotive for the country Pakistan. There are three primary components which are powered by electrical vehicle. Unlike a power source or battery source to supply the power, an inverter or controller and an electric motor are the major components. A solar car, which is made of solar PV panels that can be replenished by the power battery. The controller normally consists of power electronic devices that gives input signals from the driver and regulates the speed and torque of the electric motor.

Manivannan S *et al.* (2016) proposed a solar driven electric car. Solar power electric vehicle (SPEV) are sustained by a charging wire that fits into the car and a 230-V wall outlet. Electric vehicles afford features such as a collision detection, security system, and seatbelt detection.

8.2 Need of Electric Vehicle

As stated before, electric vehicles are the emission-free solution for the future generation since the internal combustion system is still relying on carbon-based combustion system, which is so hazardous to the human

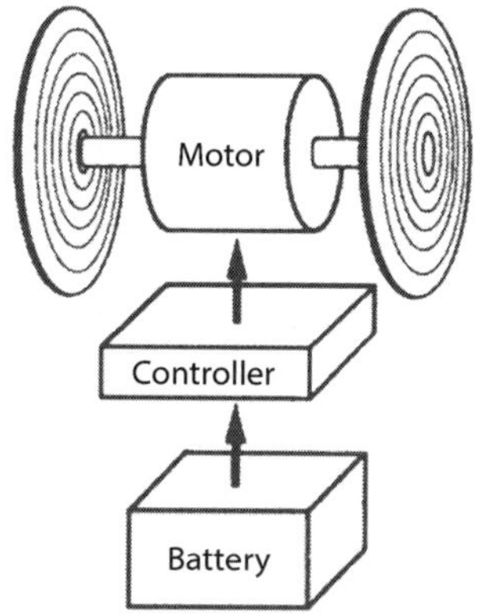

Figure 8.1 Simplified configuration of EV components.

lifestyle. Meanwhile, electric vehicles are the pollution-free solutions which do not emit greenhouse gases. Even power generation from the power plant may also generate electricity from the coal with some standard. These may require only water and electricity to generate power, and do not produce any harm to society. An electric vehicle shown in Figure 8.1 includes a battery, an electric motor, a controller, a charger, and driving system, which are enough for an electric vehicle setup.

8.2.1 Overview of Single Phase Induction Motor

In the industrial sector, single phase induction motors are frequently employed, notably in the fraction horse-power sector. In cases where a three-phase supply is not easily available, they are commonly employed for auxiliary drive for low-powered consistent speed equipment such as industrial machinery, household appliances, and farming gear. Vending machines, toys, hair dryers, and other uses employ single phase induction motors with size varying from 1/500 kW to 1/15 kW. Universal motors are widely found in kitchen equipment, transportable tools, and vacuum cleaners.

However, these devices are suitable for low-power applications and they are not utilized for high-power applications due to a number of limitations, and are only used when three-phase machines are available. Single-phase induction motors are much less efficient, have less increasing inventory, have a reduced power factor, and they are more expensive than three-phase motors with the same output. Their performance is only half that of a three-phase motor for a similar body frame and temperature increase. There is no beginning torque incorporated into these motors.

8.2.2 Objectives

- To utilize renewable energy to charge the battery to reduce emission of gases and carbon dioxide.
- To make use of the movement of the vehicle.
- To provide full torque.
- To minimize the running cost.
- To control the motor speed.

8.3 Methodology

The block diagram in Figure 8.2 portrays the proposed model of battery-powered induction motor, which runs electric vehicles. The reference speed is tracked by the actual speed. In such cases, the motor should always run under synchronism.

8.4 Simulation Results and Discussion

Figure 8.3 presents the simulation model of the solar panel which was developed in the MATLAB/Simulink software. The output voltage of the PV panel is shown in Figure 8.4 under the standard test conditions of PV module. In the system, DC-DC boost converter was used and the input of the converter waws fed as an output to PV module.

Figure 8.5 represents the simulink model of wind generation which was developed in the MATLAB/Simulink software, and the output waveform of wind generator is represented in Figure 8.6.

Figure 8.7 shows the Simulink model of a hydro generator which was developed in the MATLAB/Simulink software, and the output waveform of the hydro generator are presented in Figure 8.8.

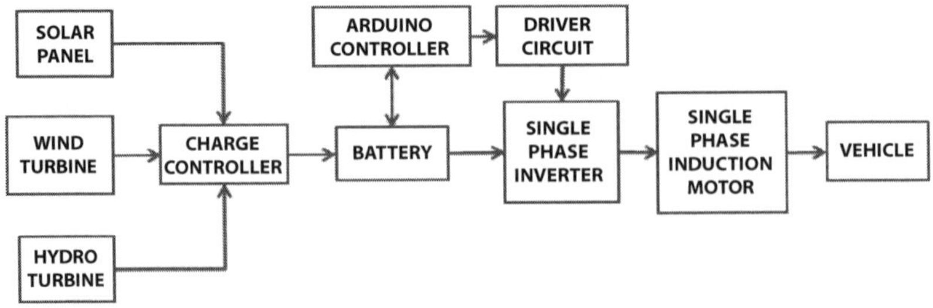

Figure 8.2 Block diagram of the proposed system.

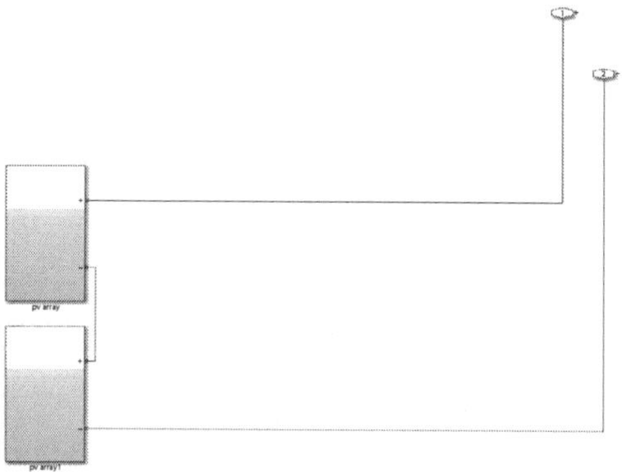

Figure 8.3 Simulation diagram of solar energy generation.

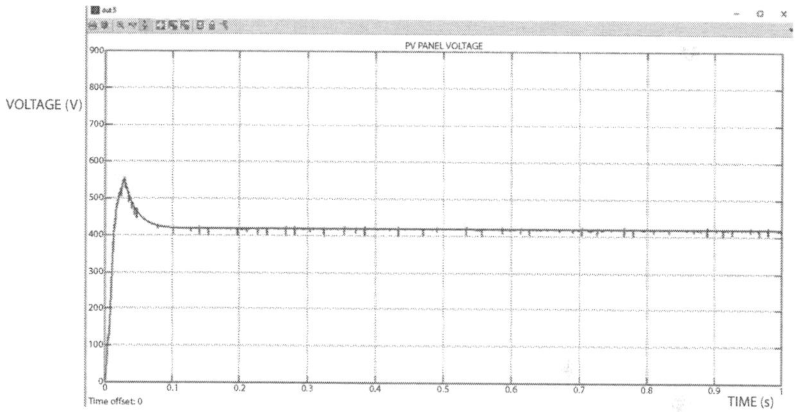

Figure 8.4 Simulation result of solar energy generation.

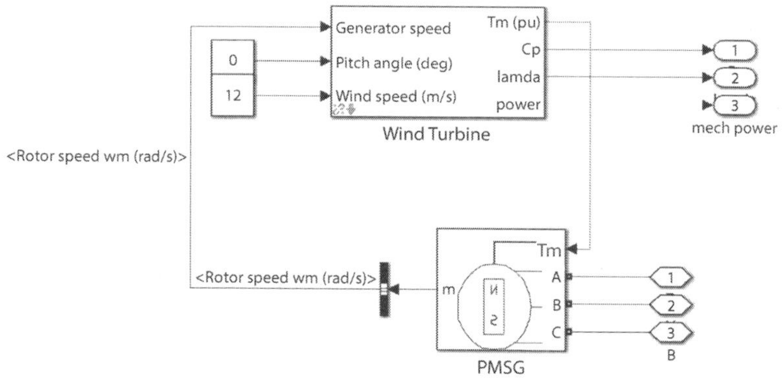

Figure 8.5 Simulation diagram of wind energy generation.

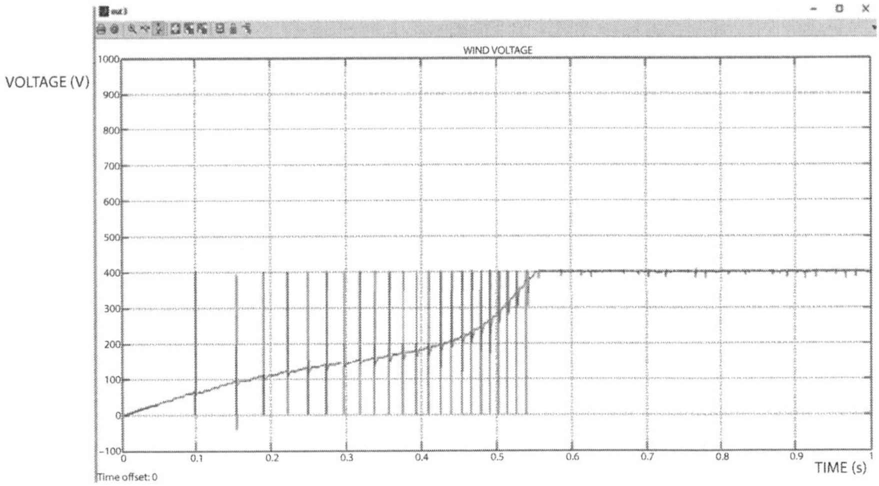

Figure 8.6 Simulation result of wind energy generation.

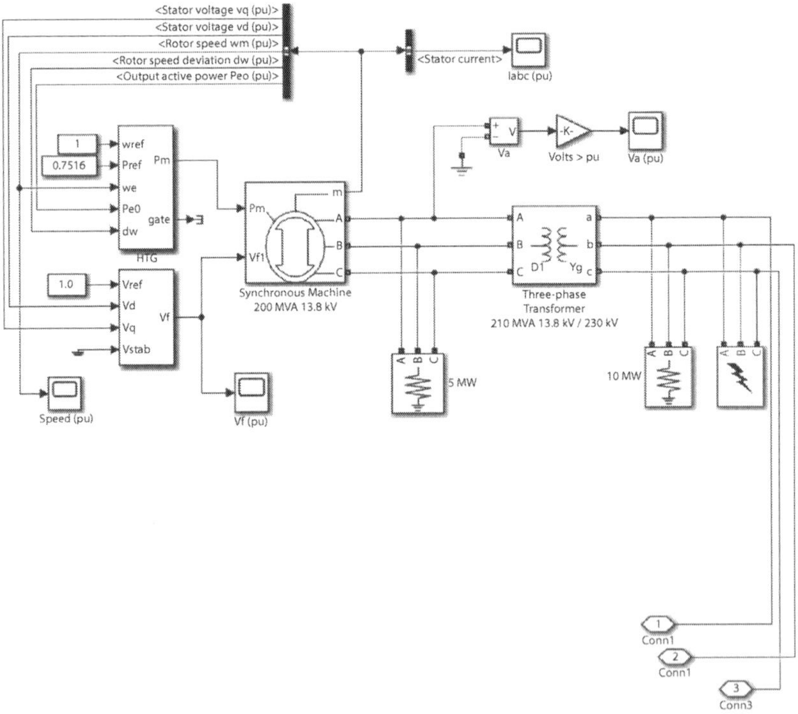

Figure 8.7 Simulation diagram of hydro energy generation.

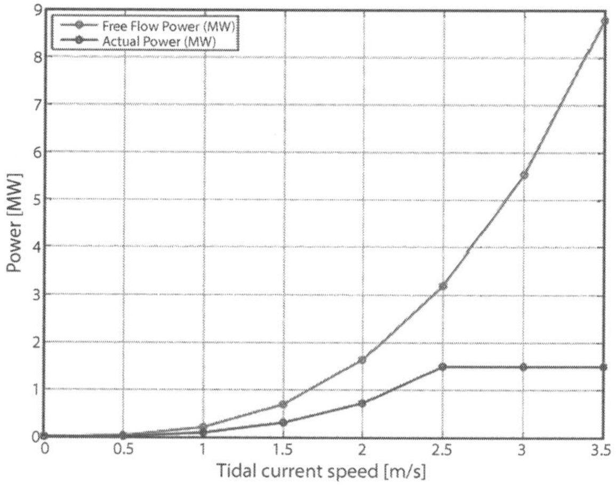

Figure 8.8 Simulation result of hydro energy generation.

Figure 8.9 shows the digital storage oscilloscope (DSO) output waveform for the single phase inverter.

Figure 8.10 shows the DSO output waveform for the 12-V DC battery which is converted into AC to run the single phase AC motor.

Figure 8.11 shows the DSO output waveform for the single phase induction motor which was used as the input for the vehicle.

Figure 8.12 indicates the Battery output power. Figure 8.13 shows input voltage and current to the vehicle.

Figure 8.14 shows the comparative speed responses with and without PWM control. It shows that PWM control technique gives a better performance in terms of speed regulations which enhance the performance of the single phase induction motor.

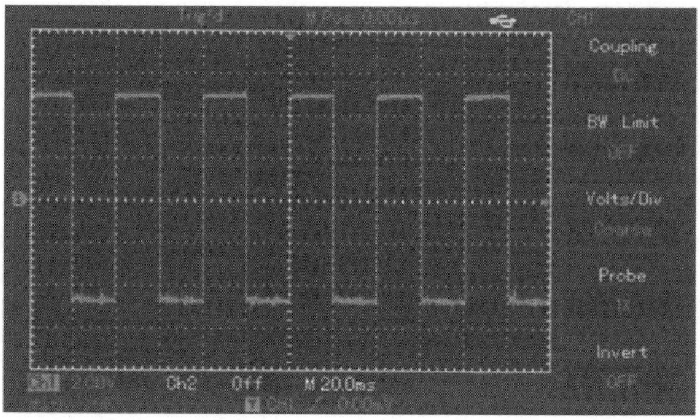

Figure 8.9 Inverter output waveform.

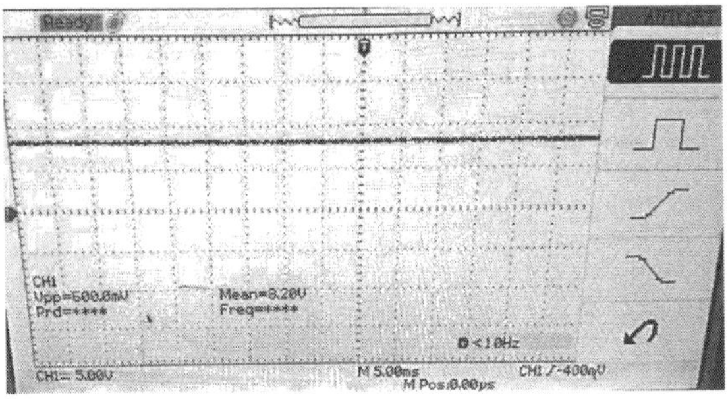

Figure 8.10 DC battery output waveform.

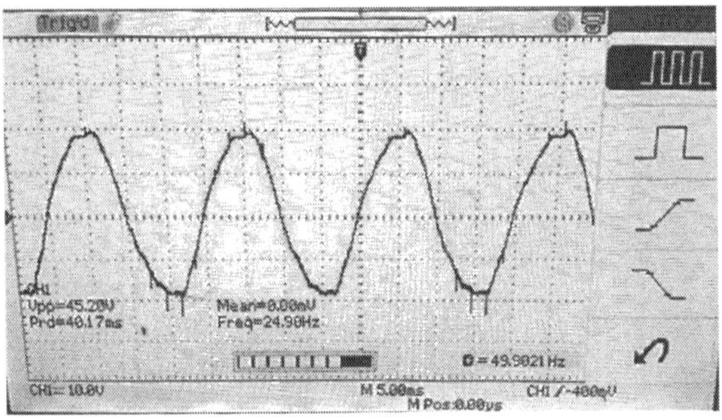

Figure 8.11 Single phase induction motor output waveform.

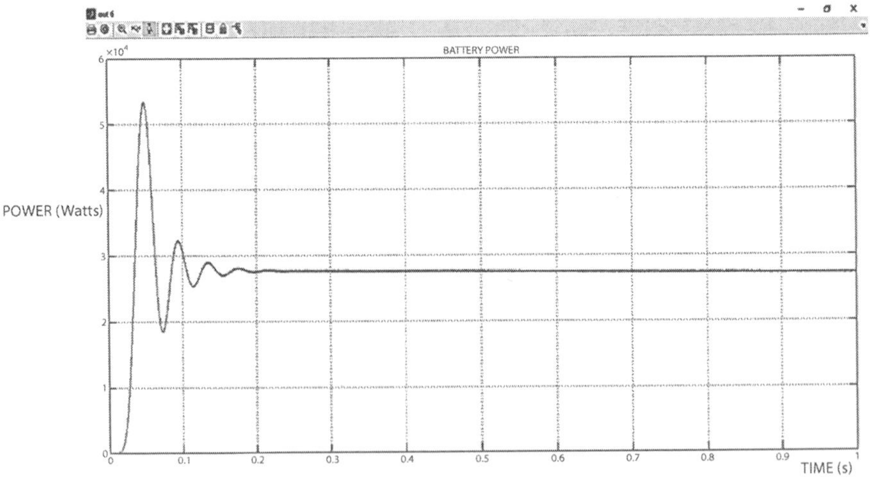

Figure 8.12 Battery output power.

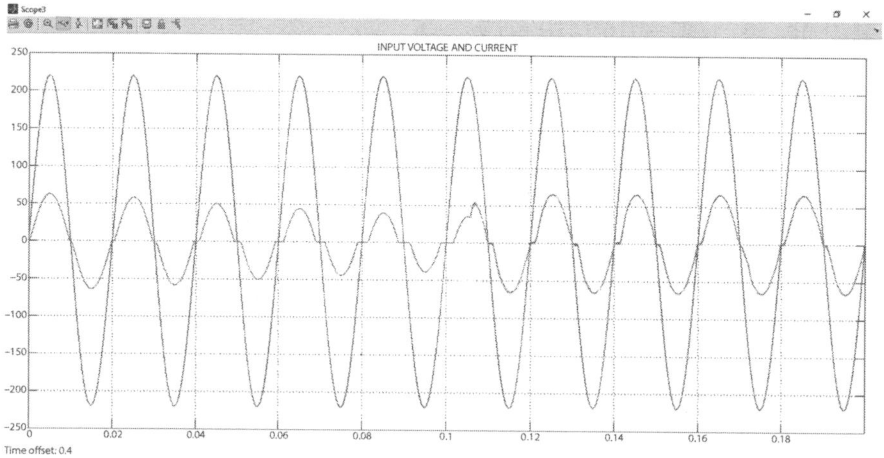

Figure 8.13 Input voltage and current for the vehicle.

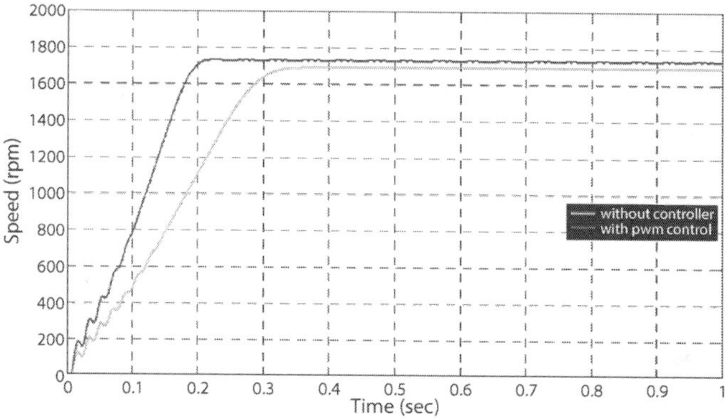

Figure 8.14 Comparison of speed responses.

Table 8.1 Comparison of SPIM without and with controller.

Constraints	Without controller	With PWM control
Speed (rpm)	1691	1728
Settling time (s)	0.34	0.28
Rise time (s)	0.31	0.21
Stator current (A)	9.5	7.78

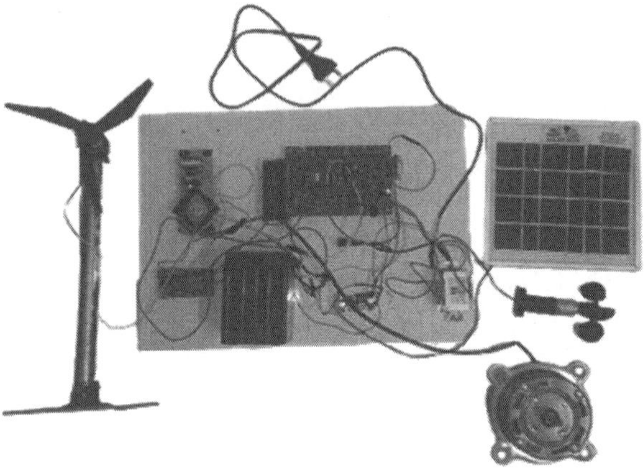

Figure 8.15 Hardware kit of the proposed system.

Table 8.1 shows the comparative results of single phase induction motor with and without controllers.

Figure 8.15 presents the experimental setup of the proposed system.

8.5 Conclusion

To meet the rising demand for gasoline and combat the severe environmental pollution caused by driving carbon-based cars, a new source of energy is required. Solar power would be a low-cost, efficient, inexhaustible, and, of course, environmentally beneficial alternative. They are zero-emission, odorless, smokeless, and noiseless cars. Wind power, like solar power, has grown in popularity as a source of renewable energy. The movement of the vehicle is also utilized to charge the battery via the hydro generator. The Arduino controller generates the PWM pulses. As a result, the suggested project was carried out, and the system fulfills the load requirements. The outcomes are scrutinized and contrasted. The comparison shows that SPIM provides superior speed response using a speed controller.

References

1. Fan Zhang, Xu Zhang, "Literature Review of Electric Vehicle Technology and its Applications", *5th International Conference on Computer Science and Network Technology*, IEEE, pp.832–837, 2016.

2. Z. Li, "High Performance Current Control for Switched Reluctance Motors Based on On-Line Estimated Parameters", *IET Electri. Power Appl.*, Vol. 4, pp. 67–74, 2010.
3. Xueqin and Wu, Yinbo and Lian, Jie and Zhang, Yangyang and Chen, Chao and Wang, Peisong and Meng, Lingzheng, "Energy management of hybrid electric vehicles: A review of energy optimization of fuel cell hybrid power system based on genetic algorithm," *Energy Convers. Manage.*, Elsevier, vol. 205, pp. 112474, 2020.
4. Tang, Xiaolin and Jia, Tong and Hu, Xiaosong and Huang, Yanjun and Deng, Zhongwei and Pu, Huayan,"Naturalistic data-driven predictive energy management for plug-in hybrid electric vehicles," *IEEE Trans. Transp. Electrif.*, vol. 07, no. 02, pp. 497–508, 2020.
5. Kanumilli, C., Singh, A., Ganesh, A. and Srinivas, M, "Plug in electric solar vehicle", *IEEE Biennial International Conference on Power and Energy Systems: Towards Sustainable Energy*, pp. 1–4, 2016.
6. Chetan Mahadik, Sumit Mahindraka, Prof. Jayashree Deka, "An Improved & Efficient Electric Bicycle system with the Power of Real-time Information Sharing", Citeseer, *Multidisciplinary Journal of Research in Engineering and Technology*, 2014, Vol: 1, pp. 215–222.
7. Yousuf.S.A., Venayagamoorthy.G.K, "Plug-in vehicles and renewable energy sources for cost and emission reductions", *IEEE Trans. Ind. Electron.* 2011, Vol. 58, pp. 1229–1238.
8. Abhinya Chaturvedi, Kirti Kushwaha, Parul Kashyap, Dr. J. P. Navani, "Solar Powered Vehicle", *Int. J. Electr. Electron. Res.*, 2015, Vol: 3, pp. 27–273.
9. Ke Bao, Shuhui Li, Huiying Zheng, "Battery Charge and Discharge control for energy management in EV and utility integration", *IEEE*, 2012, pp.1–8.
10. Zarma, Tahir Aja and Galadima, Ahmadu Adamu and Aminu, Maruf A, "Review of motors for electric vehicles", *J. Sci. Res. Rep.*, 2019, Vol: 24, pp. 1–6
11. M Saravanan, B Selvababu, Anandhu Jayan, Angith Anand and Aswin Raj, "Arduino Based Voice Controlled Robot Vehicle", *IOP Conf. Ser.: Mater. Sci. Eng.*, 2020, Vol: 993, pp. 012–125.
12. Dhumal, Amol and Naikoji, Amol and Patwa, Yutika and Shilimkar, Manali and Nighot, MK, "Vehicle tracking system using GPS and Android OS", *International Journal of Advanced Research in Computer Engineering & Technology*, 2015, Vol: 4, pp. 1220–1224.
13. G. Harper, R. Sommerville, E. Kendrick, L. Driscoll, P. Slater,R. Stolkin, A. Walton, P. Christensen, O. Heidrich, S. Lambert and A. Abbott, "Recycling lithium-ion batteries from electric vehicles", *Nature*, 2019, Vol: 575, pp. 75–86.
14. Singh, Krishna Veer and Bansal, Hari Om and Singh, Dheerendra, "A comprehensive review on hybrid electric vehicles: architectures and components," *J. Mod. Transp.*, Springer, vol. 27, no.2, pp. 77–107, 2019.

15. Xueqin and Wu, Yinbo and Lian, Jie and Zhang, Yangyang and Chen, Chao and Wang, Peisong and Meng, Lingzheng, "Energy management of hybrid electric vehicles: A review of energy optimization of fuel cell hybrid power system based on genetic algorithm," *Energy Convers. Manage., Elsevier*, vol. 205, pp. 112474, 2020.
16. Tang, Xiaolin and Jia, Tong and Hu, Xiaosong and Huang, Yanjun and Deng, Zhongwei and Pu, Huayan,"Naturalistic data-driven predictive energy management for plug-in hybrid electric vehicles," *IEEE Trans. Transp. Electrif.*, vol. 07, no. 02, pp. 497–508, 2020.

9

FPGA-Based Smart Building Access Control

Sakthi Ram T.[1], Yogesh L.[1], Vetriashwath S.[1], Nishanth G.[1] and O.V. Gnana Swathika[2*]

[1]*School of Electrical Engineering, Vellore Institute of Technology, Chennai, India*
[2]*Centre for Smart Grid Technologies, Vellore Institute of Technology, Chennai, India*

Abstract
Smart buildings use integrated and automated technologies and processes to improve efficiency and optimize usability and accessibility. And while building security has traditionally been costly and limited, intelligent building security offers many new opportunities. The recent rapid growth of smart buildings has brought many opportunities and challenges to the existing building systems arena, especially in terms of access control systems. This chapter proposes a new access control system for smart buildings using field programmable gate arrays (FPGA). We are using a sequence detector to authorize the entry through the buildings with the help of universal ashynchronous receiver transmitter (UART) serial communication and pattern detector using finite state machines.

Keywords: FPGA, smart building, serial communication, access control, finite state machine

9.1 Introduction

This chapter proposes a solution to encrypt and decrypt data in Verilog-based coding for field programmable gate arrays (FPGA). The project offers a smart contract-based access control that has multiple access control contracts (ACC), one judge contract (JC), and one register contract (RC). Each ACC provides one access control method, JC implements a

**Corresponding author*: gnanaswathika.ov@vit.ac.in

O.V. Gnana Swathika, K. Karthikeyan and Sanjeevikumar Padmanaban (eds.) Smart Grids as Cyber Physical Systems Volume 1: Artificial Intelligence, Cybersecurity, and Clean Energy for Next Generation Smart Grids, (137–144) © 2024 Scrivener Publishing LLC

misbehavior judging method, and RC records these processes [1]. The project proposes an automatic security system login lock using a finite state machine based on FPGA [2]. The project proposes a security login system where FPGA is used and when a user enters the password, Global System for Mobile communication (GSM) module is activated and an One Time Password (OTP) is sent, which changes every 3 min. Two-factor authentication is thus performed [3]. The project proposes a secure authentication protocol for wireless medical sensor networks. The XOR-based scheme is applied to FPGA for maintaining the confidentiality of data and prevention of tampering [4]. The project proposes an improved efficient hardware implementation of cipher-based message authentication code on FPGA for satellite applications, reducing FPGA consumption and time taken [5]. The project proposes an MD5 hash-producing algorithm-based FPGA block to store passwords to give communication integrity and signature authentication. Verilog is used in this project [6].

9.2 Methodology

In the past, flexible workspaces' access control system was powered by actual computers housed in server farms inside their structures. These days, smart access encompasses much more than door access. Smart access ensures standardized and streamlined procedures across all sites by integrating with a provider's whole portfolio. This enables service providers to give members access to all pertinent websites at any time. With smart access, you as an operator have flexibility in the access options you offer. In our concept, the password is input through the keyboard interface, and the data (the password) is delivered to the finite state machine via serial transmission. The given password is compared with our pre-set password using a finite state machine; if the comparison is successful (the password matches), the UART transmitter is used; if it is unsuccessful (the password is incorrect), the process loops back to the finite state machine for another password input. Bi-directional, asynchronous, and serial data transmission are all supported via UART. Issues with synchronization management between computers and external serial devices can also be handled via UART.

Issues with synchronization management between computers and external serial devices can also be handled via UART. Parallel data from the master device is converted into serial form by transmitting UART and transmitting in serial form to the receiving UART. The serial data will subsequently be changed back into parallel data for the receiver device. Since UART lacks clocks, it inserts start and stop bits to the transmission

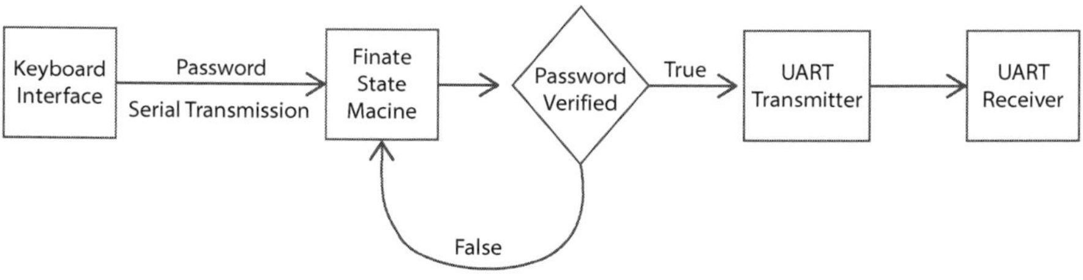

Figure 9.1 Flowchart for the components and processes involved.

to signify the beginning and conclusion of a message. The simplicity of implementing an finite state machine-based pass code detector along with the usage of UART transmission enables secure and lightweight password detection suitable for everyday low power applications. For instance, a smart door lock might make use of this proposed system.

Figure 9.1 explains the working of our proposed model. Here, the keyboard interface initially transmits the password serially to the sequence detector which checks for the inputs, if the sequence matches the passcode, *i.e.*, "102210" in this case, then it signals a high output, which is then sent to the UART transmitter which sends the data/command to the required appliances. For instance, it can used to send a command to open a garage door or turn on a specific machine enabling smart home automation with secure access. The UART transmitter runs on the FPGA's clock, hence, to get the desired baud rate, we can find the term N up to which the counter will run based on the following expression. Usually, UART transmission takes place at 11.52 KHz, so we find that the number of clock cycles it takes to transmit a single bit would be 434 clock cycles.

$$N = FPGA\ Base\ Clock\ /\ UART\ Baud\ Rate$$

9.3 FSM Sequence Detector

A sequential state machine called a sequence detector is used to identify successive bits in a binary string. When the desired sequence is found, a sequence detector creates an output 1 from a sequential state machine that accepts a string of bits as input. The output of a Mealy machine is dependent on both the current state and the external input (x). In a Moore machine, output depends only on the present state and not dependent on the input (x).

The sequence we built comes under overlapping detectors, as shown in Figure 9.2. Here, we fixed the password to be "102210" in our model. Our Finite State Machine (FSM) module therefore functions for this password, as shown in Figure 9.3. If the first input is 1, the stage S1 is reached; if the

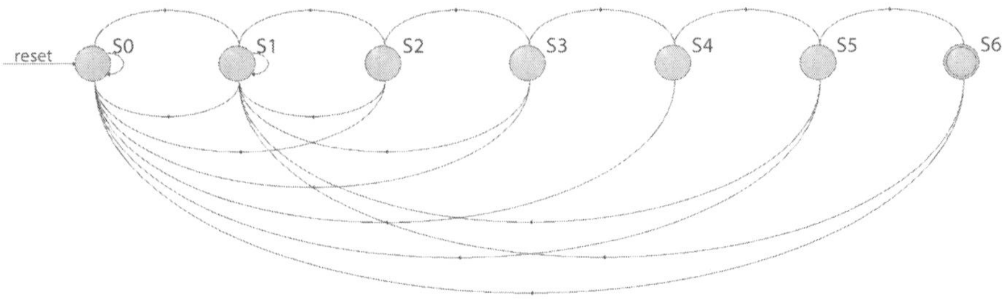

Figure 9.2 Flowchart explaining FSM loops.

	Name	S6	S5	S4	S3	S2	S1	S0
1	S0	0	0	0	0	0	0	0
2	S1	0	0	0	0	0	1	1
3	S2	0	0	0	0	1	0	1
4	S3	0	0	0	1	0	0	1
5	S4	0	0	1	0	0	0	1
6	S5	0	1	0	0	0	0	1
7	S6	1	0	0	0	0	0	1

Figure 9.3 Matrix showing input versus reach.

second input is 0, the stage S2 is reached; otherwise, the stage S0 is maintained. Then, if the next input is a 2, it advances to S3; otherwise, if it is a 1, it stays in S1 or returns to S0. If the input is once more 2, the S3 stage advances to S4, and if it is 1, the stage advances to S1 before returning to S0. If the input is 1, the S4 stage advances to the S5 stage; otherwise, S0 is reached. Our output, S6, is activated if our final input is 0, otherwise, the process loops back to the reset S0 stage.

9.4 UART Transmitter

In UART, data is transmitted serially at a fixed baud rate, which is known to both the transmitter and receiver, as shown in Figure 9.4. UART only

Figure 9.4 Bits allocation in data transmission.

requires two wires and also does not require a synchronous clock. The first bit indicates the beginning of the data transmission. Then next 8 bits are chosen as data bits which allows us to send any 8-bit ASCII character in a single transmission cycle. The parity generator is a combinational circuit at the transmitter, it takes an original message as input and generates the parity bit for that message depending on the parity scheme that is followed. We are using odd parity so the parity bit is generated such that the resultant number of bits transmitted in the given cycle is odd. The UART transmitter runs on the FPGA's clock, hence, to get the desired baud rate, we can find the term N up to which the counter will run based on the following expression. Usually, UART transmission takes place at 115.2KHz, so we find that the number of clock cycles it takes to transmit a single bit would be 434 clock cycles.

Figure 9.5 shows a flowchart that explains the data transmission. If Stage S6, which is our output, is high, following the conclusion of the FSM-based sequence detector, it signifies that the password has been verified and complies

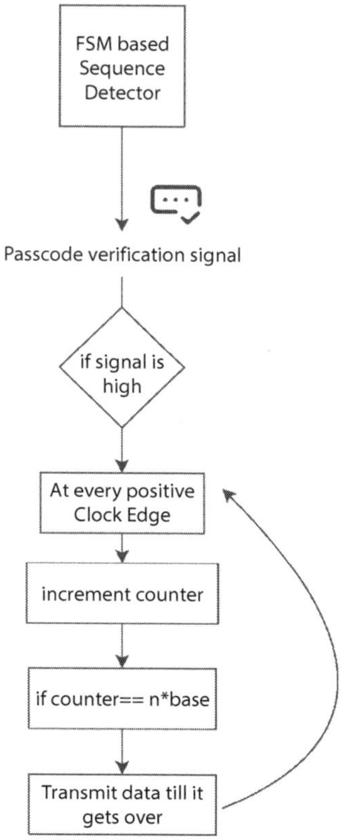

Figure 9.5 Flowchart explaining the data transmission.

with our pre-set value. The counter in our UART transmitter then advances on every positive clock edge. The transmission of data terminates and the loop returns to the clock edge if the counter reaches a maximum value of N times the base.

9.5 Results

Case 1

In this case, the right passcode sequence is sent to the finite state machine, which generates a high signal after it reaches the entire sequence, after which, the UART transmitter sends the command of "DATA," in this case, serially through at a baud rate of 115.2KHz, and also, the parity bit is generated at the end of each cycle, enabling error detection capabilities for the receiver. The data is successfully transmitted serially after verifying the authenticity by sequence detector. The output window and the waveform for Case 1 are shown in Figures 9.6 and 9.7.

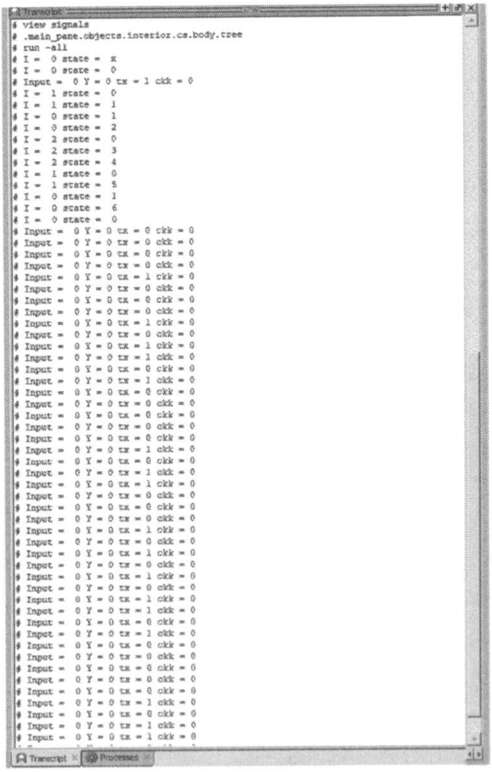

Figure 9.6 Output window.

Figure 9.7 Output graph.

Case 2:

In this case, the right passcode is not sent to the sequence detector. So the command/data is not transmitted, thus, it can be seen that this ensures secure transmission of data for intercommunication of devices in smart homes and offices. The output window and the waveform are shown in Figures 9.8 and 9.9.

```
# I =  0 state =  x
# I =  0 state =  0
# Input =  0 Y = 0 tx = 1 ckk = 0
# I =  1 state =  0
# I =  1 state =  1
# I =  0 state =  1
# I =  0 state =  2
# I =  2 state =  0
# I =  2 state =  3
# I =  1 state =  4
# I =  1 state =  1
# I =  0 state =  1
# I =  0 state =  2
# I =  0 state =  0
# Input =  0 Y = 0 tx = 1 ckk = 0
# Input =  0 Y = 0 tx = 1 ckk = 0
# Input =  0 Y = 0 tx = 1 ckk = 0
# Input =  0 Y = 0 tx = 1 ckk = 0
# Input =  0 Y = 0 tx = 1 ckk = 0
# Input =  0 Y = 0 tx = 1 ckk = 0
# Input =  0 Y = 0 tx = 1 ckk = 0
# Input =  0 Y = 0 tx = 1 ckk = 0
# Input =  0 Y = 0 tx = 1 ckk = 0
# Input =  0 Y = 0 tx = 1 ckk = 0
# Input =  0 Y = 0 tx = 1 ckk = 0
# Input =  0 Y = 0 tx = 1 ckk = 0
# Input =  0 Y = 0 tx = 1 ckk = 0
# Input =  0 Y = 0 tx = 1 ckk = 0
# Input =  0 Y = 0 tx = 1 ckk = 0
# Input =  0 Y = 0 tx = 1 ckk = 0
# Input =  0 Y = 0 tx = 1 ckk = 0
# Input =  0 Y = 0 tx = 1 ckk = 0
# Input =  0 Y = 0 tx = 1 ckk = 0
# Input =  0 Y = 0 tx = 1 ckk = 0
# Input =  0 Y = 0 tx = 1 ckk = 0
# Input =  0 Y = 0 tx = 1 ckk = 0
# Input =  0 Y = 0 tx = 1 ckk = 0
# Input =  0 Y = 0 tx = 1 ckk = 0
# Input =  0 Y = 0 tx = 1 ckk = 0
# Input =  0 Y = 0 tx = 1 ckk = 0
# Input =  0 Y = 0 tx = 1 ckk = 0
# Input =  0 Y = 0 tx = 1 ckk = 0
# Input =  0 Y = 0 tx = 1 ckk = 0
# Input =  0 Y = 0 tx = 1 ckk = 0
# Input =  0 Y = 0 tx = 1 ckk = 0
# Input =  0 Y = 0 tx = 1 ckk = 0
# Input =  0 Y = 0 tx = 1 ckk = 0
# Input =  0 Y = 0 tx = 1 ckk = 0
# Input =  0 Y = 0 tx = 1 ckk = 0
# Input =  0 Y = 0 tx = 1 ckk = 0
# Input =  0 Y = 0 tx = 1 ckk = 0
# Input =  0 Y = 0 tx = 1 ckk = 0
# Input =  0 Y = 0 tx = 1 ckk = 0
```

Figure 9.8 Output window.

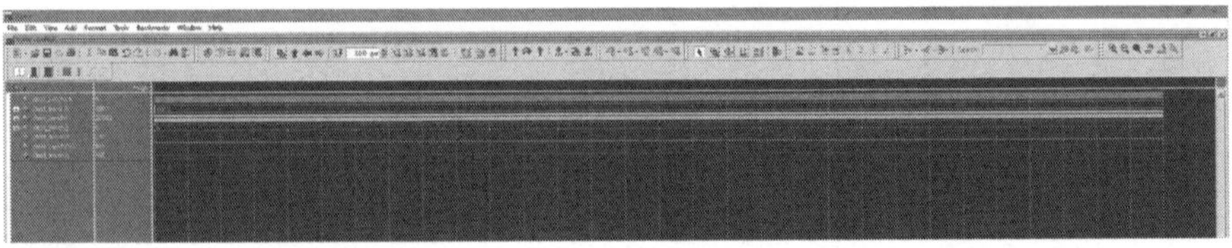

Figure 9.9 Output graph.

9.6 Conclusion

Smart access control is achieved with the help of a simple finite state machine-based password detector that is lightweight, and once the pass code is successfully verified, the command/data is transmitted serially at the required baud rate of 115.2KHz with parity bit to ensure data integrity.

References

1. Yuanyu Zhang; Shoji Kasahara; Yulong Shen; Xiaohong Jiang; Jianxiong Wan; "Smart Contract-Based Access Control for the Internet of Things"; *IEEE Internet Things J.* (Volume: 6, Issue: 2, April 2019).
2. Kavita Saroch, Abhilasha Sharma; "FPGA Based System Login Security Lock Design Using Finite State Machine"; *IOSR Journal of Electronics and Communication Engineering (IOSR-JECE)* Volume 5, Issue 3 (Mar. - Apr. 2013).
3. Samiksha Subba, Bhawana Dahal, Nirmal Rai, Rochan Banstola, Suman Das, Surya Prakash Tamang; "FPGA based Security Login System using GSM with OTP Generation"; *Int. J. Adv. Eng. Res. Sci.,* [Vol-4, Issue-2, Feb- 2017].
4. Wei-Chen Lin 1, Po-Kai Huang 2, Chung-Long Pan 2 and Yu-Jung Huang, "FPGA Implementation of Mutual Authentication Protocol for Medication Security System", *J. Low Power Electron. Appl.*, 2021.
5. Syed Jahanzeb Hussain Pirzada; Abid Murtaza; Jianwei Liu; "Implementation of CMAC Authentication Algorithm on FPGA for Satellite Communication"; *2019 IEEE 3rd Information Technology, Networking, Electronic and Automation Control Conference (ITNEC).*
6. Shadab Ahmad Khan; "FPGA Implementation of MD5 Algorithm for Password Storage"; *Int. J. Sci. Res.*

10

Artificial Hyperintelligence-Enabled Cyber-Physical System Control for Autonomous Vehicles

S. Srithar[1*], Vetrimani E.[2], Kumbala Pradeep Reddy[3], Sarangam Kodati[4] and S. Alagumuthukrishnan[5]

[1]*Department of CSE, Koneru Lakshmaiah Education Foundation, Vaddeswaram, Andhra Pradesh, India*
[2]*Department of CSE, Presidency University, Bangalore, India*
[3]*Department of CSE, CMR Institute of, Technology (Autonomous), Medchal Road, Kandlakoya, Hyderabad, India*
[4]*Department of Information Technology, CVR College of Engineering, Hyderabad, India*
[5]*Department of Computer Science and Engineering, CMR Institute of Technology, Hyderabad, India*

Abstract
The cyber-physical intelligent systems (CPIS) integrate the physical and cyber-computational devices into a point for effective communication among the connected devices. The application of the cyber-physical intelligent system is extensively used in robotics, avionics, and intelligent transportation, etc. Hyperintelligence is widely adopted into an autonomous vehicle for perception and auto driving. The major concern during cyber communication is to guarantee proper channel allocation during the packet transmission. The control and service channel utilization strongly defines the positive performance of a communication network. The cyber connection between the ad-hoc node is seriously affected by the hidden terminal problem. The cyber intelligent system is difficult to operate on the no-network areas which seriously affect the packet transfer and performance is degraded especially in an ad-hoc environment. The proposed hyperintelligence incorporates the imperial relay cohort (IRC) technique to minimize the line series problem. The frequency reuse is addressed on the Layer 4 peer-to-peer and Layer 5 networked systems to

**Corresponding author*: sss.srithar@gmail.com

O.V. Gnana Swathika, K. Karthikeyan and Sanjeevikumar Padmanaban (eds.) Smart Grids as Cyber Physical Systems Volume 1: Artificial Intelligence, Cybersecurity, and Clean Energy for Next Generation Smart Grids, (145–162) © 2024 Scrivener Publishing LLC

resolve the channel allocation issues. The connected components may be disrupted by a malicious intermediate node. At the time of network connection, the physical devices verify its genuineness among the other cohort peers to confirm authentication. The hidden channel utilization helps to avoid the access of non-genuineness connection components. The handover failure and the packet delivery success rate have been experimented with through the network simulator.

Keywords: Intelligent cyber active-physical system, hyperintelligence, autonomous intelligent vehicles, packet transmission, malicious

10.1 Introduction

The cyber-physical intelligent system (CPIS) is a package of communication, computation, and large-scale processing devices. The cyber intelligent system will communicate directly or through some intermediate devices. The CPIS is closely related with the Internet of things (IoT) because the IoT devices are connected through the CPIS devices. The commonly used CPIS is like an autonomous self-driving vehicle, intelligent transportation, smart city, smart power plant, and cloud services. The sensing devices sense the event and notify it to the cyber-connected physical components. The connectivity components are wired wireless or hybrid. The robust connectivity among the cyber world is very much essential to upgrade communication technologies [1]. The centralized communication system may not give an optimal solution. The distributed system will give an optimal solution to reduce the load balancing and response delay.

Nowadays, billions of nodes are connected and transfer the data over the communication medium. Still, the research is in line with intelligent transportation due to the behavioral obstacle of the model. There are various parameters that are needed to be considered to upgrade from a semi-autonomous environment to a fully autonomous one. The human population demands vehicle autonomy for safety benefits. The survey says billions of road accidents are reported with serious injuries and fatalities. The smart vehicle has integrated cyber-connected physical devices such as sensors, actuators, processors, controllers, and navigation devices. The cyber-connected physical systems are widely considered for car-to-car and car-to-infrastructure in intelligent transportation. The autonomous cyber-physical intelligent vehicle system receives the input from various sensing devices periodically [5, 14]. The received inputs are pre-processed to eliminate the redundant and missing values through some intelligent computing devices. The computing devices will study and analyze the sensed values for perception and classification. Finally, the actuators will incorporate the controllable output into the action. The cyber-physical intelligent devices are self-organized and have tight coordination among the nodes [9]. The service quality is measured in terms of quicker response, high precision, and distributed coordination in large-scale systems.

10.2 Analytical Framework

10.2.1 Literature Review

Vijaya G *et al.* [15] explored the great circle algorithm which distributes the messages as multi-cast to know the adjacent peer vehicle position and speed. Shaik, S *et al.* [10] explored the improved cross-layer cooperative routing (CLCR) which selects the best and the reliable node for end-to-end packet transmission. The cyber-physical internet-connected systems will predict reliable routes based on the nature of the lane and the vehicle speed. S. Shafi *et al.* [8] examined cross layer autonomous route recovery (CLARR) routing by considering the stability of the multi-hop link with the shortest relay. B. M. Reddy *et al.* [6] explored the gateway tiny message transmission approach which constantly shares the control packet to get the regular update of the vehicles.

Srithar [12] investigated the route cause analysis during overtaking. The cyber-connected physical devices will use the Cooperative Overtaking Sight Assistance (COSA) protocol standard to provide the assistant message through the connected wireless devices [13]. Basha, S *et al.* [2] investigated the relay which contains good stability and portability. The transport layer utilizes the traditional cyber protocols such as transmission control or datagram protocol. K. Karuppasamy [4] explored the several communication proactive and reactive routing protocols for easy packet dissemination. As per the analysis reactive protocol, ad hoc on-demand distance vector (AODV) will be given an optimal result over the packet success delivery rate. Srithar, S. [11] explored cyber-physical system movement intelligent prediction through the velocity, position, and communication range of the vehicle.

10.3 Layer Architecture of Cyber-Physical Intelligent Systems (CPIS)

The CIPS is divided into different categories namely physical, cyber, and human. The physical devices access the channel by frequency, time

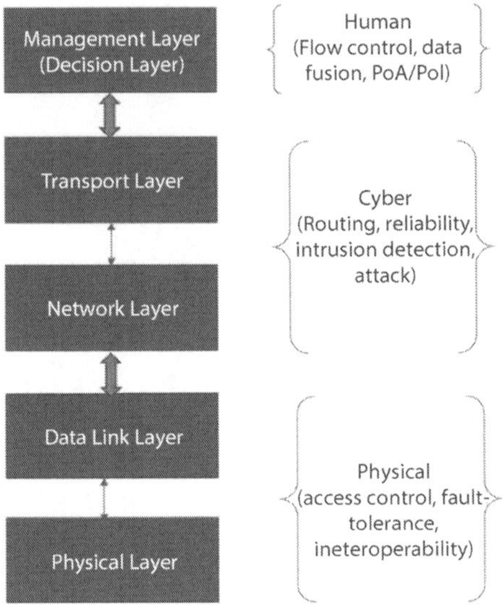

Figure 10.1 Layer interface architecture.

division, and code division. The medium access control layer ensures multi-device connectivity along with collaborative channel access. The technique named imperial relay cohort (IRC) is embedded in a cyber-physical system component to avoid the line series issues.

The cyber layer mainly deals with communication among all the devices for routing, reliability, and security. The routing will take the unique internet protocol (IP) address and Media Access Control (MAC) for cooperative communication. Finally, the cyber and physical components work together to effectively control the system. The layered architecture is shown in Figure 10.1, and the CPIS utilities are represented in Figure 10.2.

10.3.1 Layer Approach of Autonomous Vehicle Control

The cyber-physical system autonomous vehicle must pass into the five layers for load balancing and autonomy. In the level one layer, the smart vehicle will observe the surrounding through various sensor devices. The level two layer focuses on the packet transmission, brake, and steering control. The sensitive information is controlled and coordinated in layer three, whereas layer four ensures the inter- and intra-connectivity communication. The top layer is widely adopted to interact with the surrounding world through peer modeling. The layered architecture of the autonomous vehicle is represented in Figure 10.3. Technology such as the IoT, wireless technology,

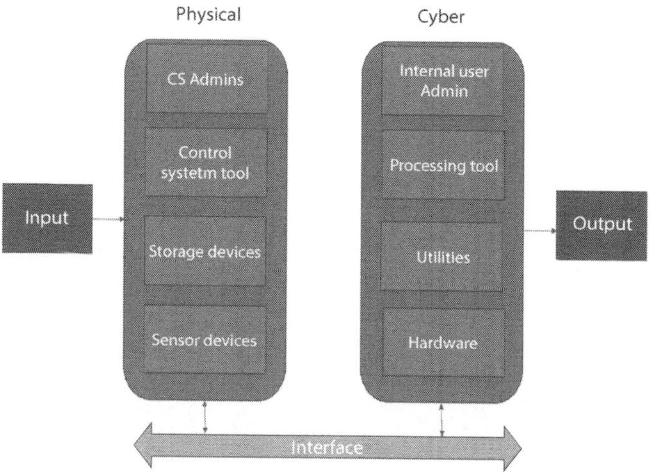

Figure 10.2 Cyber-physical intelligent system utilities.

blockchain, and cybersecurity are the heart of CPIS . The security aspects are seriously considered for the success of cyber-physical autonomous vehicle communication. The system is modeled by the discrete, finite state, and continuous-time models.

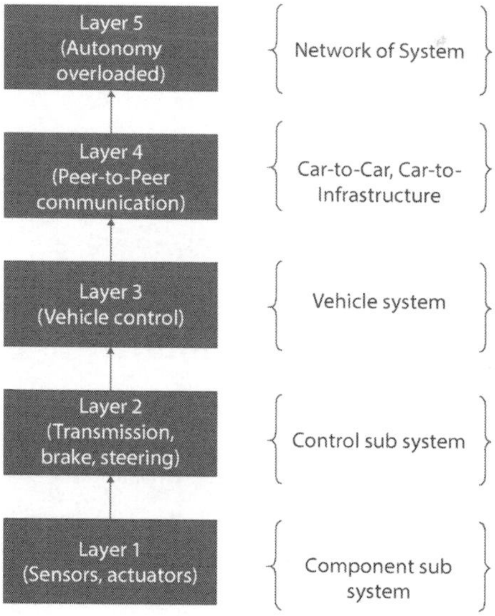

Figure 10.3 Autonomous vehicle layer approach.

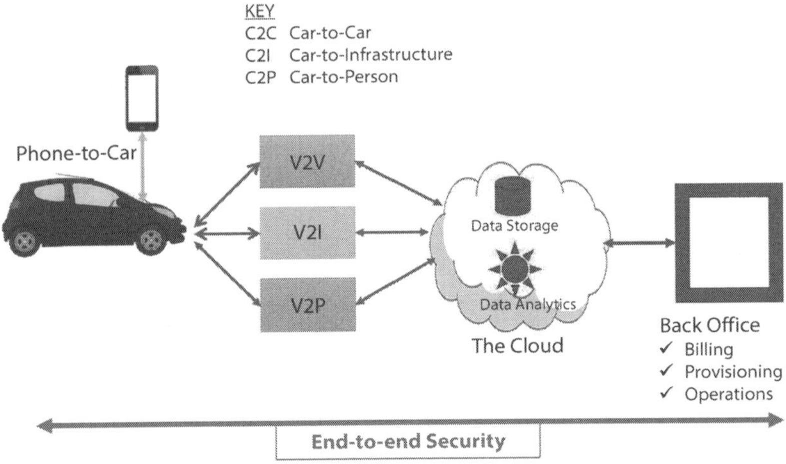

Figure 10.4 End-to-end security model.

10.3.2 End-to-End Security Parameters

The cyber-attack will be the biggest threat that could cause traffic, economic loss, and accidents. The connected systems are connected cooperatively with the centralized system through the wired or wireless channels. Some risk factors seriously affect and collapse the autonomy of the vehicle. The CPIS security model is represented in Figure 10.4. Some of the security parameters are listed below.

- Ransomware installation
- Hijacking
- Launch mobile bots
- Information theft
- Unauthorized access

10.4 Cyber-Physical Autonomous Vehicle vs. Machine Learning Systems

The traditional neural network modes are mostly used to detect the objects by machine training. The training is in the form of custom or through some predefined models like You Only Look Once (YOLO). The application of Convolutional Neural Network (CNN) is widely used in fully autonomous vehicles for pedestrians and lane detection. To detect the target objects, the system is trained with the dataset. The accuracy of the object detection is based on the number of test images and the feature extraction. The model gets trained using the cascade classification and the object is located with the bounding box. The false-positive rate of the training will be a serious impact during the decision making which lead to prediction failure. The performance of the system is depending on the following key factors.

- The volume of the training data
- Real-time environment changes
- Cooperativeness

The application of autonomy and cooperativeness reduces the cost of the traffic sign installation. The vehicle itself can take its decision during the travel safely through trustworthy cooperative behavior. There is a wide range of challenges to execute the operation especially in autonomous travel by the trained machine learning models. Because the trained model may be simulated with smaller size or training data. To enhance the effectiveness of the model machine learning is integrated with a cyber-physical autonomous system for decision making.

10.4.1 New Entry Authentication Procedure

The autonomous vehicular system is connected through cyber-physical devices. So, there may be a possibility of a security breach during the packet transmission. The secured connected devices may act as a relay between two or more devices in a network. The trustworthy verification is done by the master names roadside control unit. The master can connect with other master control units through wireless or through a wired medium. The master unit has the certificate authority to verify the genuine behavior in terms of transmission or re-transmission. The self-organization is initialized through the master unit after the certificate verification. The certificate authority will issue a session-initiated key for the secure transmission. The new entry vehicle authentication sequence is shown in Figure 10.5.

The following list contains some important characteristics which affect the performance of the ad-hoc vehicular environment.

- Stability of the network
- Connectivity strength
- Trustworthiness
- Portability and integration

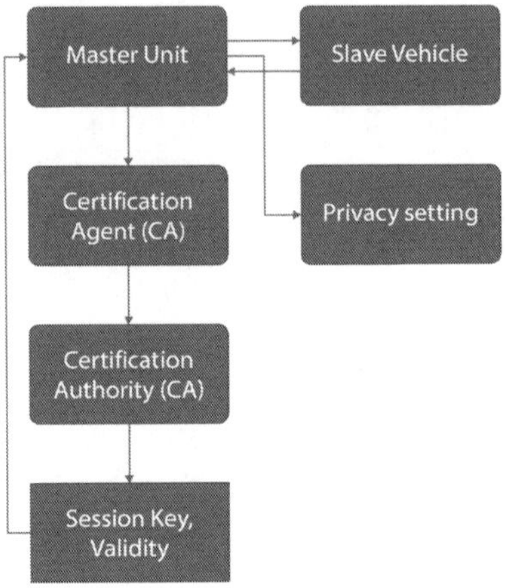

Figure 10.5 New entry authentication procedure.

10.4.2 Autonomous Vehicles Basic Requirements

There are some basic key components that are required to make the normal vehicle into a smart vehicle. Figure 10.6 shows the basic requirements of a smart autonomous CPIS vehicle.

- Global positioning system (GPS)
- Short-range communication transceiver
- Cameras
- Ultrasonic sensor
- Light detection and ranging (LIDAR)
- Radar sensor
- Server controller

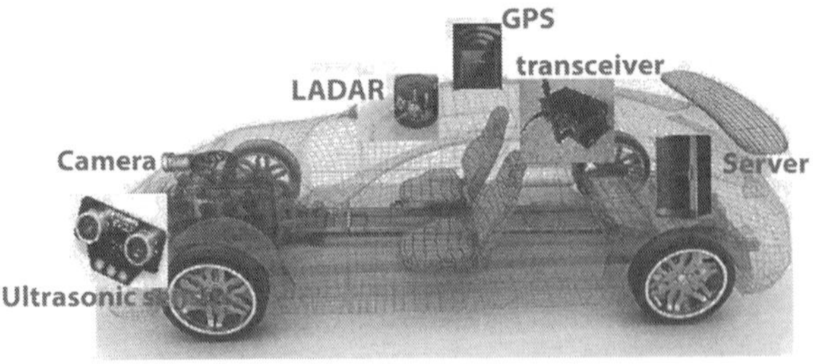

Figure 10.6 Smart vehicle basic requirements.

10.4.3 Global Positioning System (GPS)

The current location of the coordinates such as latitude and longitude value is detected from the GPS. The received GPS signals are combined with altimeter, tachometer, and gyroscopes for better accuracy.

10.4.4 Short-Range Communication Transceiver

The IEEE standard 802.11p communicates to another vehicle using Dedicated short-range communication (DSRC) for reliable data dissemination. The range of the communication is from 200 to 300 m. The communication is in the form of car-to-car, car-to-control unit, and control unit-to-control unit, etc.

10.4.5 Cameras

The camera image or video is the key factor to train the image frames for lane detection, traffic sign detection, and obstacle detection. The full-length video is converted into image frames for neural network training.

10.4.6 Ultrasonic Sensor

The coverage range of the ultrasonic sensor is very limited, which can detect the obstacle bypassing the sound waves. The response time and the send time is used to determine the distance between two objects. The sensor transmits the electromagnetic sound waves and waits for the response from the obstacle object.

10.4.7 Light Detection and Ranging (LIDAR)

To measure the obstacle object, the LIDAR will send the laser light pulse, whereas the reflection will be considered to degree the distance between two objects. To detect the lane and the road edge, LIDAR is used.

10.4.8 Radar Sensor

The radar sensor is used to identify the velocity, angle, and distance of objects. In an autonomous vehicle, speed is the very important parameter that will decide the success of the message transmission.

10.4.9 Server Controller

The controller is nothing but the server which keeps track of all the sensing output and utilizes it for perception and recommendation. It is a computer system that can control the steering, speed, and brakes.

The system is logically divided into two different layers: physical and control layers. The physical system contains the sensor devices whereas the control layer contains the driving model, interaction unit, and the services of the cyber unit. Figure 10.7 shows the layer connectivity of the cyber autonomous vehicle. Figure 10.8 shows the On-Board Unit (OBU) to actuator process flow and Figure 10.9 shows the CyPhy concerns.

10.4.10 Protocol Specification

For any kind of wireless communication over the medium, IEEE802.11 standard is preferred. The short distance prefers Bluetooth standard IEEE 802.15 whereas the IEEE802.11b is used for data transmission. The autonomous vehicle preferably takes the IEEE802.11p standard for packet dissemination for fast-moving node-to-node communication. The system will suffer to operate under no network areas and during the handover.

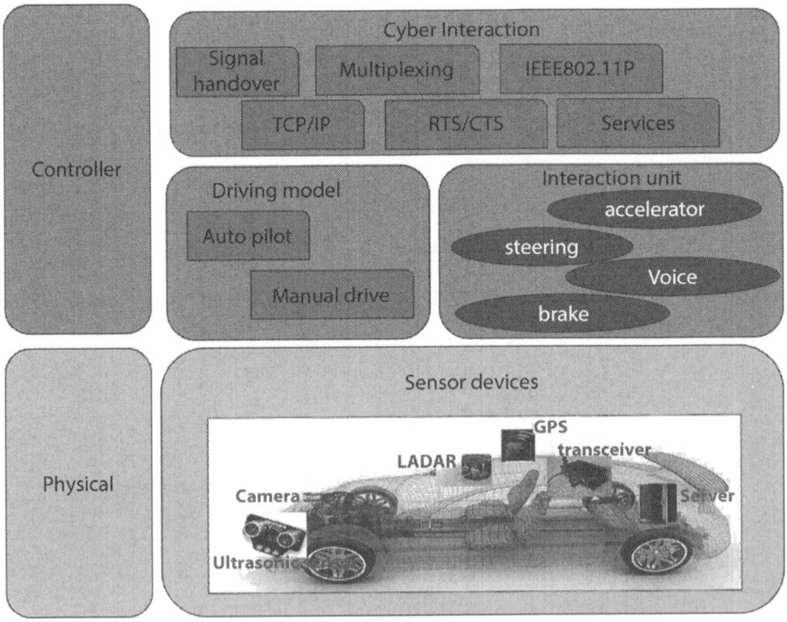

Figure 10.7 Cyber-physical layer connectivity.

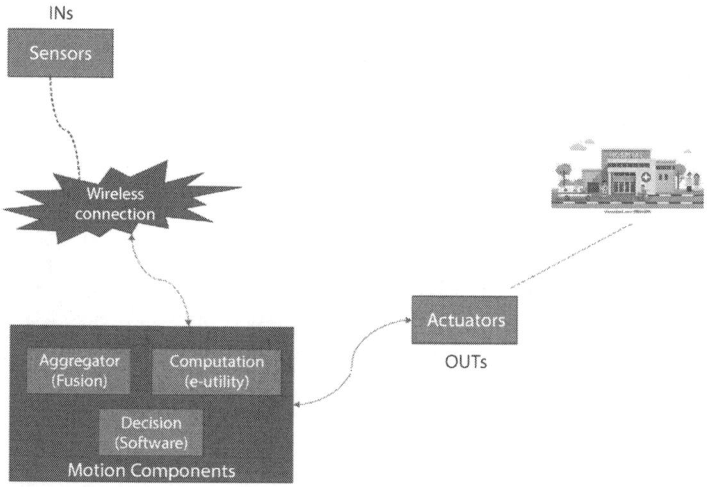

Figure 10.8 OBUs to actuator process flow.

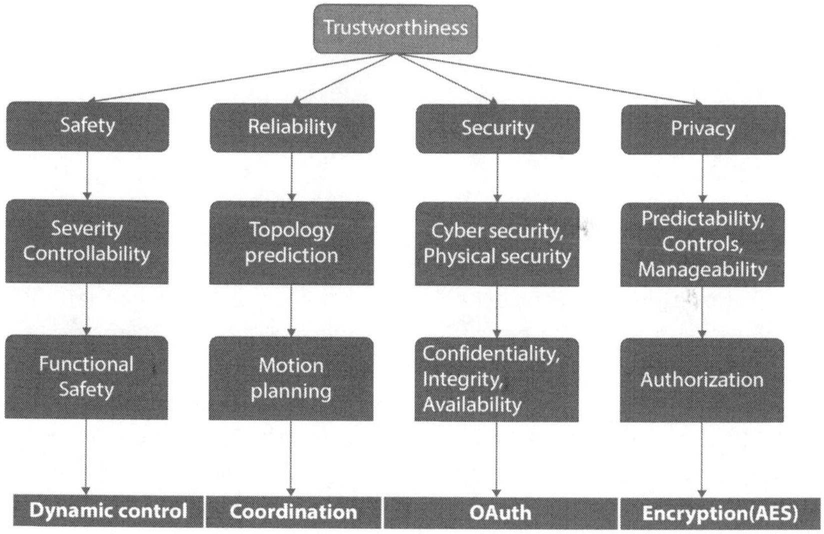

Figure 10.9 CyPhy as concerns.

The system makes use of the alternate standard named IEEE 802.11ax for faster data transmission. The cyber-physical connected system now works with other connected devices with the data rate up to 1.3 Gbps and the system will operate under 2.4 and 5 GHz of the frequency band. Whenever multiple vehicles are trying to access the wireless medium, there may be a change of connectivity issue and line series problem. The line series problem is taken over by the IRC after the trustworthy evaluation. By assigning separate frequency, time, and code for all the vehicles is difficult to manage.

The control unit will suffer from assigning and revoking the frequency bands during the multiple channel access. The technique named channel partition will give an optimal solution and will avoid bandwidth utilization. The channel allotment will be reserved based on the adjacent control unit request and provision. The new and existing entries are periodically updated by the nearby units for easy resource allocation. The reservation scheme will provide an optimal solution during the handover process. The soft and hard handover scheme sometimes takes delay at the time of access point changeover. The relay vehicle will share the same frequency but communicate the message through the mutual key of 32 bits. The original message and the mutual key are taken for converting the plain message into a cipher message.

10.4.11 Imperial Cohort Reply Procedure for Optimal Channel Selection

Autonomous vehicles disseminate the messages through the control and service channels. Whenever multiple vehicles are trying to access the channel, the control unit will allocate the channel and its frequency based on the available slot. The traditional channel allocation will not give an optimal result because of the handover frequency rate. The channel allocation is initiated whenever the vehicle is beyond the articulation point. The conventional method will distribute the request message to the control unit and waits for the new handover. Many a time, the handover fails if the channel utilization is already full. The problem could be eliminated by the early reservation scheme through the cohort reply.

The road-side control units have the proper connectivity with other control units which eliminates the line series problem. The communication channel is allocated to the vehicle based on the initial request. The control unit will detect the speed of each vehicle and estimated the average sit-back timing. The control unit will distribute the token number which is supplied by the nearest control unit. The sit-back time is intimated to the nearest control unit based on the direction of the movement. The process flow of the channel allocation is shown in Figure 10.10.

Now, the control unit will receive the identity of the vehicle and its current position. The unit will estimate the approximate entry time and do all the pre-processing works early. The channel slot reservation is done based on the sit-back timing of the vehicle. Whenever the vehicle crossed the

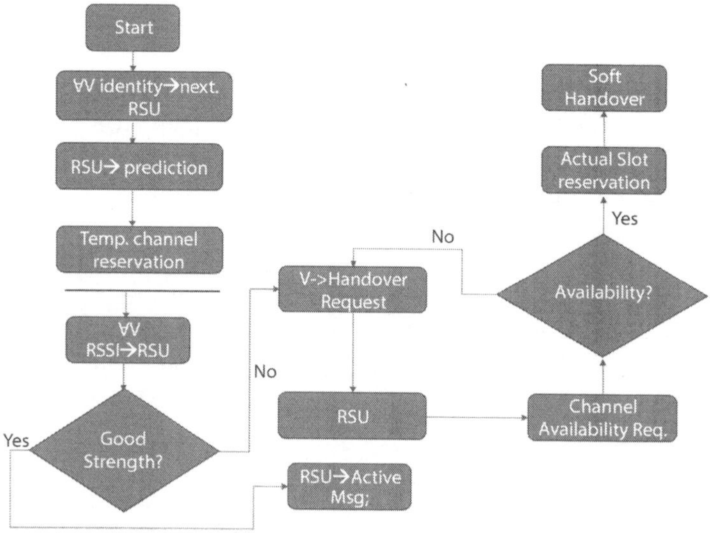

Figure 10.10 Channel allocation process flow.

articulation point, the vehicle itself will transmit a handover request communication to the control unit via the neighbor vehicle or directly. The channel allocation is based on the estimated speed of the vehicle. Initially, the channel allocation is done based on time division. Based on the vehicle density, it will change the allocation style as frequency or code division access. Figures 10.11 and 10.12 show the multi-channel route request and response packet structure, respectively.

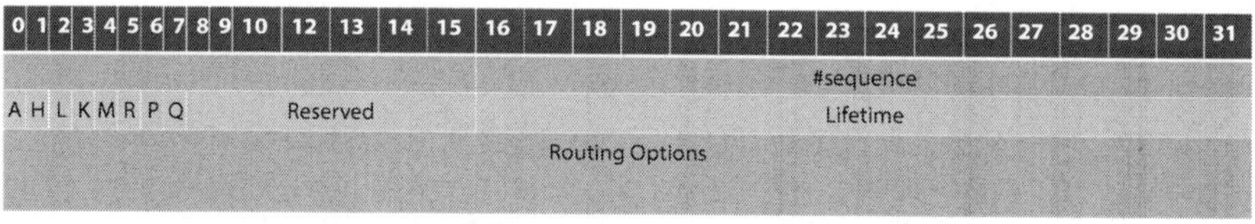

Figure 10.11 Multi-channel route request packet structure.

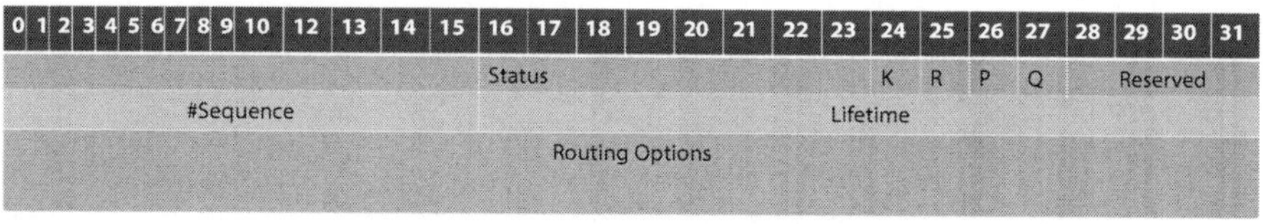

Figure 10.12 Multi-channel route reply packet structure.

10.5 Results and Discussion

The proposed model is directly incorporated into our cyber-physical intelligence with the simulated model. The model will take Sumo 1.9.2 version to build the initial configuration of the model. The interior design of the system is done in the NS2.34 simulator which utilizes the IEEE 802.11ax for faster and efficient data dissemination. The channel allocation depends on the Imbalanced Cluster-based Redirecting (ICR) which took 3 RSUs into a consideration. The proposed algorithm is examined with the existing approaches Self-organizing Multi-channel MAC (SOMMAC) [3] and dynamic Proxy MIPv6-based Mobile Internal Vertical Handover (PMIP-MIVH) [7]. The simulation constraints are tabulated in Table 10.1. The cyber and physical configurations are shown in Figures 10.13 and 10.14.

Table 10.1 Simulation parameters.

Parameters	**Value**
Simulation tool	NS2.34
Sumo version	1.9.2
Field size	3000*3000
Range of vehicle	100–350 m
Traffic	CBR
No. of lanes	3
Packets	512 bytes
Transfer rate	0.05 s
Avg. simulations	50
MAC	IEEE 802.11ax
vehicle speed (Max)	50100 km/h
vehicle speed (Min)	50 km/h
Vehicle count	50 to 250
Number of RSUs	3
Simulation time	150 s

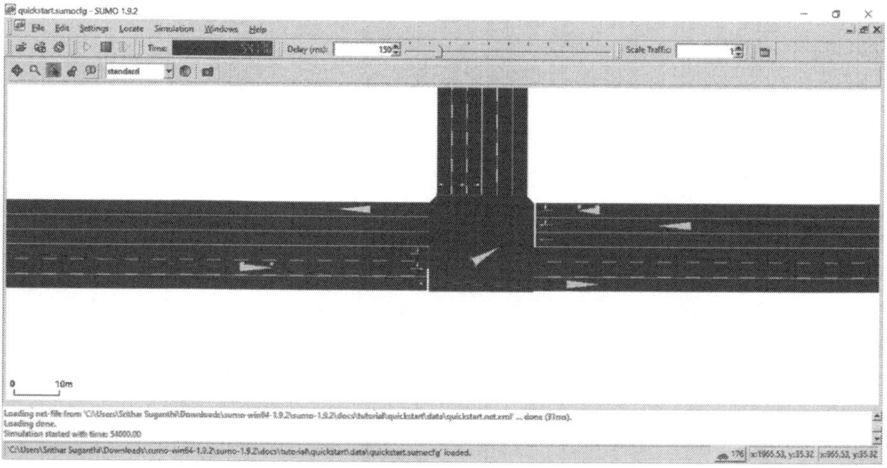

Figure 10.13 Cyber-physical system configuration.

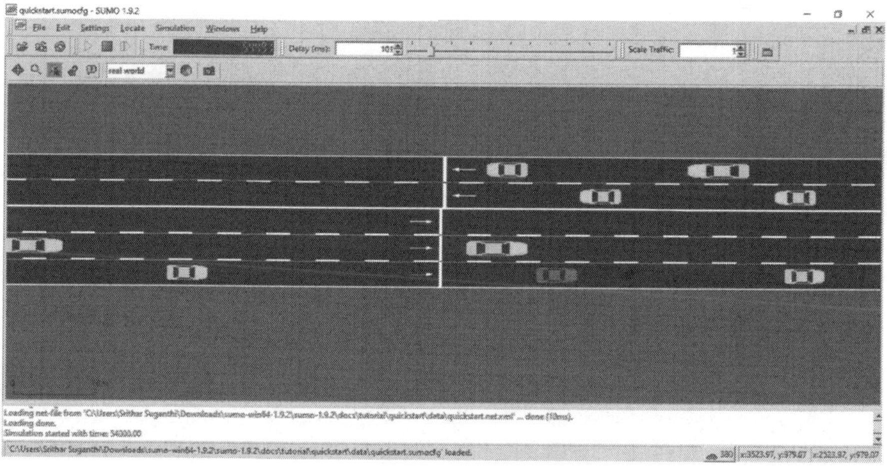

Figure 10.14 Physical components formation.

10.5.1 Handover Rate of Failure vs. Vehicles Count

The failure rate of the handover is eliminated by our hyperintelligence technique which constantly utilizes the early reservation to avoid the hidden and exposed terminal problem. The handover scheme is implemented to enhance the packet transmission rate. The traditional approaches will not be given an optimal result during the fast movement. The early channel allocation is given the best result in failure rate and channel waiting time.

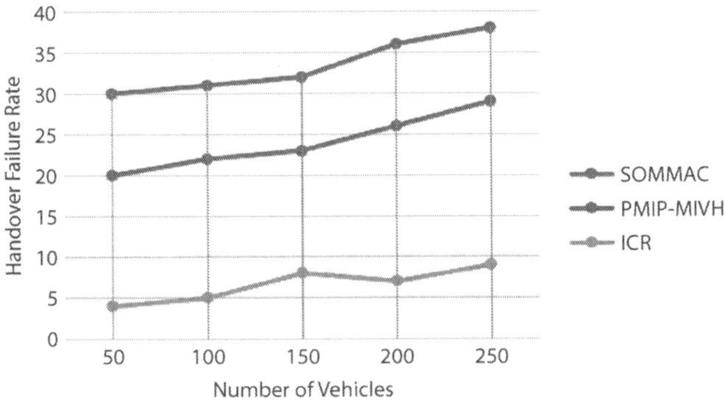

Figure 10.15 Handover rate of failure vs. vehicle count.

The proposed ICR is outperformed over the SOMMAC with 23% and PMIP-MIVH with 17%. The handover rate of failure is denoted in Figure 10.15.

10.5.2 Packet Delivery Rate (PDR) vs. Vehicle Count

The proposed cyber active-physical intelligent system will effectively be taken over the handover problem which can provide optimal results in terms of packet transfer hit rate. The optimal channel allocation reduces the waiting time during the handover. The proposed ICR is outperformed over SOMMAC with 19.2% and dynamic PMIP-MIVH with 11.3%. The packet delivery hit rate is represented in Figure 10.16.

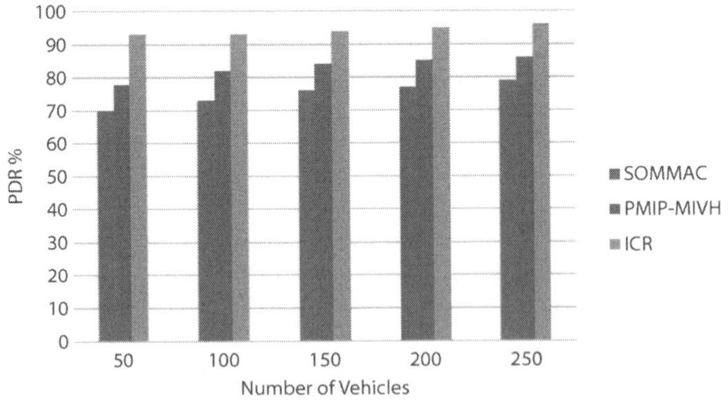

Figure 10.16 PDR vs. vehicle count.

10.6 Conclusion

The cyber active-physical intelligent system is widely adopted to improve the performance of the autonomous system. The traditional cyber-connected system suffers a lot with the line series, channel allocation, and hidden terminal problem. In addition, the malicious node will impact seriously during the message transmission. The proposed approach effectively handles the line series problem by considering the early channel reservation technique. The early channel reservation scheme will avoid the queuing delay during the vehicle movement. The channel waiting time is mostly avoided which also gives a positive result in terms of the hidden terminal problem. The proposed approach will give the best result in terms of packet delivery hit rate and error rate during handover.

References

1. Azeem & Syed, Notification of Data Congestion Intimation for IEEE 802.11 Adhoc Network with Power Save Mode, *Indo. J. of Elect. Engg. and Comp. Sci.*, Vol.5, No. 2, 2017, pp. 317-320, 2017.
2. Basha, S. & Shankar, T, Fuzzy logic based forwarder selection for efficient data dissemination in VANETs, *Wirel. Netw.*, Issue 3, 2021. DOI. 27. 10.1007/s11276-021-02548-8.
3. Fikri,A, Musa,Ç, Shafqat,Ur-Rehman, A new Self-Organizing Multi-channel MAC schema for RSU-centric VANETs, *Physica A*, Volume 551, 124098, 2020. https://doi.org/10.1016/j.physa.2019.124098
4. K. Karuppasamy, Comparative Study of Traditional Routing Protocols in Mobile Wireless Networks. *Asian J. Inf. Technol.*, 15: 359-364, 2016
5. Livinus,T, *et al.*, A mobile internal vertical handover mechanism for distributed mobility management in VANETs, *Veh. Commun.*, Vol.26, No. 100277, 2020. https://doi.org/10.1016/j.vehcom.2020.100277
6. Reddy,BM, Anumandla,KK and Tiwari,VK , Optimization of smart vehicle ad hoc network (SVANET) communication for traffic related issues with a security, *2017 Int. Conf. on ICCCI*, pp. 1-4, 2017.
7. Sadakale,R, Bhosale,A and Ramesh,NVK, Performance Analysis of Traffic Types in AODV Routing Protocol for VANETs, *10th Int. Conf. on Computing, Communication and Networking Technologies (ICCCNT)*, pp. 1-5, 2019, doi: 10.1109/ICCCNT45670.2019.8944448.
8. Shafi,S, Bhandari,BN, and Ratnam, DV, A cross layer design for efficient multimedia message dissemination with an adaptive relay nodes selection in VANETs, *Conf. on SPACES*, pp. 81-84, 2018, doi: 10.1109/SPACES.2018.8316321.

9. Shaik, MJ, Syed Umar, 2016, Proposed Potential Security Infrastructure in VANETS using Tamper Registered ', *Int. J. Electr. Comput. Eng.*, Vol. 6, No. 1, pp. 344~348, 2016.
10. Shaik, S., *et al.* An Efficient Cross Layer Routing Protocol for Safety Message Dissemination in VANETS with Reduced Routing Cost and Delay Using IEEE 802.11p, *Wirel. Pers. Commun.*, Springer, 100, 1765–1774, 2018. https://doi.org/10.1007/s11277-018-5671-z
11. Srithar, S *et al.* An Improved Heterogeneous Motion Prediction Algorithm for Vehicle-to-Vehicle Message Ferrying in VANETs. *Wirel. Pers. Commun.*, 113, 695–714, 2020. https://doi.org/10.1007/s11277-020-07248-8
12. Srithar S *et al.*, Comfort and danger zone boundary prediction for flying car safe overtaking in highways, *IJRTE*, 8(5): 1155-1159, 2020.
13. Srithar, S & Saravanakumar, NM, An Awareness Show: Vehicle Accident Root Cause Analysis in Tamilnadu, *Research Review International Journal of Multidisciplinary*, vol. 4, Issue 1, pp. 341-345, 2019, ISSN:2455-3085 (Online), https://doi.org/10.5281/zenodo.2544761
14. Suresh Babu, E, Nagaraju, C, Krishna Prasad, MHM, A Secure Routing Protocol against Heterogeneous Attacks in Wireless Adhoc Networks', pp. 339–344, 2015.
15. Vijaya *et al.* Controlling the Traffic Interactions with High Mobility and Constant Network Connectivity by Vanets, *2nd Int. conf. on MEET, Lecture Notes in Electrical Engineering*, vol 434. 2018. https://doi.org/10.1007/978-981-10-4280-5_62

11

FPGA-Based Smart Delivery Bot

Sakthi Ram T.[1], Yogesh L.[1], Vetriashwath S.[1], Nishanth G.[1] and O.V. Gnana Swathika[2]*

[1]*School of Electrical Engineering, Vellore Institute of Technology, Chennai, India*
[2]*Centre for Smart Grid Technologies, Vellore Institute of Technology, Chennai, India*

Abstract

Smart delivery is an autonomous bot that enables the delivery of goods from one node to another while taking the shortest possible time to minimize time and fuel, and hence, increasing the productivity and efficiency of the delivery system. In India, the majority of deliveries are handled by humans. Compared with other delivery systems, robotic systems offer additional advantages as they are large transportable vehicles. Automation, such as unmanned robots, is the result of advances in technology. This proposed model represents an automated robotic system that moves from point A to point B without the need to move and also without human involvement. Sensors are used to detect obstacles. This bot uses Dijkstra algorithm to find the shortest distance between two nodes or delivery points. This has several applications in growing fields such as food delivery systems in large restaurants or automating the delivery of goods in railways stations. Our model does all the computation such as finding the shortest path and sensor interfacing all within a field programmable gate array (FPGA)-minimizing space and optimizing performance.

Keywords: Smart delivery, FPGA, optimized performance, robotic systems, Dijkstras algorithm

11.1 Introduction

We propose a completely self-functioning autonomous delivery bot for contactless delivery of packages with tampering prevention using contact

**Corresponding author*: gnanaswathika.ov@vit.ac.in

O.V. Gnana Swathika, K. Karthikeyan and Sanjeevikumar Padmanaban (eds.) Smart Grids as Cyber Physical Systems Volume 1: Artificial Intelligence, Cybersecurity, and Clean Energy for Next Generation Smart Grids, (163–172) © 2024 Scrivener Publishing LLC

sensors powered by FPGA. This chapter proposes a contactless delivery bot which is a four wheeler that can deliver the package to the destination and return to the starting node with the help of Global Positioning System (GPS) although this is efficient and suitable for dynamic paths but the need for GPS connectivity can be eliminated in static paths, saving space and power, and making the bot more robust as it works independent of network connectivity. Our proposed model can even work in remote areas where there is poor connectivity and is more power efficient [1]. While GPS-based navigation is only suitable for long arenas, this can be scaled very effectively for smaller areas like a restaurant delivery system for example, hence, making our proposed model more robust. This chapter discusses the various robotic systems implemented to fight against the pandemic, how it uses connected technologies along with data collections and Big Data to improve the efficiency of the delivery system [2]. This chapter proposes a delivery aiding system where there is a detector at the parking system's entrance that is turned on to look for a vehicle coming back. A secret code is needed to open the gate after the detector has been set off. If the entered secret code is accurate, the gate would graciously permit the vehicle to enter. Without it, the gate remains closed. If the exit detector on this vehicle detects it entering the park, and once the car arrives, the door will swing open quickly to allow the oncoming vehicles to enter passwords using FPGA [3]. In these papers, they have implemented the Dijkstra algorithm on an FPGA or slightly modified version of it for finding the shortest paths in long range roads [4–7]. This project in multi-processor systems and network on chip allows finding the shortest paths in the presence of destructive factors (deadlocks, live locks, starvation, failures). It also does not use routing tables but uses relative addresses of nearby nodes [8]. In these papers, the authors implemented other advanced algorithms like A* and Bellman-Ford and Floyd Marshall algorithms [9–11]. We propose a Dijkstra-based algorithm to find the shortest distance between two nodes for static arenas.

11.2 Methodology

Figure 11.1 shows the flowchart for the components and processes involved. The smart delivery system was designed using Verilog coding in FPGA to deliver in the shortest route across two points. The sensor gives an analog signal for end points which is converted to digital for processing. It works on Dijkstra's algorithm for the shortest route mapping by the addition of neighboring points from start point to end point and comparing for the shortest path among those in an array form. Additionally, extra measures have been taken to return the package if opened/tampered with in between

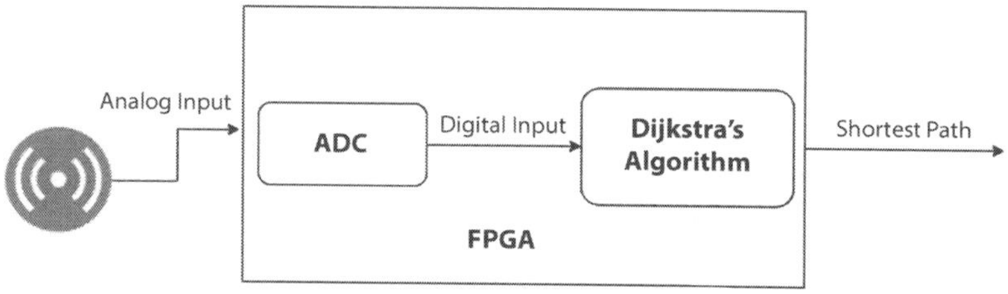

Figure 11.1 Flowchart for the components and processes involved.

delivery by sensing at every point in the path chosen, if damage is detected, a return gets initiated back through the same path from the detected point itself. This part of the supply chain can be implemented in the logistics industry to optimize delivery time, improve profits, and trust, and mainly ensuring minimal carbon footprint. We propose an autonomous contact-less delivery system for packages powered using a low power and highly efficient FPGAs along with Analog to Digital Converter (ADC) which is coupled with sensors for data acquisition. Our system consists of a robotic bot which is powered by an FPGA. The packages have a starting node and an ending node location-printed on them which are then scanned and sent to the FPGA through the means of the FPGA, and once it receives it, computes the final path and the path is sent serially to through the output port of FPGA to the robot for navigation.

In our approach, road networks could be represented as nodes in a graph using Dijkstra's algorithm, which determines the shortest pathways between nodes in a graph. The algorithm determines the shortest path between a given source node and each of the other nodes in the graph. Labels for the Dijkstra technique are fully ordered positive integers or real values. It may be generalized to use any partially ordered labels as long as they are monotonically non-decreasing because a following label is formed while traversing an edge.

The shortest path between vertices A and D and any subpath B -> D of that path are both the shortest paths between vertices B and D, according to Dijkstra's algorithm. Djikstra applied this characteristic the other way around, *i.e.*, we overstate the distances between each vertex and the origin. After that, we go to each node and its neighbors to determine the shortest subpath there. In order to identify the next best answer, the algorithm takes a greedy approach in the hopes that the final result will be the best solution for the entire problem.

We provide a clever delivery bot in this suggested model shown in Figure 11.2.

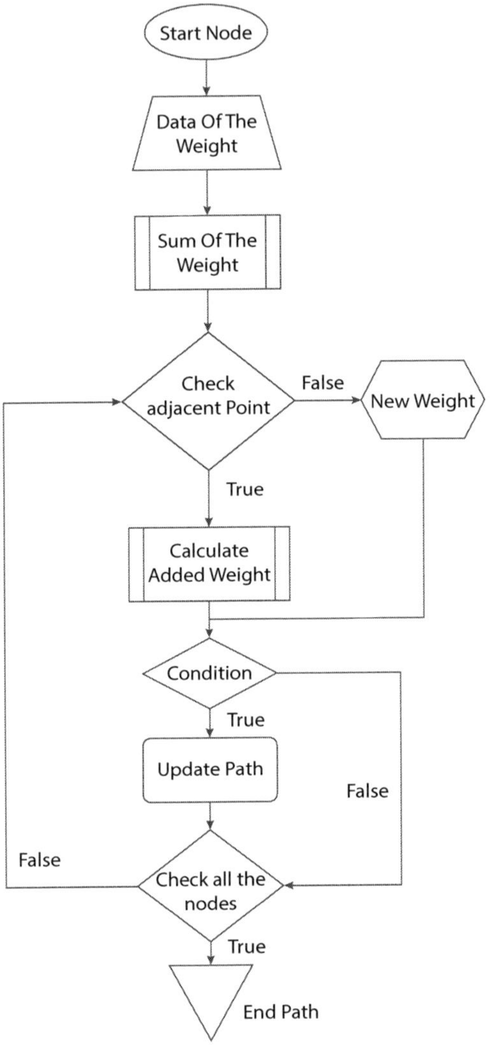

Figure 11.2 Flowchart of data transmission.

- At the early part of our algorithm, we first provide the start and end nodes to our model as input. The algorithm then gets the data for weight, calculates the sum of weight using the Dijkstra algorithm, checks the neighbor point to see if it is true, and then uses our model to calculate the weight of the added route if it is false. The amount is added up, and if the total is higher, we update the route so that we get the shortest path. If the total is lower, we check all the nodes in a loop to determine the shortest way between the start and end nodes.

- Basically, Dijkstra's algorithm starts at the node you select and examines the graph to determine the shortest path between that node and every other node in the graph.
- The algorithm keeps track of the shortest paths that are currently known to exist between each node and the source node, and it updates these values when a shorter path is discovered.
- The other node is added to the path and designated as "visited" once the algorithm has determined the shortest path between the source node and that other node.
- Up until every node in the graph is added to the path, the process is repeated. In this manner, a path that connects the source node to every other node and takes the shortest route to each node is created.

11.3 Test Graph

Figure 11.3 shows a simple graph with weighted edges representing the distance between the delivery stations. If the package is tampered with, then the bot returns to the start point that is 0. We are using a simple graph to demonstrate the working of this algorithm. The shortest path between node 0 and node 1 is 0->5->1 and the distance between them is 3 km in this test graph. The edge list of the graphs is given as follows:

- 0->2 5 km
- 0 ->5 1 km

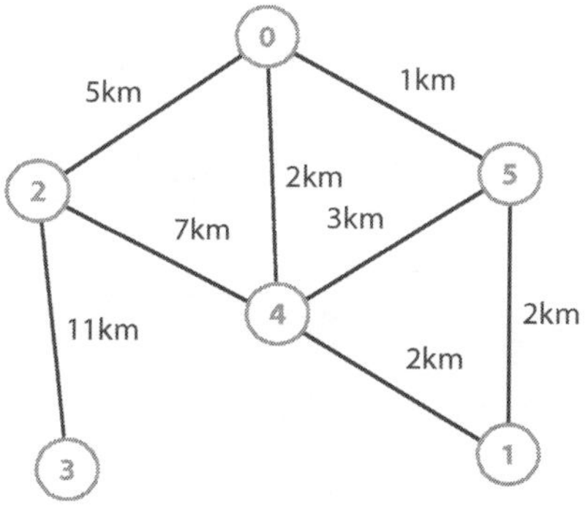

Figure 11.3 Nodes pointing sample locations and distances between them.

- 1->4 2 km
- 1 ->5 2 km
- 2-> 3 11 km
- 2 ->4 7 km
- 4 ->5 3 km

11.4 Results and Discussion

ADC Interfacing with FPGA

Our model uses ADC128S022 12-bit ADC which can allow us to interface up to eight different sensors. Here we use three different inputs, the first input gives the starting point, the second input gives the ending point, and the last input checks if the package is tampered with or opened using contact sensors. The ADC operates at a clock of 2.5 MHz which is generated from 50-MHz clock signal of the Cyclone IV FPGA. The ADC communicates with Serial Peripheral Interface (SPI) interface which provides full-duplex communication, and enables simultaneous data transmission and reception for improved signal integrity, enabling high-speed applications. We tested this by interfacing three sensors in three different channels and we used finite state machine to cycle between the three different sensors sequentially to demonstrate the working of this ADC, and once the data is read, it is stored in registers where data can be extracted in parallel. Then, according to the signals, if the package is not tampered with, then the algorithm computes the shortest path between the provided starting and ending node. If the device is tampered with, then it computes the shortest distance between the node 0 and current node to deliver back to the station.

The 12-bit ADC is interfaced with up to three different analog sensors. In Figure 11.4, the waveforms relating to the ADC interfacing are showcased.

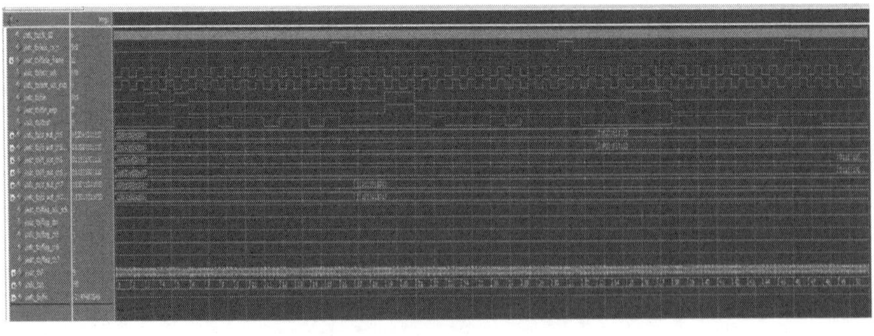

Figure 11.4 Output graph of ADC interfacing.

The ADC operates on SPI protocol so a low frequency 2.5-MHz clock signal is produced by a PWM generator with the help of the FPGA through the register adc_sck. The chip select or adc_cs_n signal is generated to select the slave connected through the SPI interface in this case, *i.e.*, to select the ADC connected to FPGA. The data input is simulated through the data_frame registers. The ADC is made to cycle between acquiring inputs from three different channels, *i.e.*, ch5–ch7. The din register represent the signal produced by the user to specify the address of the sensor to be accessed that is chosen from the eight available channels present in the ADC. The Data Output (DOUT) signal provides the 12-bit converted value of the selected channel. On power-up, channel 0 is selected by default, while subsequent reads will use the address provided in the previous operational frame. The data bits are transmitted in descending order, such that the highest-order bit is delivered first. It is captured by the user on the rising edge of adc_sclk.

The Data Input (DIN) signal is used to select the channel to be converted in the following frame. It is delivered in descending order, and is captured by the ADC128S022 on the positive edges of adc_sclk. d_out refers to the serialized signal from the ADC, the data is acquired from channel 5 in the first cycle and it is de-serialized and stored in a 12-bit register in the next cycle when the data for the next sensor in queue that is channel 6 data is being read by the ADC. d_out_ch5, d_out_ch6 , d_out_ch7 represent the data acquired from the sensors connected to the respective channels during the previous cycle.

The planning module takes less than 1500 clock cycles to compute the shortest path. In Figure 11.5, the d_out (final path) is generated before d_out_exp (final path exp), and dn (done) is generated before dn_exp. Actual outputs and expected outputs are compared only when the check signal is high. When the check signal is high for the first time, acknowledgement to the start condition is evaluated. When the check signal is high for the second time, the shortest path and done signals are evaluated. The resulting path from the starting node to the end node is computed and stored in the final path, as shown in Figures 11.6 and 11.7, which is a 50-bit register, which also acts as the output port; each node is represented by a 5-bit number from the ending node to the starting node. And to indicate the completion of the path, we used a dummy node 11111, which indicates that the path is completed.

Figure 11.5 Output window of the path planner.

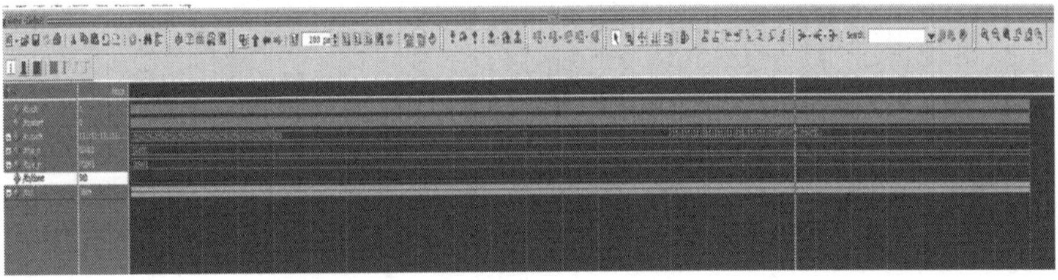

Figure 11.6 Output waveforms of the path planner.

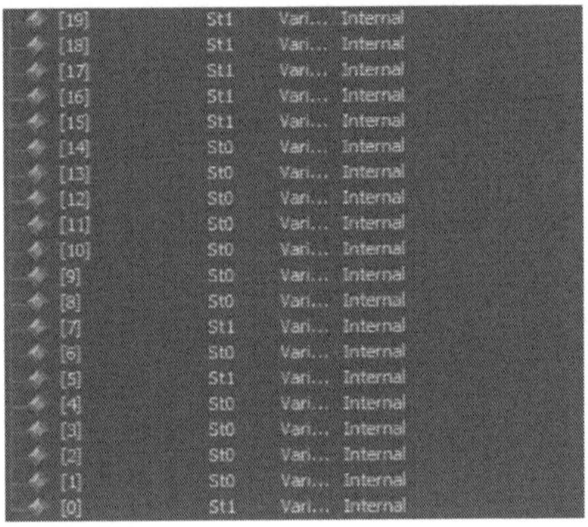

Figure 11.7 The 50-bit final path register.

We can see that we get the expected path as output from the given graph, also, a high signal is generated to indicate that the computation is completed, so immediately after which, the path can be accessed in parallel in just one cycle by the robot.

References

1. M. M. Abrar, R. Islam and M. A. H. Shanto, "An Autonomous Delivery Robot to Prevent the Spread of Coronavirus in Product Delivery System," *2020 11th IEEE Annual Ubiquitous Computing, Electronics & Mobile Communication Conference (UEMCON)*, 2020, pp. 0461-0466, doi: 10.1109/UEMCON51285.2020.9298108.
2. Khamis, Alaa and Meng, Jun and Wang, Jin and Azar, Ahmad Taher and Prestes, Edson and Takács, Árpád and Rudas, Imre J. and Haidegger, Tamás (2021) Robotics and Intelligent Systems Against a Pandemic. *ACTA*

POLYTECHNICA HUNGARICA, 18 (5). pp. 13-35. ISSN 1785-8860 (print); 2064-2687.

3. Anitha, K. and Srija, V., Implementation of Automatic Car Parking System Using Verilog HDL.
4. Guoqing Lei; Yong Dou; Rongchun Li; Fei Xia; "An FPGA Implementation for Solving the Large Single-Source-Shortest-Path Problem"; *IEEE Transactions on Circuits and Systems II: Express Briefs* (Volume: 63, Issue: 5, May 2016).
5. Matti Tommiska and Jorma Skytt¨a; "Dijkstra's Shortest Path Routing Algorithm in Reconfigurable Hardware"; *Field-Programmable Logic and Applications, 11th International Conference*, FPL 2001, Belfast, Northern Ireland, UK, August 27-29, 2001, Proceedings.
6. "Mohammad Tanzil Idrisi; Enhancing the Field Programmable Gate Array (FPGA) Use by Shortest Path Algorithm; *International Advanced Research Journal in Science, Engineering and Technology (IARJSET),* Vol. 7, Issue 11, November 2020".
7. Yasuhiro Takei, Masanori Hariyama and Michitaka Kameyama; Evaluation of an FPGA-Based Shortest-Path-Search Accelerator; *International Conference on Parallel and Distributed Processing Techniques and Applications (PDPTA)* pp.613-617 (2015-07-29, Las Vegas, Nevada, USA).
8. OLEG G. MONAKHOV, EMILIA A. MONAKHOVA, ALEKSANDR YU. ROMANOV, ANDREJ M. SUKHOV, AND EVGENII V. LEZHNEV; "Adaptive Dynamic Shortest Path Search Algorithm in Networks-on-Chip Based on Circulant Topologies"; *IEEE Access* (Volume: 9).
9. Yuzhi Zhou,1Xi Jin,1and Tianqi Wang1; "FPGA Implementation of A* Algorithm for Real-Time Path Planning"; *Int. J. Reconfigurable Comput.*, vol. 2020.
10. Michael J Alexaner; Gabriel Robins; "New Performance-Driven FPGA routing algorithms"; *IEEE Transactions on Computer-Aided Design of Integrated Circuits and Systems*, Vol. 15, No. 12, December 1996.
11. Uday Bondhugula, Ananth Devulapalli, Joseph Fernando, Pete Wyckoff, and P. Sadayappan, "Parallel FPGA-based All-Pairs Shortest-Paths in a Directed Graph"; *20th IEEE International Parallel & Distributed Processing Symposium (IPDPS '06)*, Apr 2006.

12

Cabin Cooling System for Heavy Commercial Load Vehicle

Aditya Burde, Samridhi and P. Sriramalakshmi*

School of Electrical Engineering, Vellore Institute of Technology, Chennai, India

Abstract

In this chapter, the idea of a solar energy-powered cooling system for trucks to maintain the cabin temperature for long journeys (without utilizing power from the engines) is proposed. Beginning with a review of the problem the drivers are facing due to the lack of a proper cooling system in the cabin. The impact it has on the environment, with the present resources, and its monetary effects. The design of the proposed model is elaborately discussed in this chapter. Practical implementation details from the various laboratory results and outdoor tests are also included.

Keywords: Solar energy, sustainability, cooling system, cabin temperature control

12.1 Introduction

In this fast-paced world, a large number of trucks can be seen traveling endlessly to deliver goods on time. According to a survey conducted in 2019, there are over 295 million registered vehicles in India. And due to the scorching heat and the issues stated before, the level of pollution does not look like they are going to be reduced anytime soon. Trucks are a significant part of the transportation of products in street transport. They are used to go over significant distances fluctuating from district to area under various climatic circumstances. A study says that 66% of the air pollution deaths in India is caused by diesel vehicles [1].

**Corresponding author*: sriramalakshmi.p@vit.ac.in

O.V. Gnana Swathika, K. Karthikeyan and Sanjeevikumar Padmanaban (eds.) Smart Grids as Cyber Physical Systems Volume 1: Artificial Intelligence, Cybersecurity, and Clean Energy for Next Generation Smart Grids, (173–184) © 2024 Scrivener Publishing LLC

The examinations show that lodge temperature surpasses 55°C in summer and bone-dry regions. It is incredibly hard for the driver to work under such circumstances. It additionally decreases the work effectiveness of the transport drivers. So, lessening the temperatures is essential inside the lodge of the truck and to give solace to the driver.

The Indian government, through a change to the Motor Vehicle Act of 1988, has made it obligatory for producers to fit forced air systems in the lodge of all-new trucks from April 2017. But the system cost and maintenance directly put a big load on truck owners' pockets, resulting in higher transportation charges. To resolve this issue, a solar-powered Peltier cooling unit is used, which is low in maintenance and almost one-tenth of the price of an air conditioning unit [2, 3]. The system is not as efficient as that of an air conditioning unit but still attains a minimum temperature of 19°C.

12.2 Literature Survey

Trucks stand by for extensive stretches to keep the vehicle's temperature suitable for the weather. That is on the grounds that nearby, state and local specialists the nation over have sanctioned the violators of sitting guidelines to assist with decreasing outflows from substantial trucks. Obviously, drivers actually need an environment-controlled climate when the truck is left for a short-term visit, or while they are stacking and dumping. One warming/cooling strategy that is acquiring notoriety is battery-controlled assistant HVAC frameworks. These independent systems can work with the truck's current battery or a committed bank of batteries that are charged with the energy from the alternator when the motor is working. When the motor is switched off, the put-away energy controls the helper regarding HVAC framework. Volvo Trucks recently introduced its own battery-powered HVAC system. The parking cooler was also introduced for all sleeper models. It's likewise accessible as an additional arrangement for professional drivers that may need to keep up with vehicle cabin temperature without sitting. Evaluated to save the vehicle inside at 78°F for 10 h, the parking cooler promises a harmless to the ecosystem answer for environment control.

It is necessary to keep in mind that, alongside cutting outflows from sitting, a battery-controlled HVAC framework can likewise set aside one's cash. Since sitting (where allowed) utilizes liters of fuel per hour, the reserve funds could sum up rapidly when the motor never again should be on for taxi solace. Running high drive motors constantly at low revolution per minute (RPM) adds pointless mileage on basic motor parts, bringing about higher support costs and more limited motor life. Contrasted with motor-terminated assistant power source units, battery-fueled frameworks function more calmly and require less support as they require fewer moving parts. A couple of battery-controlled systems, including the new Volvo trucks parking cooler, gives the decision to chip away at shore power for added flexibility.

These points can assist with guaranteeing maximized execution from the battery-controlled helper HVAC framework:

12.2.1 Beginning With the Principal Warmer or A/C

A helper HVAC unit works best in the event that it is utilized to decisively keep up with the current temperature inside a taxi, not intensify or cool it. The Program to utilize the vehicle's primary warmer or climate control system no less than 30 min prior to closing off the motor so you begin at an open setting. Then, at that point, utilize the helper framework to keep up with the thermoregulator.

12.2.2 Additional Protection

The better the in-taxi protection, the simpler it will be to assist HVAC to go about its business. Consider including froth away from plain view or introducing additional protection in the taxi and sleeper. Weighty shades outwardly windows and between the sleeper and taxi can likewise assist with keeping the temperature constant as per the requirement which reduces the heap in the helper section.

12.2.3 Utilizing Genuine Profound Cycle Batteries

Battery-fueled helper HVAC frameworks pull low units of current into the battery plates for a long time, releasing a battery from 80° or more north of a long reprieve interval. Not at all like beginning or double reason batteries, profound cycle batteries are exceptionally intended to deal with the intensity and stress of the continued releasing and charging. While they cost more, it is found that profound cycle batteries provide superior worth in light of cost per cycle.

For loosening up rest times: with the MAN genuine helper cooling frameworks, the transporter can have a loosening up break in the taxi without leaving the motor running for a really long time and consuming a lot of fuel. A very much refreshed driver likewise increases wellbeing. Due to their profoundly effective innovation, they give fast and dependable temperature control for up to 12 h at minimal working expenses.

The conservative, battery-operated rooftop-mounted cooling framework for heavy-duty vehicles can, without much of a stretch, be introduced into the current rooftop hatch.

It is advised that the vehicles are outfitted with a 225-Ah battery and with a better presentation Longlife ecological power source. The framework's preset low voltage security should not be adjusted. In the wake of involving the framework for a lengthy period, it is suggested to fire up the vehicle prior to applying one more power buyer to try not to deplete the battery.

12.2.4 Roof-Mounted Air-Conditioning System RTX 1000

COST: 1500 USD = 118635 INR

The auxiliary cooling framework is intended for loosening up rest periods at halt. Because of its profoundly efficient innovation, it cools rapidly and dependably for as long as 12 h. The compact, battery-worked rooftop-mounted cooling framework for trucks can be effortlessly introduced into the current rooftop hatch.

12.2.5 Roof-Mounted Air-Conditioning System RTX 2000

COST: 2244 USD = 177477.96 INR

Due to its high-tech performance, it cools rapidly and dependably for as long as 6 h and longer. The minimized, battery-worked rooftop mounted cooling framework for trucks can be effortlessly introduced into the current rooftop hatch.

MAN SPX 1200 auxiliary air-conditioning system for roof and rear wall mounting:

COST: 2999 USD = 237190.91 INR

The auxiliary cooling framework is intended for loosening up rest periods during a halt. Because of its efficient technology, it cools rapidly and dependably. The additional level rooftop unit saves space and barely influences the general level of the rooftop.

Consequently, it is likewise appropriate for TGX vehicles with GM or XLX vehicles and rooftop spoilers. The evaporator unit SPX 1200T (in the rooftop hatch) and the condenser unit SPX 1200C (on the back wall) is introduced independently (split mounted).

12.2.6 Cooltronic G2.5 Auxiliary Air-Conditioning System

COST: 2495 EURO = 205889.98 INR

MAN genuine rooftop-mounted cooling frameworks guarantee to loosen up breaks and rest periods in the vehicle under ideal circumstances. They likewise increase the driver's consideration and in this manner, guarantee expanded wellbeing out and about. They are free-cooling frameworks with electric air blowers and a shut circuit that consequently changes its cooling result to the least conceivable power utilization.

On account of the extraordinary two-section plan, the truck rooftop-mounted cooling framework can be mounted underneath the rooftop spoiler and is subsequently especially appropriate for MAN TGX GM or XLX vehicles with the cabin of medium level and rooftop spoiler. However, the MAN Cooltronic G2.5 can likewise be utilized with any remaining MAN TG vehicles with long cabins.

The proposed project unit contains a manual temperature setting panel from which the user can set the required temperature. And a smart temperature sensor to detect any increase in temperature, if any. Some of the major elements that are part of the system are the Peltier cooling unit and the thermal insulators. Their working principles are stated below.

12.3 Working Principle of Peltier Cooler

Thermoelectric coolers work as indicated by the Peltier impact. The impact makes a difference in temperature by moving heat across the nodes. The required amplitude of the voltage is applied across the electrodes to regulate the flow of current. At the point when the ongoing courses cross the intersections of both the conductors, heat is drained at one of the nodes in the junction and cooling happens. The heat energy released is stored across the second intersection.

12.3.1 Elements of Peltier Cooler [4–7]

The thermoelectric coolers during II-VI go about as a solid-state heat sink. Each includes different subbing n-and p-type semiconductors. The semiconductors of various categories have compared Peltier coefficients. Various components are affixed in between a pair of pottery plates. These are connected in series which are found to be thermally equal. Strong arrangements of bismuth telluride, antimony telluride, and bismuth selenide are the favored materials for Peltier impact gadgets since they provide the best exhibition from 180 to 400 K and constructed using both n-type and p-type. The cooling effect between as many numbers of thermoelectric coolers is compared with a number of coolers used. Normally, different thermoelectric coolers are very closely related to the one placed next and kept between two metal plates. There are three different types of unique thermoelectric coolers, namely, thermocyclers, single-stage coolers, and multi-stage coolers.

12.3.2 Heat Absorption

When the current passes through one bunch of portions from n- to p-type, cooling happens. Moreover, there is a reduction in temperature at the convergence ("cold side"), achieving the ingestion of hotness from the encompassing. The hotness is conveyed alongside the parts by electron transport and followed through on the reverse ("hot") side as the electrons crosses from a high-energy state to low-energy state.

The Peltier heat holding capacity is expressed as,

$$Q = P\,I\,t$$

Where P is Peltier coefficient, I is current, and t is time. At this stage, thermoelectric cooler can convey a most outrageous difference in the temperature of around 70°C. Nonetheless, II-VI's Triton ICE thermoelectric cooler cools the equipment at the maximum of 2°C under stream market commitments.

12.3.3 Thermal Insulation

Vehicle thermal insulation will keep you cool in the late spring by reflecting undesirable motor, header, exhaust, and summer heat, yet in addition, keep you warm in the colder time of year. It is incredibly adaptable and exceptionally simple to cut, handle, and introduce. It is utilized by proficient race groups, auto reclamation shops, and many clients worldwide.

12.4 Proposed Idea

To solve this problem, it was planned to make a semi-automated cooling system powered by solar panels [8–12]. Place a small solar-powered cooling unit inside the car's cabin which will be connected to an external battery. A switch in the cabin will control the entire system and users can adjust according to their needs. Alongside the main system, we will be using thermal insulation material and reflective sheets on the doors, windows, and roof to overcome the extra temperature rise of the cabin. For this project design, the design of the most selling car that has been taken into consideration is Swift Dzire. Swift Dzire was observed and studied thoroughly on the basis of dimensions, weights, battery capacity, engine, volume of air required per person, and other aspects.

Dimensions: $1.735*2.154\ m^2$

According to the study, it is needed to install solar panels of rating up to 40 W or more as per the charging capacity of the battery [13–15].

Figure 12.1 shows the block diagram of the proposed idea.

Figure 12.2 depicts the circuit diagram of the power supply set up which is connected with the solar panel and through which the cooling unit is powered up.

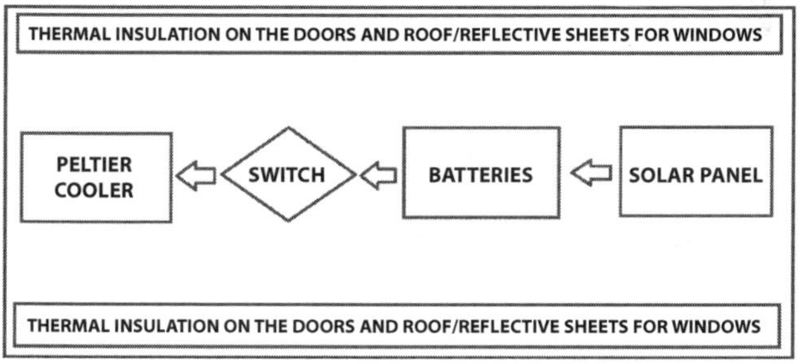

Figure 12.1 Block diagram of proposed idea.

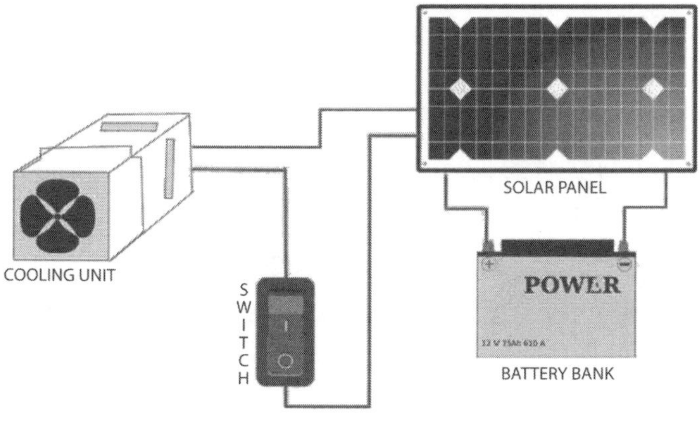

Figure 12.2 Circuit diagram.

12.5 Design Specifications

Specifications for various components required to complete one unit are as stated :

- Thermoelectric Peltier cooling system
  Voltage: 12 V
  Operating current: 6 A
  Dimensions in mm: 220*140*40
  Weight in grams: 390 g
- Battery Requirement
  Lithium-Ion: 12 V 8 AH
- Insul-Layer 6 mm roof lining barrier
  Dimensions:1200 mm*900 mm
- Solar panel
  Maximum power: 60 W
  Open circuit voltage: 22 V
  Voltage at maximum power: 18 V
  Short circuit current: 3 A
  Current at maximum power: 2.8 A
  Tolerance: ±5%
- Switch: push-button switch
- Reflective sheets
- Wire: 2.5-mm copper wire

12.6 Prototyping

SOLIDWORKS software was used to design the layout of the commercial vehicle with a cabin length of 1534.21 mm, cabin width of 1960.18 mm, and cabin height of 1760 mm. The dimension of the product that is attached on the rear is 1500 x 300 x 250 mm and here, the solar panel dimension is 1518 x 732.33 x 36 mm. Figures 12.3.1 and 12.3.2 show the lateral and top view of the panel placement set up.

Figures 12.4.1 and 12.4.2 show the back and lateral view of the product placement. The proposed idea is initially tested in a car.

The idea is proposed for truck or heavy vehicles. For the testing purpose, the prototype of the cooling system is kept inside the car. The prototype is placed in a car, as shown in Figure 12.4.3, with a bigger cabin dimension than the truck cabin, and is successfully able to cool the surroundings in several minutes.

Figure 12.3.1 Panel placement (lateral view).

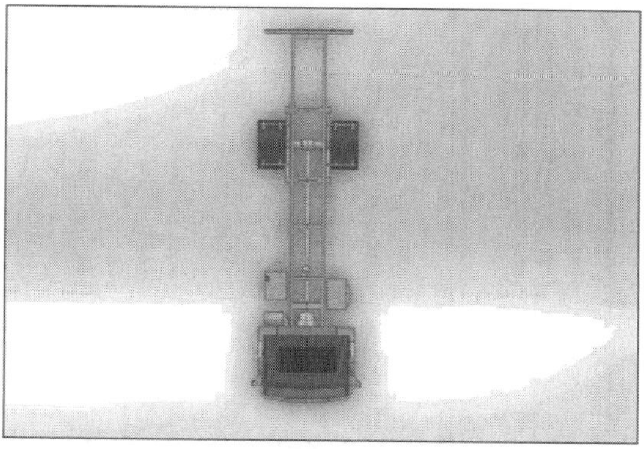

Figure 12.3.2 Panel placement (top view).

Figure 12.4.1 Product placement (back view).

Figure 12.4.2 Product placement (lateral view).

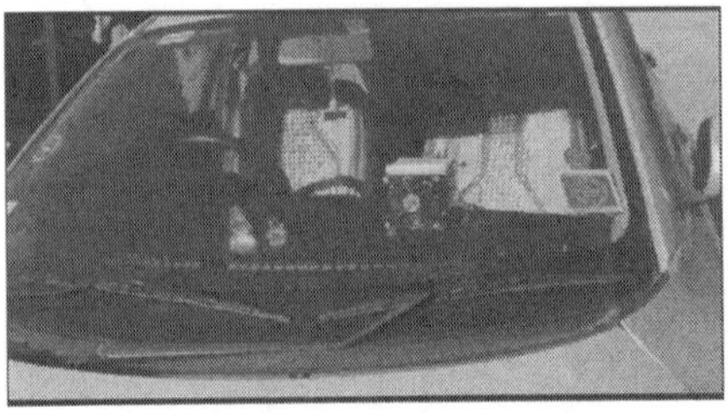

Figure 12.4.3 Product placement in a car.

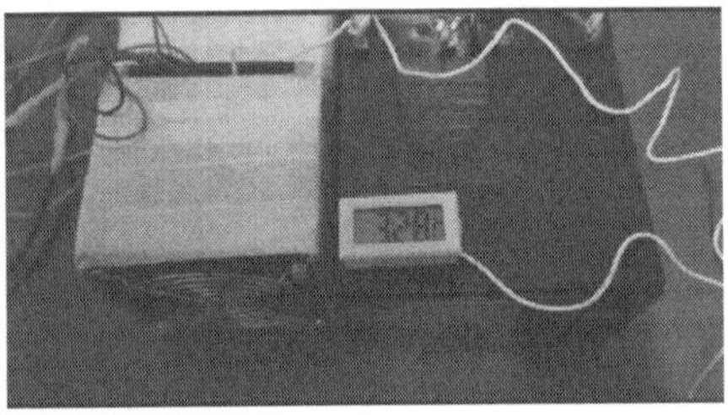

Figure 12.5 Surrounding maximum temperature.

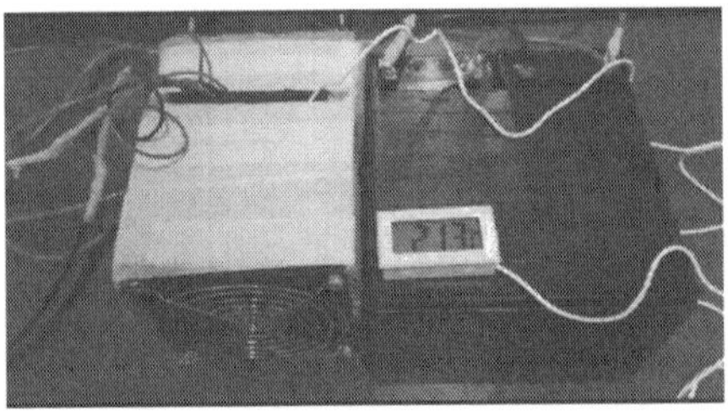

Figure 12.6 Minimum temperature attained.

After testing the product, a temperature drop of 11 degrees was noted in an open space and when tested in the car, a minimum of 17°C temperature was noted. The product can run without any charge for 45 min with the battery backup and can maintain the temperature eventually.

Figure 12.5 shows the temperature display and the minimum temperature attained after few minutes is shown in Figure 12.6.

12.7 Advantages of Proposed Idea

The proposed idea enables the user to enjoy the comfort of a cool cabin while stuck in traffic without having to turn on the engine or wasting money on the extra fuel.

It is the first of its kind as no such relevant product is present in the market which will allow the customers to maintain the cabin temperature without turning on the engine. In addition, this proposed idea will also help to reduce the carbon footprint in the environment, hence decreasing pollution, majorly bringing down the pollution rates. Some high-end trucks do come with a temperature regulation module but that needs the engine to be turned on and is very expensive to be used by common people. The installation of these modules is not easy as well. On the other hand, the project is easy to install as well as easy on the pockets of the customers, which is unique and can be used by anyone with any kind of truck.

12.8 Conclusion

At the end of the testing, favorable results are attained. The temperature drop is noticeable significantly. Several in-laboratory and outdoor tests are conducted on the prototype to optimize the product results. The flow of air in the cabin is able to cool the cabin and the sensors are working fine to regulate the temperature according to the user settings.

References

1. https://swachhindia.ndtv.com/india-air-pollution-related-deaths-due-to-diesel-vehicles-says-a-new-study-31853/#
2. Kreith, F. Solar energy handbook. United States: N. p., 1981. Web.
3. T.T. Chow, A review on photovoltaic/thermal hybrid solar technology, *Appl. Energy*, (2010).
4. K. G. B. Arjun, B. G. Pruthviraj, K. Y. K. Chethan, and P. Rashmi, "Design and implementation of Peltier based solar powered portable refrigeration unit," 2017 *2nd IEEE International Conference on Recent Trends in Electronics, Information & Communication Technology (RTEICT)*, 2017, pp. 1971-1974, DOI: 10.1109/RTEICT.2017.8256943.
5. R. Sharma *et al.*, "Peltier Effect Based Solar Powered Air Conditioning System," *2009 International Conference on Computational Intelligence, Modelling and Simulation*, 2009, pp. 288-292, doi: 10.1109/CSSim.2009.40.
6. Hirabayashi, Timothy, *et al.* "Application of Noise Control and Heat Insulation Materials and Devices in the Automotive Industry." *SAE Transac.*, vol. 104, 1995, pp. 2529–46.
7. Cai, Z.; Al Faruque, M.A.; Kiziltas, A.; Mielewski, D.; Naebe, M. Sustainable Lightweight Insulation Materials from Textile-Based Waste for the Automobile Industry. *Materials* 2021, 14, 1241

8. Mario Pagliaro Dr., Rosaria Ciriminna Dr., Giovanni Palmisano," Flexible Solar Cells", 14/09/2008, Citation no:224.
9. Gao, Ew., Zhang, Xp. & Yao, Zq. Simulation and analysis of flexible solar panels' deployment and locking processes. *J. Shanghai Jiaotong Univ. (Sci.)* 13, 275–279 (2008).
10. M. Parans ParanthamanWinnie Wong-NgRaghu N. Bhattacharya, Semiconductor Materials for Solar Photovoltaic Cells, 2020 Springer Nature Switzerland.
11. L.C. Hirst, Principles of Solar Energy Conversion, *Comprehensive Renew. Energy*, 10.1016/B978-0-12-819727-1.00112-6, (234-255), (2022).
12. Solar energy storage in the rechargeable batteries panel QiLi YangLiu Shaohua Guo Haoshen Zhou.
13. Mantell, C L. Batteries and energy systems. United States: N. p., 1982. Web.
14. Operation conditions of batteries in PV applications, Andreas Jossen, Juergen Garchea, Dirk UweSauerb.
15. Drebushchak, V.A. The Peltier effect. *J. Therm. Anal. Calorim.*, 91, 311 (2008).

13
Renewable Energy and Its Dynamic Value

Abhinav Koushik and Milind Shrinivas Dangate*

Chemistry Division, School of Advanced Sciences, Vellore Institute of Technology, Chennai, India

Abstract
Intermittent renewable energy sources significantly reduce local air pollution and mitigate climate change. However, because of their erratic generation, it is difficult to incorporate these resources into the electricity system. In order to value wind and solar energy, it is necessary to address both of these problems, particularly in light of the laws and incentives designed to encourage their widespread use. The utility of wind and solar energy is assessed in this dissertation by simulating their daily intermittency and interactions with hydropower in California and storage in Texas. Here, we estimate how much the state of California's hourly wind and solar power cut carbon dioxide, sulfur dioxide, and nitrogen oxide emissions. So this offsets depend on direct replacement of expensive natural gas producers and the reallocation of hydropower to times when the growth in renewable energy is least. Due to the intermittent nature of solar power, hydro generation shifts from the afternoon to the evening. This increases the amount of carbon offsets needed because the gas generators that take their place in the evening are worse than those that are left running in the day. In contrast, as wind causes the replacement of generators with comparable emissions intensities, wind offsets are less susceptible to the reallocation of hydropower.

Keywords: Renewable energy, climate change, wind, solar

**Corresponding author*: milind.shrinivas@vit.ac.in

O.V. Gnana Swathika, K. Karthikeyan and Sanjeevikumar Padmanaban (eds.) Smart Grids as Cyber Physical Systems Volume 1: Artificial Intelligence, Cybersecurity, and Clean Energy for Next Generation Smart Grids, (185–198) © 2024 Scrivener Publishing LLC

13.1 Introduction

1. What is the value of renewable sources, similar as wind and sun energy? These resources can lower emigrations and the cost of generating electricity for the grid. But integrating them into traditional power grids can be difficult because of their cyclical and random intermittency. By accepting electricity to use when it is most precious, electricity storage facilities could facilitate this integration. But simply increasing grid storage could actually decrease welfare. If intra-day storehouse arbitrage results in a significant rise in out-peak coal power rather than cleaner, on-peak creation, then determining the entire value of intermittent renewables calls for a dynamic framework that accounts for all internal and external advantages and costs. The value of intermittent energy in California will be evaluated in this debate, which have embraced renewable energy early. Investigated in the context of how the relationships between hydro, wind, and solar power affect the favorable effects of the final two. Estimated how significantly they lower emissions of carbon dioxide, sulfur dioxide, and nitrogen oxides in California using arbitrary oscillations in hourly wind and solar power. These equipoises rely on the direct demotion of expensive natural gas producers and the relocation of hydropower to the times when the amount of renewable energy is at its lowest. Addressed the shift in hydro with a dynamic econometric model that accounts for the relations between the three renewable sources. Wind, solar, and hydro generation trends in California for 2011–2015 show that redistribution follows diurnal patterns. Hence, these data clearly shows that accounting for the dynamic relationships among the three renewable energy sources is required for determining the correct external benefits and value of intermittent renewables. In California, hydropower improves the exports of solar electricity, whose external benefits range between 9.78 and 30.08 USD/MWh, whereas the range for wind is 8.31 to 1.00 USD. One of a crucial perceptivity from this environment is that hydropower in California is formerly helping to smooth the integration of wind and solar generation. Since implicit hydroelectric generation (*i.e.*, water behind the levee) can be erected when prices are low and drawn down when prices are high, hydroelectric capacity acts much like storehouse capacity in this regard. As the context makes clear, the impact of intermittent renewables on costs and emigrations will probably depend on the presence of storehouse capacity. Thus, in the third environment, pretend the impact of mileage-scale storehouse in the Texas electricity request [1]. While in Chapter 2, they used literal data and time series econometrics to value the environmental benefits of wind and solar

power in the environment, and developed an electricity request model that accounts for environmental benefits and energy cost reductions in assessing the value of wind power and storehouse.

2. With a confidence interval of 95% built using the Newey West standard errors of the dynamic offset estimates.
A theoretical model and an empirical model of welfare maximization in the quotidian electricity request are used to examine how the value of wind, storage, and (dis)charging allocations depend on emigrations price. The model's key insight, consistent with the Texas grid, is that wind power is higher during off-peak hours and adversely correlated with demand. Demonstrate that the value of wind generating is diminishing because as its capacity grows, new generation will take the place of the most valuable archconservative power, resulting in lower wind value and lower equilibrium pricing, leading to a weaker wind value. Since the endogenous arbitrage decisions take into account both private costs and emissions damages, emissions taxes raise the value of storage. At current wind capacity (providing 12% of load) and without storage, wind's economic value equals its levelized cost in all scenarios (60.00 USD/MWh). However, disregarding wind's ability to reduce CO_2 and air pollution results in private surplus gains that are barely half the levelized cost (26.30 to 27.40 USD/MWh on average).

3. Using the literature's social cost of carbon for a discount rate of 3%, measured in 2015 US dollars (40.00 USD/tCO2).
The profitable value of storehouse nearly doubles its private fat earnings. Indeed, under the large intermittency of the 30 wind script, ERCOT's planned 324 MWh of storehouse has an estimated profitable value lower than its levelized cost of 420.00–950.00 USD/kWh [2]. Still, the enforced model only accounts for hourly intermittency smoothing. The value is likely to be larger if we include minute position intermittency smoothing (1 min, 5 min) and frequency regulation services. It showed that storehouse moves power from off- peak to peak. This arbitrage is grounded on the differences in electricity demand, wind power generation, and emigrations intensities throughout the ages. Trying emigrations reduces storehouse and arbitrage by counting for the advanced emigrations intensity of out-peak creators. Simulation results show that simply trying emigrations leads to a larger weal gain than enforcing ERCOT's planned storehouse of 324 MWh with no environmental levies. Both low and unconstrained storehouse have less than a 50% utilization rate since hourly arbitrage mostly charges during several hours in the morning to release power in the afternoon and evening.

13.2 Is a Wetter Grid a Greener Grid? Estimating Emigration Equipoises for Wind and Solar Power in the Presence of Larger Hydroelectric Power

In sequence to achieve CO_2 and original adulterants reduction pretensions, an electric grid powered by a significant share of renewable energy is one of the given alternatives [3]. Nonetheless, the intermittent nature of solar and wind power and the incipient mileage scale electricity storehouse make it gruelling to assess the emigrations equipoises of these power sources. Prostrating those methodological challenges is needed to approximate the profitable value of energy and to estimate affiliated programs similar to feed-in tariffs, renewable portfolio norms, and subventions. Using literal facts on the randomness of hourly cosmic and air creation from 2013 to 2015, it was estimated how important CO_2, SO_2, and NO_x are to these sources abate in the electric grid in California, a worldwide leader in renewable energy relinquishment. The new challenge lies in relating the borderline generation and emigration equipoises caused by adding two intermittent sources to a grid that has a significant share of hydropower. Former literature, centered on grids with small fragments of hydro generation, has linked emigration equipoises related to the immediate relegation of the loftiest borderline cost reactionary creators by zero borderline cost renewables through the merit order effect [4].

13.2.1 Data

For the econometric models, this is used in per hour on level load and all generation types (fossil, hydro, nuclear, wind, solar, imports and few sources of energy, which combines solar, natural mass, and biogas) from 2013 to 2015 in California. Scraped data from the OASIS server in order to get 26,280 observations. Missing observations represent 0.11%. Used hourly wind and solar power hour-ahead forecasts from OASIS but this data was only publicly available from December 2014 onwards.

Also used hourly average temperature data from several California rainfall stations and hourly carbon dioxide, sulfur dioxide, and nitrogen oxide emigrations data only mingles information for reactionary energy powered units whose capacity is lesser than 25 MW. Nonetheless, utmost power shops in California Independent System Operator (CAISO) report their emigrations to Continuous Emission Monitoring System (CEMS). To regard for the carbon equipoises coming from electricity significances, GHG emigrations force of all power shops outside California that vend power to this grid were used [5].

13.3 Wind, Solar, and Hydropower Trends in CAISO

13.3.1 Power Generation Trends

From 2013 to 2015 in the California independent system operator, the share of air and cosmic capability and creation was summed up to the column. Therefore, there are lower than 1 to 7, and also it dropped to 5 in 2015. It reached 22 shares of renewable in the column. As stated in the theoretical model, this growing share of power does not indeed make a generation donation throughout the day. During these times, the hydropower wind exhibits an elaboration towards the "duck wind," a hollow pace. Notice that the goods of the California failure push the hydro generation angles down throughout the times. Nonetheless, the appearance of the convex belly at noon is entirely a redistribution effect (Figure 13.1) [6].

Hydropower reallocation can be better depicted by using the relative change in the hourly generation shares of each day. In 2011, solar power was only 1% of the total generation while wind was 3%. In the following years, solar capacity outgrew wind power capacity which determined three reallocation patterns. First, from midnight to 7 AM, when the sun is gone and wind peaks, there has been a drop in hydro generation which follows a wave pattern: reductions are larger from 1 to 3 AM and as wind power dwindles and solar kicks in, they decrease. The second wave goes from 8 AM to 3 PM: reductions increase and decrease with solar generation, peaking at midday. Finally, the displaced hydro is reallocated during the

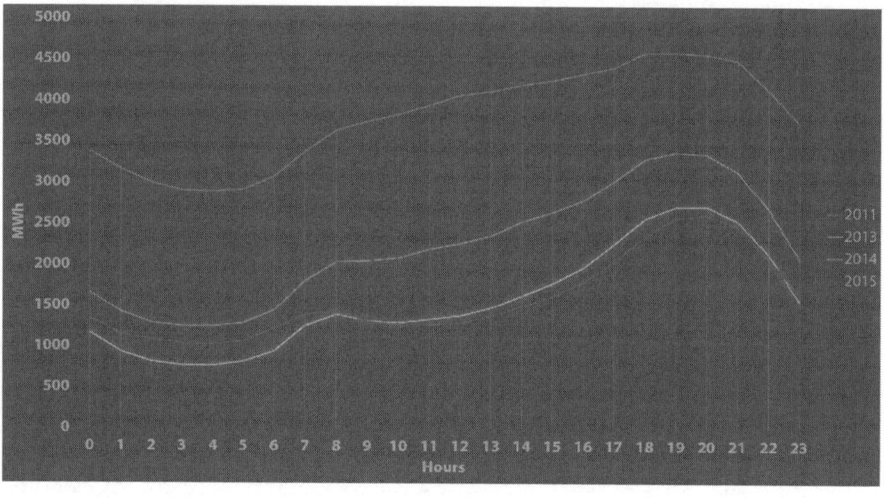

Figure 13.1 CAISO average aggregate hydropower hourly generation.

third wave (4 PM – 11 PM) as the sun sets, and these generation increases dwindle as wind ramps up.

13.4 Identification

The assumption and randomness of air and cosmic power to find their reactionary generation equipoises and affiliated CO_2, SO_2, and NO_x emigrations equipoises during 2013–2015. Since air and cosmic affair are decided by the environment, which are external to the profitable determination way that helps electricity generation in the non-commercial request, their arbitrary time difference can find differences in the electric patch and posterior emigrations. This identification strategy is the same as the one used by Cullen (2013) and Novan (2015) to identify wind power borderline equipoises in Texas. Likewise, the ultimate uses wind speed as an instrument to control for possible endogeneity, chancing that the equipoises estimates are analogous using wind generation or speed. This suggests that using hourly arbitrary generation does not lead to prejudiced estimates. Given this: these vaticinations are grounded on exogenous rainfall data which makes them uncorrelated with any curtailment opinions that could affect the factual renewable generation. In the proposed, fixed goods can address the impact of failure and snow pack loss on hydropower generation. Grounded on the stoked, none of those time series view the presence of roots. This allows for the identification of unproductive goods without using any cointegration procedures. Likewise, carbon emigrations equipoises peak during summer due to the large electricity demand in those months, which lead to a larger use of reactionary generation and advanced emigrations than the rest of the time. On the other hand, sulfur and nitrogen oxides emigration equipoises show lower seasonal variability and show a slight drop during the summer. The seasonality of emigrations equipoises depends substantially on cargo than on hydro force situations and generation (Figure 13.2). Nonetheless, controlling for the dynamics of hydropower redistribution is still necessary to estimate equipoises adequately.

As shown in Figure 13.2, dynamic average carbon emissions offsets and load throughout seasons. In principle, a price on emissions, through a tax or cap and trade, should automatically send the correct signals on the value of renewables. Nevertheless, if this price is lower than the estimated marginal damage of the externality, it could lead to undervaluation and underinvestment. Average carbon prices in California for 2015 were 12.00 USD/tCO_2, which significantly reduces the value of the external benefits of renewables and undermines the rationale for justifying their incentives (Climate Policy

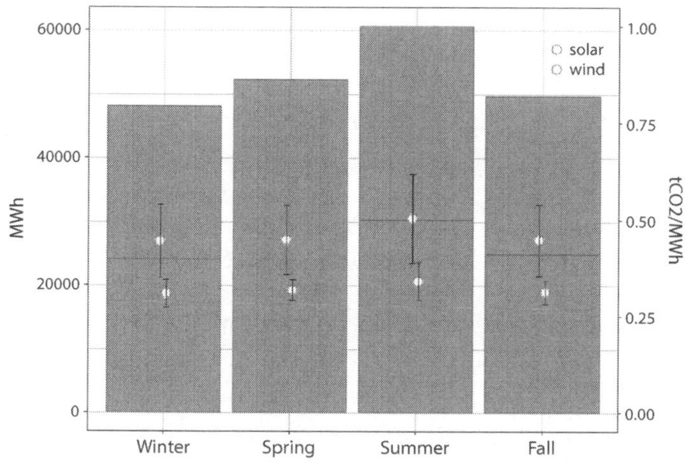

Figure 13.2 Dynamic average carbon emissions offsets and load throughout seasons.

Initiative, 2017). Reference [7] argue that these low carbon prices are caused partially by overlapping environmental policies and incentives. Harmonizing these different incentives to provide clear signals on the value of wind and solar is a pending challenge.

Finally, the current policies and incentives should reflect the varying level of benefits of renewables throughout seasons and years. This is difficult to implement with the renewable portfolio standard (RPS) and tax credits but can be more easily implemented in the case of feed-in-tariff contracts. Therefore, the above lessons can shed some light to new contracts and schemes that want to take into account the heterogeneity of benefits among technologies and throughout time.

13.5 Electricity Storehouse, Emissions Levies, and Value of Renewable Energy

13.5.1 Introduction

What use do intermittent renewable energy sources like air and sun serve? These resources can lower emigrations and the cost of generating electricity for the grid. But integrating them into traditional power grids can be tricky because of their cyclical and random intermittency. By allowing renewable electricity to be used when it is most valuable, electricity storage facilities could facilitate this integration. However, merely adding more storage space to the grid can cause more emigration. Intra-day storehouse arbitrage results in a significant rise in out-peak coal power rather than cleaner, on-peak natural gas generation. As a result, determining the entire value of intermittent renewables calls for a dynamic framework that accounts for all internal and external gains and costs. In this investigation, the worth of wind generation and storehouse capacity was evaluated by developing theoretical and empirical models of the Texas Electric Reliability Council of Texas (ERCOT) electricity grid. The models regard for the relations between wind, storehouse, and emigrations levies. The main theoretical perceptivity show that the value of wind is exogenously determined by electricity demand and its cycles while the value of storehouse depends on the timing of charging and discharging events. The empirical results show that wind and storehouse are complementary since wind's intermittency raises arbitrage benefits, which leads storers to reallocate power grounded on wind cycles, adding its value. Emigrations levies increase net weal and the value of storehouse since they give the correct impulses to dispatch generation that maximizes social benefits. At 2015 wind-power situations in ERCOT, it is borderline values. The model shows that the borderline value of wind is the increase in generation, from fresh capacity, valued at equilibrium prices. This value entails the benefits of avoided reactionary generation and the value of emigrations equipoises. Since wind capacity generates electricity grounded on exogenous and stochastic patterns, once a turbine is erected, its value depends on the correlation between wind cycles and the electricity demand patterns.

13.5.2 Literature Review

Gowrisankaran *et al.* (2016) assessed the additional intermittency costs of incorporating solar power at different integration levels (0%, 10%, 15%, and 20%) with detailed operational data for southeastern Arizona. The main intuition behind their research states "solar generators produce only when the sun is shining, adding to social costs and requiring electricity system operators to reoptimize key decisions" [8]. The authors documented the additional grid level costs, such as increasing operating reserves to support the intermittent solar power, and demand curtailment costs or paying consumers not to use electricity at peak hours in order to reduce the risk of failing to meet load.

De Sisternes *et al.* [9] explores the long-run impact and value of storage in the ERCOT grid by using a detailed capacity expansion model with unit commitment constraints of the power grid. The authors determined that the optimal fossil and renewable energy generation mix for meeting the projected 2035 electricity demand by exogenously varying the storage capacity (0–30 GW). This research accounts for a chronological or cyclic intermittency of demand and renewables as well as system requirements for operating reserves. Rather than accounting for the external damages from emissions for choosing the optimal allocation, the authors constrain the optimal solution to comply with emissions intensity targets in the range of 0.02–0.05 tCO_2/MWh, which are much lower than those for the average combined cycle unit in 2015 (0.47 tCO_2/MWh) [10].

The model results show in great detail hourly dispatch for weekly timeframes of all resources, including the charging and discharging of storage. It incorporates technical details about storage such as the 80% round-trip efficiency and power ratios for maximum hourly discharge. The authors conclude that storage increases the cost-effective penetration of renewables, reducing capacity investments in nuclear and gas peaking units. Its adoption ultimately will depend on reducing technology costs and improving power ratios.

Sioshansi [11] was one of the first papers to model the value of wind power in the presence of storage under market power. This paper used a Stackelberg model in which wind producers are the leaders since they operate a monopoly of storage capacity. Thus, wind generators choose storage allocations and then fossil fuel producers react and set their generation profiles. Nevertheless, the model did not consider the external costs of fossil generation and, by modelling a market power setting, it could not render the Pareto optimal allocation and the first-best value of wind power.

Lamadrid [12] simulated the impact of wind power and energy storage in ERCOT with a detailed model of the grid that accounts for carbon dioxide emissions. The model takes an engineering approach with a very detailed account of ramping and reserve costs as well as hourly wind stochasticity. Using hourly data for 2011, the authors found that taxing carbon increases welfare, and that the net benefits of wind and storage are much larger than their impact on reserve and ramping costs. While the authors modelled fossil generation, storage allocation, and emissions, they did not explicitly model the marginal value of increasing wind capacity or storage.

13.5.3 Emissions Functions

Using the ordering from the marginal private and social costs, the emissions functions for both cases was computed by regressing the total market emissions on their corresponding generation quantities. Model aggregate emissions using a fifth order polynomial for the case of CO_2 emissions and sixth order polynomials for NOx and SO_2 (Figure 13.3). Thus, we have two emissions functions for each pollutant in each hour, one corresponding to the ordering of the plants based on the private marginal cost and another corresponding to ordering of the plants based on the social marginal cost.

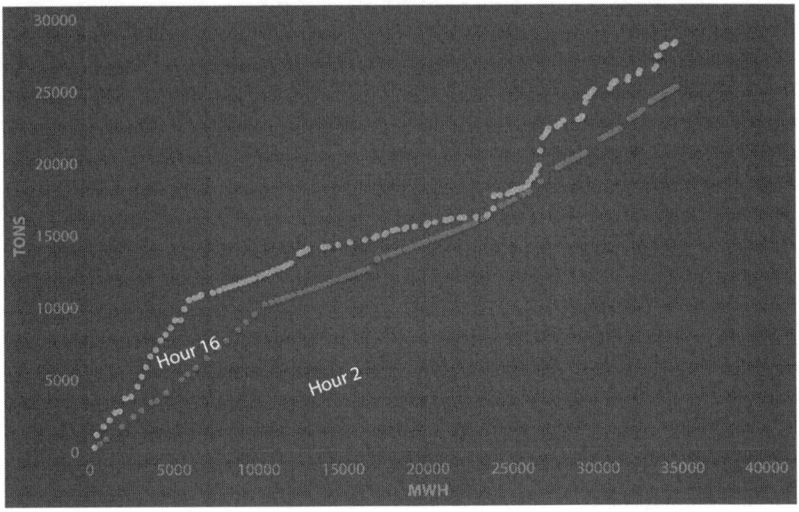

Figure 13.3 CO_2 emissions function for a market dispatch based on the private Motor Carrier (MC) in ERCOT.

13.5.4 Wind Power and Storage Parameters

We use efficiency and cost parameters from Byrne *et al.*, 2018. The storage round-trip efficiency parameter (η) is 90%, the battery lifetime is 10 y, and levelized cost of storage ranges from 420.00 to 950.00 USD/kWh.

Monte Carlo simulations were implemented by randomly drawing 24-hour-wind capacity factor profiles from all 365 days during 2015. To simulate the value of wind under different integration levels (non-marginal increases in wind capacity), the optimistic case that the hourly capacity factors will be kept at larger wind capacities was assumed. Hence, the wind generation profiles were computed by multiplying projected capacity by these hourly factors. We computed the value of wind for capacity levels that generate wind power that represents 12% to 30% of load in ERCOT, on average.

13.5.5 Policy Scenarios and Monte Carlo Simulations

We implement the model for four scenes mingling storage and taxes: no storage no taxes, v_0 (baseline), no storage and taxes, v_1, storage and no taxes, v_2, and storage and taxes, v_3. We compute average allocations, welfare and emissions (CO_2, NO_x, and SO_2) using Monte Carlo simulations (1000 draws) for each scenario.

13.5.6 Welfare and Allocations

Implementing storage and emissions taxes jointly (v_3) leads to the largest welfare increase compared to the current baseline of neither (v_0) since energy is allocated precious.

13.5.7 Emissions Offsets

Increasing wind capacity leads to decreasing CO_2, NO_x and SO_2 emissions offsets in the scenarios with no tax on the externalities and to increasing SO_2 offsets when pollution is taxed. Without a tax, the merit order based on the private marginal cost sets gas turbines as the marginal generators during peak hours. Incoming wind generation replaces these plants and offsets their emissions. As wind capacity increases, the cleaner combined cycle generators become marginal and the additional wind power that replaces them offsets less emissions than with lower capacities.

On the other hand, with environmental taxes, dispatch is based on the social marginal cost, and the merit order of power plants changes. Thus, gas turbines are still the marginal generators for current wind levels but coal plants are on the margin for large adoption levels (30%). Hence, as wind generation increases, it replaces coal, delivering increasing sulfur emissions offsets. In spite of these increasing external benefits, the total economic value of wind is decreasing in the storage and tax scenario since fossil generation costs, which drive equilibrium prices, decline at a faster rate. These previous results hold with low levels of storage and without storage since limited arbitrage does not move enough generation to alter average marginal emissions.

13.5.8 Accounting for Regulating Reserves Costs

Once we account for the additional regulating reserves costs due to wind power intermittency, it was found that the average value of wind power with the unconstrained storage and taxes is 60.44 USD/MWh for the current 12% share. This value is similar to the main specification and the levelized cost of wind. The marginal value of wind still has a downward but less steep slope since equilibrium prices decline at lower rate when accounting for additional reserves costs.

The value of storage decreases in all scenarios compared to those of the main specification. This implies that the effect of wind's intermittency on increasing reserve costs significantly reduces the gains in value from a huge space in middle of low and high rates. It is worth noting that extrapolating the regulating reserves formula from the National Renewable Energy Laboratory (NREL) study (Ela *et al.*, 2011) to compute aggregate reserves costs at the wholesale level could lead to overestimation of these expenses. To sum up, the main insights on the marginal value of wind and storage (and their complementarity) from the theoretical model and the main empirical specification still hold valid in the reserves scenarios.

13.6 Conclusion

This chapter tackled the new challenge of relating the borderline generation and emigrations equipoises caused by adding two energies which are renewable sources to an electric grid.

Hence, we control for hydropower reallocation using the standard module which adds up on hourly basis and on a daily basis. The results allow us to infer the correct marginal external benefits of wind and solar due to CO_2, SO_2 and NO_X emissions reductions. In California, hydropower dynamics particularly enhance the external benefits of solar power.

In a wider view, the given dynamic modelling is crucial for understanding electricity generation and grid position emigrations in systems with adding relinquishment of storehouse technologies since hydro is formerly an amiss form of storehouse whose charging functions by keeping water in the force. The deduced perceptivity about hydropower arbitrage and redistribution would also apply to profit maximizing storers. Therefore, in unborn backed storehouse grids, profitable dynamics will also play a crucial part in determining renewable energy emigrations equipoises, external benefits, and eventually, its profitable value.

Large scale relinquishment of wind and solar power is one of the crucial programs for a cleaner grid that improves original air quality and mitigates climate change. As renewable energy inventories a larger share of power and its intermittency brings new integration challenges, a significant storehouse capacity will be needed to smooth its cycles and force energy when it is most precious. This exploration results show that when wind serves a 30 of the cargo, its average value declines to a bit lower than 50.00 USD/MWh. In an ideal case, carbon prices should rise at the rate of the reduction rate as the Hoteling model requires a limited atmospheric Gomorrah. It is worth noting that the reckoned wind values are an upper bound since they assume no ramp-up, start-up, or transmissions costs. Likewise, these specialized constraints plus transmission constraints make it delicate to the project wind's value at advanced relinquishment situations (> 50).

Throughout this discussion, we have argued and shown substantiation that the value of wind and solar power depend on having storehouse that smooths their intermittency. The environment showed how force hydropower is formerly acting as storehouse in California by shifting water generation from the hours with the largest solar and wind affair to those with the smallest. This hydro redistribution helps to integrate solar and wind into the grid by handling their intermittency. As a result of the redistribution, the carbon equipoises and the profitable value of solar power increases relative to those of wind. This exploration has also exfoliated light on the value of storehouse. While environment shows with literal data how hydro accommodates wind and solar, environment explicitly quantifies the value of power arbitrage. The value of storehouse increases for larger wind capacities since the intermittency of wind raises the benefits of arbitrage. This value also increases with a duty on emigrations since stores can respond to prices reflecting all social costs and allocate energy efficiently. Accounting only for hourly intermittency smoothing, it was found that the value of storehouse is lower than its current levelized cost. Eventually, it shows that counting for a daily rather than diurnal planning horizon slightly increases the value of storehouse for capacity situations larger than 11,000 MWh. In both, weekly an diurnal horizon models, the average battery application situations are less than 50. While the equal rate of storehouse remains high, the assiduity is likely to see limited deployment of batteries aimed at arbitraging short term intraday differences.

References

1. ERCOT, 2016b. Wind Power Production - Hourly Averaged Actual and Forecasted Values http://www.ercot.com/gridinfo/generation. (accessed 11. 2016)
2. Lazard. 2016. Lazard's levelized cost of Storage-Version 2.0. December 2016.
3. Cardell, J., Anderson, C., 2015. Targeting existing power plants: EPA emission reduction with wind and demand response. *Energy Policy*. 80: 11-23.
4. Low Carbon Fuel Standard and Alternative Diesel Fuels Regulation 2018
5. Borenstein. S., Bushnell. J., Wolak. F., and Zaragoza-Watkins. M. (2016). "Expecting the Unexpected: Emissions Uncertainty and Environmental Market Design". Energy Institute at HAAS Working Paper 274. August 2016.
6. California Independent System Operator. CAISO. (2016a). Daily renewables watch. http://www.caiso.com/green/renewableswatch.html (accessed 1.16 several downloads)

7. Climate Policy Initiative. 2017. Carbon Price in California http://calcarbondash.org/?gclid=EAIaIQobChMI3oW-yoqZ1QIVTDqBCh17fQMlEAAYAiAAEgJsRfD_BwE. (accessed 7. 2017)
8. Gowrisankaran. G., Reynolds. S., and Samano. M. 2016. Intermittency and the Value of Renewable Energy. *J. Polit. Econ.* 124: 1187–1234.
9. De Sisternes. F., Jenkins. J., and Botterud. A. 2016. The value of energy storage in decarbonizing the electricity sector. *Appl. Energy*. 175: 368–379.
10. Energy Information Administration. EIA. 2017. Fuel prices. https://www.eia.gov/naturalgas/data.php#prices (accessed 1. 2017) Environmental Protection Agency (EPA), 2017. FTP database of the Continuous Emission Monitoring System CEMS. (accessed 8. 2016)
11. Sioshansi. R. 2011. Increasing the Value of Wind with Energy Storage. *Energy Journal*. 32: 1-29.
12. Lamadrid. A.J. , D.L. Shawhan, C. Edmundo, M. Sanchez, R.D. Zimmerman, Y. Zhu, D.J. Tylavsky, A.G. Kindle, Z. Dar. 2015. Stochasticall y optimized carbon-reducing dispatch of storage generation and loads. *IEEE Trans. Power Syst.*, 30(2): 1064-1075.

14

Energy Resources and Reliability Assessments

Kavinkumar Ravikumar and Milind Shrinivas Dangate*

Chemistry Division, School of Advanced Sciences, Vellore Institute of Technology, Chennai, India

Abstract

The dependability assessment of power systems has drawn more attention in recent years. The continuing changes in generating portfolios and ecological restrictions are primarily to blame for this. Reliability approaches linked with enhanced modelling and related phenomena play a crucial role in the calculation of rising levels of penetrating the variable energy resources (VERs), like PV (photovoltaic) and WTGs (wind turbine generators) systems. Measurement of the input of such properties to the sufficiency of the power grid is a crucial step in the development of wind power and photovoltaic system works. The intermittent nature of VERs for power systems adds a new degree of difficulty to modelling such resources using typical generator reliability methodologies. Variable energy resources (VERs) are different from traditional generators in terms of their dispatchability and capabilities. Additionally, it is not always possible to regulate and make accessible the VERs' fuel supply. This is because VERs' outputs exhibit a strong degree of correlation and depend mostly on convenience of the input sources (wind or sun sources). Therefore, in the probabilistic reliability studies, the result of separate generators within the same place cannot be simulated separately. In this study given here, a method for modelling the output of massive wind farms and solar power plants is established for evaluating the dependability of the power system. An approach presented here is basically on expanding, generating margin's cumulative distribution function (CDF) to incorporate, output power of VERs before adding VERs. In analytical, the energy storage value is computed without repetitions from increased CDF of the generating margin. The approach that is being given lessens the computational effort and modelling complexity involved in determining VERs' capacity values.

**Corresponding author*: milind.shrinivas@vit.ac.in

O.V. Gnana Swathika, K. Karthikeyan and Sanjeevikumar Padmanaban (eds.) Smart Grids as Cyber Physical Systems Volume 1: Artificial Intelligence, Cybersecurity, and Clean Energy for Next Generation Smart Grids, (199–214) © 2024 Scrivener Publishing LLC

Keywords: Power systems, photovoltaic, wind turbine generators (WTGs), variable energy resources (VERs), cumulative distribution function (CDF)

14.1 Motivation

The transition to renewable energy sources has gathered significant pace in response to the rapidly expanding global economy, ecological issues, financial incentives for fossil fuel powered generators, and awareness about global climate change. Despite the benefits of such input sources, planning and oversight of power systems have become even more complicated due to the new realities of input uncertainty and output variation. The main issue and obstacle to efficiently utilizing intermittent resources from a dependability perspective so these are unpredictable in nature. For an example, a wind farm is made up of several WTGs (wind turbine generators) dispersed throughout a region. In terms of ability, the operational subjects of wind turbines are differing from thermal plants (conventional power plants). The varying input and output correlation of wind generators affects output, which complicates modelling. Failure of WTGs increases the difficulty and complexity of using reliability evaluation techniques that are applicable to conventional power generators, as well as the dependence on the input of the system related between the generating units.

Power electrical circuits and solar panels are only a couple of the many parts that make up grid-connected photovoltaic (PV) systems. The operational and ambient circumstances have a big impact on how well these components work. Overall decrease in efficiency and the breakdown of parts of the same system may be caused by the operational circumstances and ecological elements. Photovoltaic (PV) systems have a big number of components, and because the source (solar) is intermittent, design of those systems is really hard to find and evaluation studies. This is because the output power depends on the presence of raw materials in a large quantity. Additionally, since PV panels are installed next to one another in a PV farm, all of the panels will get the same amount of solar radiation and will be affected by the same operational and ecological conditions, which might alter the farm's production. The amount of electricity that is available is significantly impacted by the failure of solar and wind power producing equipment. Numerous studies demonstrate that various variable energy resources (VER) failure mechanisms may be caused by operational and ecological conditions [1, 4].

Furthermore, generators would not be able to produce electricity if input was absent. Forced outage rates, which are substantially greater when taking into account the input's sporadic nature, might be seen as the result of an input shortage in this situation. Because it is the major technique that is used in the modelling detail and system interdependence, Monte Carlo simulation is the technique that is widely used for dependability model of variable energy resources.

Monte Carlo simulation, however, takes a while to run, especially when used in a big, complicated system. In addition, the correctness of the answer is primarily based on the quantity of samples taken. Making a model for a cluster of wind farms and solar power systems' output power is the aim of this study. Assessment of the suitability of the power system. Monte Carlo simulation is often not regarded as acceptable for model creation in power system adequacy evaluation. For creating dependability models, a direct analytical approach is seen to be preferable. In order to incorporate a group of variable resources is the only equivalent to capture the output variability and forced outputs while remaining compatible with conventional probabilistic planning techniques, this work develops a model that combines a bunch of variable resources to a single way to measure sublevel state.

14.1.1 Objections

The following are some succinct difficulties that were addressed in the context of this book.

A. How to simulate output power of such resources while taking into account the operational features and generation unit failures is one of the major issues because of the increase in the penetration of the photovoltaic and wind generating systems.
B. The alternating nature of the system and unstable input sources from the different energy resources are made known to find the complexity of applying the relevant method, which is used before, to the thermal power generators (conventional generators).
C. Based on capacity and dispatchability, an operational characteristic of nature of different energy sources differ from conventional power plants.
D. Different energy sources are contracted from the different areas and different energy farms. These farms are made up of several generation units (wind turbines or solar panels), which are dispersed across a large region, and their output is correlated. Therefore, it is impossible to represent the output of a single unit as a set of independent random variables.

E. When designing the capacity of the electricity system, the volume of storing is used to quantify the capacity input of intermittent sources. To ensuring the system dependability is used to calculate and maintain the VERs are added to electrical web, an assessment method is used in this instance. When assessing the VERs' capacity value, modelling complexity must take into account the input variability, VER failure, and correlation. Therefore, PV system is regarded as a load modifier from the evaluation of such possible inputs or resources in the conventional effective output bearing capability calculation approach. It is possible that the capacity credit in this instance isn't accurately reported.
F. The VERs' capacity values are usually calculated using iterative approaches. For big and complicated systems, iterative approaches need a lot of computing time and resources. Therefore, a more computationally efficient replacement for the system planner would be to develop a straight, non-iterative technique of that takes into account, impacts of input sporadicity, forced VER outages, and correlation of VER output with system load.
G. While incorporating wind energy offers many benefits, it also presents a number of technical obstacles, such as the fluctuation and unpredictability of the speed of the wind and the failure of WTGs, that might reduce dependability of fuel cells. Utilizing energy storage devices is one of the most viable options to reduce these downsides. However, in addition to impairing the power system's dependability, wind speed fluctuation and uncertainty as well as WTG failures make it more difficult to choose the right size energy storage system to meet a certain reliability objective.

14.2 Photovoltaic (PV) Systems

In the literature, many PV system topologies were discussed [7, 8]. Photovoltaic (PV) system designs are often categorized depending on how the DC/DC and DC/AC converters are placed in relation to PV panels. These configurations are typically made to increase the system's dependability and efficiency. Grid-connected PV systems are often built from a variety of parts, including solar panels and power electronic circuits. The operating and ecological factors have a significant impact on the dependability of such components. Both operational and ecological conditions may cause system component failure and a general decrease in output power [9, 10]. In the literature, many PV system topologies were discussed [7]. Designs for PV systems are frequently divided into groups based on where the DC/DC and AC/AC converters are situated in respect to PV panels. Usually, the purpose of these configurations is to improve the system's dependability and effectiveness. Solar panels and power electronic circuits are dependability and effectiveness. Solar panels and power electronic circuits are only a couple of the components that are frequently used to construct grid-connected PV systems. The dependability of that component is significantly influenced by operating and ecological conditions. Failure of a system component and a general drop in output power might result from operational and ecological issues [8, 9].

14.2.1 Attributes of PV System

Power electronic components are regarded as significant parts of PV systems. These parts are crucial to power conversion and offer very effective solutions. However, number of the high energy consuming electronic components and the frequent breakdown of electronic circuits make it more difficult to assess the entire systems' dependability. Power semiconductor components and electrolytic capacitors are among the dependability issues since they are the weakest connections in a system. Most converters for power electronics lack redundant components. The system will thus shut down if any faults are found in any of its parts or subsystems. When assessing the suitability of PV systems, such components should be taken into account from a reliability standpoint. Only the dependability of important components has been taken into consideration because PV systems have a significant number of system components [9, 10]. When PV electricity is added to the power network, it's important to look into how all system elements will affect power system efficiency.

14.2.2 Grid Level PV Farm Structure

Photovoltaic (PV) parks typically consist of a large number of PV panels set up in an array. A DC/AC inverter connects a PV array, which is made up of PV panels, to the main grid. A subarray is a collection of N panels within an array. The output of the PV system was linked to the grids at the higher-level power using transformers [16, 17]. The feed-ins (mostly solar radiation) and system component availability both affect the power output of a PV park at a certain site. Since PV panels are installed side by side in PV farms, it is believed that all of the PV panels has same solar irradiation then they are subject to the same operational and ecological conditions that might impact farm performance.

14.2.2.1 Output Power of PV Systems

For PV systems, the manufacturer provides the highest output power ratings, which are frequently expressed in peak-watts (WP). The current and voltage characteristics (I V features) may be calculated using the following equations [11] under the normal test conditions (the emitted light of 1 kW/m^2 is specified for a temperature of 25°C), where s is the energy level,

$$I = s\,[I_{sc} + K_I\,(Tc - 25)] \tag{14.1}$$

$$V = V_{oc} - K_V\,Tc, \tag{14.2}$$

The terms "Isc" stands for "short circuit current," "KI" stands for "short circuit current temperature coefficient in A/°C," "Voc" stands for "open circuit voltage," "KV" stands for "open circuit voltage temperature coefficient in V/°C," and "Tc" stands for "cell temperature in C", [11]

$$T_C = T_A + s\,N_{OT} - 20/0.8 \tag{14.3}$$

here, *Ta* is the ambient temperature and (*Tno*) is the nominal operating temperature of the cell.

The following equations [11] can be used to compute the output (Ppv) for a given radiation level, ambient temperature, and I and V characteristics

$$Ppv = N \times FF \times I \times V, \tag{14.4}$$

here, N is the number of panels and fill factor denoted in FF, that may be defined as follows [11] and relies on the module characteristics:

$$FF = VmppImpp/\ VocIsc \tag{14.5}$$

here, *Vmpp* and *Impp* are the maximum power point of current and voltage.

14.2.2.2 Attributes of PV System Components

Power electronic equipment is crucial in a PV farm. Photovoltaic (PV) panel output power is sent to a load or the grid using power electronic circuits [10, 12] discuss reliability evaluation of power circuits in large scale grid connected PV systems for various configurations. Power electrical circuits are the main contributors to system failure. The mean time to failure (MTTF) has been utilizing the highly associated signal in the literature; the overall circuit reliability is obtained by adding the constituent error rates (the failure rate is the inverse of the MTTF) [2, 3].

However, this is accurate if a component failure results in the complete breakdown of the power integrated device. Durability occurs in various power electronic designs, such as multilevel inverters, so that they may still perform as intended even if a component fails [13]. There are a number of potential approaches that might be taken in this situation and yet result in an output voltage level that is acceptable. By adding the component failure rates in such setups, the overall failure cannot be predicted.

System dependability may be estimated using analytical methods such state enumeration approaches, minimal cut sets, minimal paths, and reliability network analysis [13]. R(*t*), the reliability function, often shows the likelihood that a system will perform correctly across a time range [0, t]. Typically, it is believed that a system component's failure rate during its useful lifetime is independent of time. The reliability of system components can be expressed as an exponential function of their continuous failure rate throughout their useful life.

$$R(t) = e^{-\lambda_i t}, \tag{14.6}$$

here, λi is the stable rate of the failure of a power system.

The mean time to repair (MTTR) was deemed to be insignificant enough to be disregarded. For a small system, where the duration needed to modify or change a system is short compared to MTTF, this assumption could be true. The consequences of ignoring the repair time, however, can be encouraging for big systems. Therefore, while estimating an output of photovoltaic systems, the MTTR should be taken into account. System availability and unavailability are characterized by Equations 14.7 and 14.8 and their functions of MTTR and MTTF (Equation 14.8).

$$p = \frac{\mu}{\lambda + \mu} \tag{14.7}$$

$$q = 1 - p = \frac{\lambda}{\lambda + \mu} \tag{14.8}$$

here, p is the probability of availability of the system, q is the probability of system failure, and μ is the repair percentage (reciprocal of $MTTR$).

In a system with number of systems (N) connected in the series connection, system dependability ($p^{s}{}_{sys}$) can be calculated by (Equation 14.9):

$$p^{s}{}_{\text{sys}} = \prod_{i=1}^{N} pi \tag{14.9}$$

here, pi is the probability of availability of component i.

From the expected outcome in a parallel connection, and the device's dependability is the likelihood that all potential successful pathways will lead to the required output. Device accessibility ($p^{S}{}_{SYS}$) can be calculated by (Equation 14.10):

$$p^{s}{}_{sys} = 1 - \prod_{i=1}^{N} pi \tag{14.10}$$

Utilizing network reduction strategies in series-parallel reliability configurations is beneficial Norm-Referenced Test (NRT). Once all series and parallel configurations have been combined into an equivalent unit using (2.9), and the availability and unavailability can then be determined (Equation 14.10).

14.2.3 Reliability Modelling of Major Photovoltaic System Components' Reliability

Power electronic circuits are the primary elements in PV systems that cause system failure. Insulated-gate bipolar transistor (IGBT)/ metal-oxide-semiconductor field-effect transistor (MOSFET), diodes, and capacitors are the main contributors to system failure in power electronic systems. Ecological and stress variables have an impact on the reliability performance of these components. The military handbook MILHDBK-217F, [14] is one of the papers that provide modelling and dependability data for power electronic components. The MILHDBK-217F gives the following breakdown of these components' failure rates.

14.2.3.1 Power Electronic Circuit Components

Since the MILHDBK-217F military manual does not include statistics for IGBTs, it was anticipated that IGBT accident rates would be similar to MOSFET efficiency.

The rate of failure of switches (MOSFET/IGBTs) is defined as:

$$\lambda_{IGBT} = \lambda_{IGBTb}\,\pi_T\,\pi_Q\pi_E \quad \text{Failures}/10^6\ \text{hours,} \tag{14.11}$$

here, $\lambda IGBT_{b\ is}$ rate failure of the IGBT, $\pi_{T\ is}$ temperature aspect, $\pi_{Q\ quality}$ aspect and πE ecological aspect.

Similarly, failure rates in diodes are given by:

$$\lambda_{diode} = \lambda_{diodeb}\pi_T\,\pi_S\pi_C\pi_Q\pi_E \quad \text{Failures}/10^6\ \text{hours,} \tag{14.12}$$

here, $\lambda diode_b$ is failure rate of the photo diode, πS voltage stress aspect, πC contact construction aspect, and πT, πQ, and πE are for IGBTs.

Capacitor failure rate is given by

$$\lambda_{cp} = \lambda_{cpb}\,\pi_T\,\pi_C\pi_V \quad \pi_{SR}\pi_Q\pi_E\ \text{Failures}/10^6\ \text{hours,} \tag{14.13}$$

here, λcp_b is probability of the failure rate and V denotes voltage, SR denotes resistance of the series, and T, C, Q, and E are for IGBTs.

14.2.3.2 Reliability of PV Panels

Large systems with many system components, like PV systems, may have several system states as a result of reliability modelling, which adds to the complexity of the model. Therefore, the way to an identical failure percentage and modify percentage for subsystems, an MTTF and MTTR must be equal. Markovian models with "up" and "down" states and known failure and repair rates can be used to represent the availability of system components. The n parallel branches of the PV panels are built by joining m series solar cells along each branch.

By adding the failure attempts of the solar cells connected in the m series, it is possible to get the corresponding failure rate of a branch. The PV panel malfunctions as a result of a failed solar branch. Thus, using Equation 14.7, it is possible to identify the availability and unavailability of photovoltaic panels (Equation 14.8). Thus, using Equation 14.7, it is possible to determine the availability and unavailability of photovoltaic panels (Equation 14.8).

14.3 Reliability Modelling of PV System

Two modelling processes are used to create a working model for a full PV system, taking into account for the impact of input from different sources and a model malfunction: (1) modelling the output's dependence on solar radiation variability, and (2) the outage caused by system equipment damage (failure rate). The result of combining these two models is a single generation model, which may be thought of as a specific source with non - linear and linear state. The panels in Figure 14.1 shows the system components failure and input sources different.

An element of PV system is expected to fail independently, and their availability may alternatively be described as Markovian components with the terms "up" and "down." The MTTF and MTTR for M components for each array may be calculated and corresponding MTTFs and MTTRs for each array may be used to represent it. By iteratively merging some new units in the table based on their size or bulk nature, construction technique described in section 14.1, the volume of output probability and rate of recurrence table for a photovoltaic farm with N number of collections may be created. The capacity outage probability and frequency table (COPAFT) of photovoltaic system may constructed using the addition technique based on the forced outage rate (FOR) of system arrays. In this recursive procedure, the supply of one unit is the starting point, and the distributions of the other units are sequentially convoluted with it, one unit at a time.

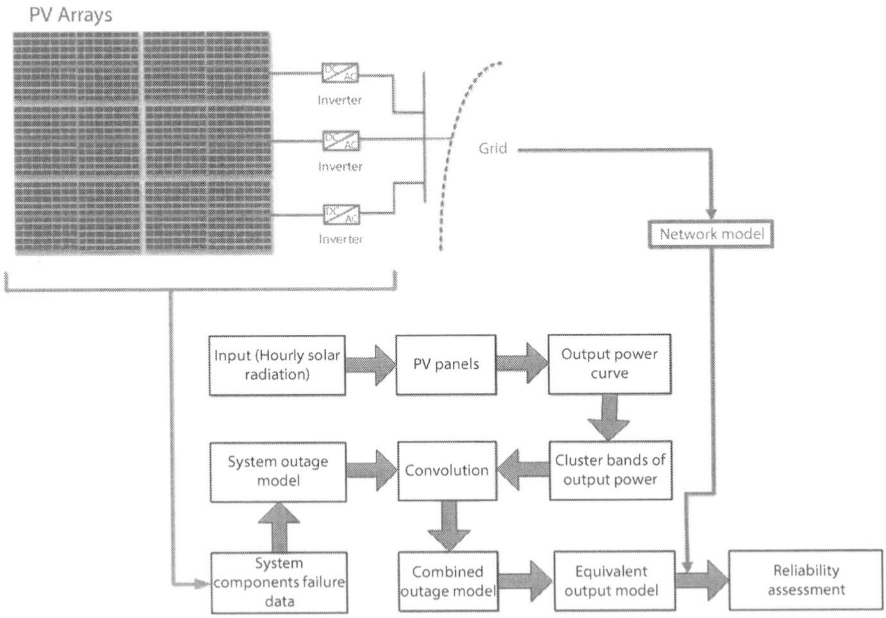

Figure 14.1 Block diagram of the failures of system component and different input sources.

Table 14.1 Reliability indices of the IEEE RTS with 100-MW PV farm.

Method	Loss of load probability	Expected demand not supplied (MW/y)	Loss of energy expectation (MWh/y)	Loss of load frequency occ./y	Loss of load expectation (h/y)
Base case					
Direct current	0.001	0.13	1181.19	2.01	9.36
PV power only					
Proposed	0.0008	0.1	849.55	1.73	6.94
Photovoltaic power with inclusion of (FOR)					
Proposed	0.0008	0.10	900.29	1.74	7.22
Monte Carlo simulation	0.0008	0.10	900.70	1.74	7.3

DC: Discrete convolution; MCS: Monte Carlo simulation.

14.4 Case Studies

From those, suggested approaches are used with the IEEE Reliability Test System (RTS) [14]. The PV system's specifications and construction were obtained [5], from the statistics on solar radiation for the Santa Maria, California location used in this investigation were acquired from the output power of the PV system was estimated using data from for a single year.

From this study, failure rates of the system were predicated to be as: 0.000027 failures per hour for DC/AC inverters and DC/DC converters, and 0.000011 leakage output per hour for solar panels [15]. As the same conversion maintenance costs was taken to be 0.1 per hour. Repair rate for PV panel was taken to be 0.2 repairs per hour, to validating results and indices of IEEE RTS are designed to calculate by MCS and the outcomes are shown in Table 14.1.

14.5 Conclusion

In order to build dependability models of VERs for planning studies, including the integration of wind and PV systems, an analytical approach for doing so was described in this study. The suggested approach employs a strategy to create an effective model that takes into account input variability, connection between VER outputs, and forced VERS outputs. Real input data and real were used to build an illustrated dependability model. From the system, which is known well, the proposed method's efficacy was shown. Several dependability indices (loss of load probability - LOLP, loss of load frequency - LOLF, expected demand not supplied - EDNS, loss of load expectation - LOLE, and loss of energy expectation - LOEE) were compared in each scenario with and without taking into account VER forced outages. The suggested method's findings were verified using Monte Carlo simulation. For the reasons outlined in section 4.2.1.3, this work submitted here concentrated on building of dependability working model for wind farms and PV systems rather than attempting to find the connections with other variables like system load. The advantages of a reliable framework that can be used to analyze methodologies for sufficiency evaluation or capability appraisal, though, could overcome these mistakes.

The method described may be employed in situations where it is judged important to take this correlation into account. However, it was demonstrated that the suggested strategy worked well for creating a stable of dependability model of a wind farm and PVs [6].

Additionally, this study has developed a straightforward way to determine the VERs' capacity value without iterations. The foundation of the suggested approach is expanding the generating reserve margin to incorporate VERs' output power. Both forced outages and the connection between WTGs on the farm were taken into account. The results of applying the suggested approach with the IEEE RTS were given. The computational cost and modelling complexity involved in iteratively determining the capacity value of VERs are lessened by the suggested strategy. Iterative assessments of capacity values backed the procedure and the outcome of the suggested strategy.

This chapter also shows a method for calculating the percentage of energy storage capability needed to stabilize wind power. To assessing dependability indices and choose the place for the energy storage systems, a number of case studies were conducted. By contrasting the improvement of system dependability indices at the candidate buses, ideal places for wind injection were also assessed. When computing reliability indices and defining the sizes of energy storage systems, sequential Monte Carlo simulations were utilized to simulate the stochastic behaviour of wind power and forced outages of WTGs. Three set of data were used in this study. These set of the data was used to investigate how temporal resolutions and the link between load demand and wind output affected the size of the energy storage. Utilizing the three data sets, the arrangement of dependability induces and the power of the energy is determining from the data what we have. These case studies simply show the impact of wind speed resolving power and relationship with the workload on network dependability and defining the size of the energy storage unit, they were not meant to compare the findings. The outcomes confirmed the suggested technique and demonstrated how wind conditions and wind power station location affect the sizes of storage devices that are needed. For instance, building a wind farm at bus 7 would not improve system dependability since the connection that connects bus 7 to the rest of the grid has a capacity constraint that prevents it from carrying all of the potential electricity.

However, some buses, such as the 13, 19, and 18, are connected to the system via a number of transmission lines and carry more loads than other load buses. The system's dependability indices for various wind farm and energy storage system locations are compared in this regard. These indices depend on the properties of the wind farm and energy storage systems, as well as how they are connected to the rest of the system. These results also show that while determining the degree of energy storage in a project design, it is crucial to take into account the influence of wind speed measurement resolutions and their connection to the load.

Also evaluated are the system reliability indices (SRI), both before and after the installation of the energy storage. From the EDNS indices, it appears that bus 18 is the best location to site the wind farm.

14.6 Future Works

This work created the framework of this book and might be expanded upon in light of the following considerations:

1. The approach described in the context can be extended to set up a uniform failure model for a bunch of output signals from various sources.
2. The approach described in this context uses a discrete convolution technique to determine the capacity value of VERs in generation sufficiency evaluation. However, the approach may also be improved for reinforced composites.
3. The combination of energy storage devices has become more important in the power system as a result of continuing changes in producing portfolios since it considerably aids in resolving the issues caused by the combination of better reliability. The strategy suggested in this study may

be developed to include variable energy supplies and data storage devices in power system adequacy assessment.

4. The assessment of variable resource penetration levels is one of the most crucial elements in power system planning projects. The best VER penetration levels can be found using the approach described in the context.

References

1. K. Fischer, F. Besnard, and L. Bertling, "Reliability-centered maintenance for wind turbines based on statistical analysis and practical experience," *IEEE Trans. Energy Convers.*, **2012** vol. 27, 184–195.
2. D. Hirschmann, D. Tissen, S. Schroder, and R. W. De Doncker, "Reliability prediction for inverters in hybrid electrical vehicles," *IEEE Trans. Power Electron.*, **2007,** vol. 22, 2511–2517.
3. G. Chen, R. Burgos, Z. Liang, F. Lacaux, F. Wang, J. D. van W. y. k., W. G. Odendaal, and D. Boroyevich, "Reliability-oriented design considerations for high-power converter modules," in *IEEE Proc. Power Electronics Specialists Conf.*, **2004,** vol. 1, 419–425.
4. F. Chan and H. Calleja, "Reliability estimation of three single-phase topologies in grid-connected PV systems," *IEEE Trans. Ind. Electron.*, **2011**, vol. 58, 2683–2689.
5. A. Ghaedi, A. Abbaspour, M. Fotuhi-Friuzabad, and M. Parvania, "Incorporating large photovoltaic farms in power generation system adequacy assessment," *Journal of Science Iran*, vol. 21, **2014**, 924–934.
6. R. Billinton and D. Huang, "Incorporating wind power in generating capacity reliability evaluation using different models," *IEEE Trans Power Syst.*, vol. 26, **2011**, 2509–2517.
7. R. Carbone and A. Tomaselli, "Recent advances on AC PV-modules for grid-connected photovoltaic plants," in *Proc. International Conf. on Clean Electrical Power (ICCEP) IEEE*, **2011**, 124–129.
8. D. Picault, B. Raison, and S. Bacha, "Guidelines for evaluating grid connected PV system topologies," in *Proc. International Conf. Industrial Technology. IEEE*, **2009**, page no. 1–5.
9. S. Forman, "Performance of experimental terrestrial photovoltaic modules," *IEEE Trans. Rel.*, **1982**, vol. 31, 235–245.
10. P. Zhang, W. Li, S. Li, Y. Wang, and W. Xiao, "Reliability assessment of photovoltaic power systems: review of current status and future perspectives," *Appl. Energy*, **2013**, vol. 104, 822–833.
11. D. K. Khatod, V. Pant, and J. Sharma, "Analytical approach for well-being assessment of small autonomous power systems with solar and wind energy sources," *IEEE Trans. Energy Convers.*, **2010** vol. 25, 535–545.

12. C. R. S. Reinoso, D. H. Milone, and R. H. Buitrago, "Simulation of photovoltaic centrals with dynamic shading," *Applied Energy*, **2013**, vol. 103, pp. 278–289.
13. M. Benidris, S. Elsaiah, and J. Mitra, "Sensitivity analysis of reliability performance of multi-level converters," in *Proc. North American Power Symposium (NAPS). IEEE*, **2012**, 1–6.
14. P. Subcommittee, "IEEE reliability test system," *IEEE Trans. Power App. Syst.*, **1979**, vol. 6, 2047–2054.
15. A. Garro and F. Barrara, "Reliability analysis of residential photovoltaic systems," *in Proc. Int. Conf. Renewable Energies Power Quality*, **2011,** vol. 1, page no. 9.
16. L. Soder, L. Hofmann, A. Orths, H. Holttinen, Y.-h. Wan, and A. Tuohy, "Experience from wind integration in some high penetration areas," *IEEE Trans. Energy Convers.*, **2007**, vol. 22, 4–12.
17. C. Singh and A. Lago-Gonzalez, "Reliability modeling of generation systems including unconventional energy sources," *IEEE Trans. Power App. Syst.*, **1985**, page no. 1049–1056.

15

Electric Vehicle Charging Stations Effect on Battery Storage Technology

Sathya Narayanan and Milind Shrinivas Dangate*

Chemistry Division, School of Advanced Sciences, Vellore Institute of Technology, Chennai, India

Abstract

The infrastructure for charging electric vehicles (EVs) now depends heavily on battery storage. Uncertainty exists on whether and how battery storage might play a significant role in enabling fast-charging stations (FCSs) to meet customer demand and provide a profit for charging station operators. This research advances our knowledge of how to build FCSs with integrated battery storage systems. Three tasks make up the work. An extensive analysis of the literature is done first, which covers EVs, requirements for charging stations, and battery technology evaluation. According to the review, lead-acid, vanadium-redox, and lithium-ion batteries are the most promising for use in charging stations. The second step involves conducting the procedures and considerations for installing an FCS and a battery storage unit. Based on battery size, type, and vendor, the findings offer a cost estimation for various FCS setups and battery storage costs. Third, a model for discrete event simulation (DES) is created to assess the costs and properties of battery storage for a network of FCSs in southeast Michigan. According to the simulation, there is no energy exchange when network expenditures (such as the price of installing a new distribution line) are taken into account, and each FCS requires more than one battery. Two-thirds of the FCSs need a battery when network expenses are not taken into account, and there is a reduction in energy exchange. In this network, zinc-air batteries are the least expensive and lithium-ion batteries are the most expensive. Due to high network expenses, a highly condensed FCS network through battery units coupled to a microgrid would provide more benefit and result in lower total cost. For example, charging station operators, battery producers, and vendors can use this model to assess the costs and battery specifications that best suit their FCS network.

**Corresponding author*: milind.shrinivas@vit.ac.in

O.V. Gnana Swathika, K. Karthikeyan and Sanjeevikumar Padmanaban (eds.) Smart Grids as Cyber Physical Systems Volume 1: Artificial Intelligence, Cybersecurity, and Clean Energy for Next Generation Smart Grids, (215–226) © 2024 Scrivener Publishing LLC

Keywords: Battery, electric vehicle (EV), fast-charging stations (FCSs), discrete event simulation (DES), network costs

15.1 Introduction

15.1.1 Background

Many nations have started electrifying their transportation infrastructure because of greenhouse gas emissions. Around 27% of all U.S. greenhouse gas (GHG) emissions and 63% of petroleum usage are produced by the transportation sector [1]. The change from fossil fuel dependent vehicles to electric vehicles (EVs) will reduce emissions [2]. The development of appropriate charging infrastructure has become crucial as more EVs hit the market. Electricity is provided by an EV charging station for recharging plug-in hybrid electric vehicles (PHEVs) and battery electric vehicles (BEVs). When an EV's battery power runs out, its range as well as the availability of recharging stations are crucial factors [3]. As of now, there are three types of charging stations available: Level 1 (residential), Level 2 (public), and Level 3 (DC fast). There are many problems and concerns about the strain that is exerted on power grid owing to short charging time and heavy loads. Numerous studies were done to determine how EV charging would affect the electrical grid in the future. A working fast-charging station (FCS) must be able to meet peak load requirements and keep a consistent power supply all day. By leveraging battery storage technology to meet the power requirements of a charging station, this may be proven. Having energy storage systems installed in FCSs can lessen the load that EV charging places on the electric grid [4].

15.1.2 Problem Statement

Problem 1: It is unknown how battery storage technology will affect how each particular EV FCS is built.
For power grid operators, storage producers, and regulators, the economic effects of grid-scale energy storage technologies are unclear [5]. It is crucial from a utility's perspective to determine whether installing a battery storage system would be less expensive than upgrading the electric infrastructure to accept FCSs. The step-by-step process of battery storage and FCS installation as well as the roles of relevant stakeholders is not clearly defined [6]. For utilization of these technologies with regard to FCSs, a thorough techno-economic analysis with up-to-date cost information and a comprehensive installation process framework is necessary.

Problem 2: It is uncertain what effect battery storage technology has on the construction of EV FCSs in a microgrid network.
The energy savings and financial benefits of using battery technology to power an EV charging network are unknown. Understanding if battery storage deployed at specific stations can help the entire network is critical as the charging network owners rapidly install hundreds of DC fast chargers on their open network [7]. To assess the changes in cost and power supply using battery storage in a network of FCSs, a detailed analysis must be done.

15.1.3 Research Objectives

The long-term goal of this research is to make it easier for people to buy electric vehicles by better equipping charging stations to fulfill their demands. Customers are given with high-power charging, using battery storage as a buffer, allowing them to recharge an EV in a limited timeframe [8], thus minimizing range anxiety. This study investigates battery systems to satisfy future FCS application needs. This research will help electric utilities, charging station network operators, battery makers, and vendors understand the financial risks and benefits, as well as potential energy savings, of deploying energy storage for a network of fast chargers. There are three main objectives:

- To provide a better understanding of battery technology and its associations with FCSs for EVs.
- To calculate the price and process of building an FCS with battery storage.
- To identify the best battery storage system for a specific FCS network.

15.1.3.1 Research Plan and Strategy

Three stages were used to divide this study. The first stage gives a general overview of battery storage and electric charging station technologies, weighs their benefits and drawbacks, and decides which ones work best with an FCS. Project activities and costs are listed in the second phase. Utilizing cost reports and installation manuals from various battery manufacturers and charging-station network operators, information for the installation of an FCS and a battery storage system. The third phase uses a case study of the greater Detroit area in Michigan to model the energy supply in a network of battery assisted FCSs. To determine the ideal battery

Table 15.1 Research plan for each phase.

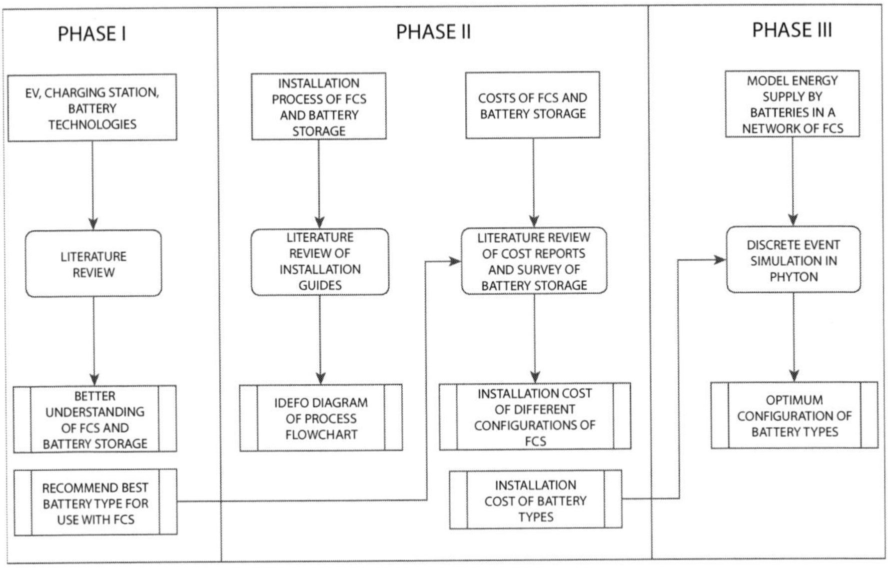

configuration that yields the cheapest energy supply, a discrete even simulation (DES) model was implemented in Python. The research plan for each phase is displayed below in Table 15.1.

15.1.3.2 Methods for Phase 1

To give readers a better understanding of EV charging and battery storage technology, a thorough literature review was carried out. These reviews were part of this phase: types of EVs, charging station types, and standards, along with their benefits and drawbacks. This also includes a discussion of the various battery types used in utility scale applications, such as lithium ion, lead acid, and vanadium redox; a comparison of different battery types based on their lifecycles, how temperature affects them, and the quantity of prior projects. Also, the security of every type of battery as well as any environmental effects from siting to end-of-life disposal.

15.1.3.3 Methods for Objective 2

This section of the study examined FCS and battery storage building costs and installation procedures. The U.S. Department of Energy, Electric Power Research Institute, battery manufacturers and installers, the Clean Energy

Council, charging station operators and installers, and these organizations' recent installation guides, handbooks, technical reports, and cost reports were examined as part of a qualitative analysis of the literature sources. The setup procedure, price of battery and FCS technology, IDEF0 diagram and function modelling were also a part of this method.

15.1.3.4 Methods for Objective 3

In this stage, an FCS based on DES model was built in Python 2.7.13 by altering the energy requirement and using several battery types. The model inputs (*i.e.*, location and power output) are based qualitatively on the network of fast chargers close to Detroit, and the installation cost range of each battery type and vendor. Battery units that can power neighboring FCSs and exchange stored energy are introduced to close the energy gap in FCSs. The model simulates these network configurations for various battery sizes, kinds, and exchange ranges in order to determine the ideal configuration of battery units needed to fill the network's energy deficit at the lowest possible cost.

15.1.3.5 Discrete Event Simulation (DES)

It models a system as a timeline of discrete events, where each event represents a moment when a substantial state change in the system takes place . Project planning is one of the many issues that DES was utilized to address. Discrete event simulation has four main parts: entities, events, time, and resources.

15.2 Literature Review

15.2.1 EVs

The term "EVs" refers to a broad group of vehicles that use electricity as part of their propulsion system. The following are the many types of EVs available today.

Hybrid electric vehicles (HEVs): do not require a plug-in power source and have an electric and internal combustion engine propulsion system. Hybrid electric vehicles convert the braking energy into electricity, which is then stored in a battery. The engine can then use the electricity to maximize fuel efficiency and reduce running expenses [9]. Therefore, regenerative braking and gasoline or diesel are required for HEVs to function.

Battery electric vehicles (BEVs) are vehicles without internal combustion engines that are propelled by electricity stored in a battery and utilized by an electric motor. By plugging them into an electricity source, usually the power grid, they are charged. A sizable DC motor and sizable battery pack are necessary for BEVs. They are also known as "pure EVs," and their range is usually between 100 and 200 miles [10].

Plug-in HEVs (PHEVs): utilize both electricity and gasoline to power the vehicle. They mostly run on a battery that is recharged by plugging into the power grid [10].

15.2.2 EV Charging

Three current and voltage levels are typically used when charging electric vehicles: AC Level 1, AC Level 2, and Level 3 DC charging. The power grid, either at home or at a charging station, must be frequently used to replenish the batteries in EVs. The bulk (more than 5000) of the country's electric charging stations, which number between 6000 and 7000, are privately owned, according to the US Department of Energy. Furthermore, almost 80% of all currently installed charging stations are Level 2.

Level 1 charging is designed to be utilized at home or other locations where the car can sit for a long time and normally takes a lot of time to complete. The maximum charging power for Level 2 charging is 10 kW and it needs a single-phase 208-V to 240-V outlet. Low installation costs and little influence on peak-demand fees for electric utilities are benefits of such pricing. Fast charging is another name for Level 3 DC charging. 480-V, 400-A, three-phase electrical supply is used for EV fast charging. The primary benefit of Level 3 charging is the significantly shorter charge time. Fast charging may put a strain on the system or result in hefty "demand charges" for drawing electricity during peak hours.

15.2.3 DC Fast-Charging Standards

Combined charging system (CCS): a higher potential rate of charging is possible with CCS chargers. The maximum theoretical power allowed by the standard through the port is 350 kW, which is more than twice as fast as a Tesla Supercharger [3]. BMW, Volkswagen, Chevrolet, and all forthcoming electric vehicles in the United States are all compatible with CCS.

CHAdeMO: the CHAdeMO standard specifies a maximum charging power of 62 kW. Cars must have two charging ports, one for AC Level 2 and the other for CHAdeMO, as the ports do not support AC charging [11]. Nissan, Mitsubishi, and Kia compatibility is also required.

Tesla Supercharger: the Tesla Supercharger system kW. Additionally, Tesla sells an add-on adapter that enables CHAdeMO or J1772 charging for Tesla users. As of 2017, there are 5339 Superchargers and 828 Supercharger stations [12].

15.2.4 Battery Technologies

Numerous battery technologies, including lithium ion, sodium sulfur, lead acid, vanadium redox flow, nickel cadmium, and others, have uses in the production and delivery of utility power. Iron-chromium, zinc air, zinc bromine, sodium nickel chloride, nickel metal hydride, and nickel metal. Recent advancements in battery technology have opened new avenues for use in electric utility transmission and distribution, renewable energy, smart grids, and electric vehicles. Installing energy storage also makes it possible to charge more EVs without having to pay for expensive utility upgrades.

Lithium-ion battery: due to its high energy density, high voltage ratings, and minimal self-discharge. lithium-ion batteries, which were initially commercialized in the 1990s, have emerged as one of the most popular storage technologies in many applications. They do, however, have several drawbacks, including limited operating temperatures, high cost, and short cycle life. More than 300 energy storage projects are currently operational worldwide, with the majority of them being deployed in projects with rated power below 5 MW.

Sodium sulfur battery: Ford Motor Company developed the first NaS battery in the 1960s. In the U.S., the first megawatt-class system of NaS batteries was installed in 2006. Around 30 NaS initiatives are currently in operation worldwide [1]. The majority of operational utility-scale installations have rated power of more than 1 MW (varying from 100 kW to 50 MW).

Lead-acid (LA) battery: in both stationary and mobile applications, lead-acid batteries are widely employed [13]. Although more expensive, stationary LA batteries are more effective than starter batteries. A LA battery has a cycle life of 1500 cycles and a normal service life of 6 to 15 years.

Vanadium redox flow battery: various stationary applications use vanadium redox flow batteries (VRFB), a relatively established form of flow battery. These batteries are naturally safe; they do not experience thermal runaway consistently stable performance. The electrolyte is aqueous and inert; they're also recyclable and environmentally beneficial. There are more than 30 operational VRFB plants globally, the majority of which have rated powers under 1 MW.

Nickel cadmium battery: the two primary types of nickel-based batteries are nickel cadmium (Ni-Cd) and nickel metal hydride (NiMH) batteries. Low temperatures (in the range of –20°C to –40°C) have no effect on this type of battery's performance. Ni-Cd batteries have a number of drawbacks as well. For instance, because it is extremely poisonous and hazardous to the environment, cadmium cannot be used by consumers.

Nickel metal hydride battery: NiMH batteries were created as an alternative to NiCd batteries because cadmium is harmful. They charge less quickly and are not as resistant to extremely low working temperatures. There is only one NiMH project (in Japan) that is now in operation and has a 300 kW rated power.

Sodium nickel chloride battery: around 10 ZEBRA projects were announced or are currently being built in North America, and there are about 20 active ZEBRA projects worldwide. The majority of utility-scale operating projects range in size from 20 kW to 5 MW, with rated power run times of 42 min to 5 h. The positive electrode in sodium nickel chloride (NaNiCl) batteries is made of nickel chloride rather than sulfur, making them high-temperature (HT) batteries. It was demonstrated that this battery type has comparatively low inherent dangers when used normally. It requires no maintenance and emits no ambient emissions.

Zinc bromine battery: for utility applications, field deployment and demonstration trials of zinc-bromine batteries (ZBBs) are still in their early phases. Zinc-bromine batteries have 20-year predicted lifespan because the active ingredients do not deteriorate on their own. In addition, they have operating AC-to-AC efficiencies of about 65%, extended cycle lives, and no shelf life. The presence of potentially contaminating elements in ZBBs is a drawback.

Zinc-air battery: zinc-air batteries provide up to three times the energy density of Li-ion, the most competitive battery technology, and are more stable and safer than other battery technologies. There are currently no zinc-air projects operational, although there are two contracts for such projects in California. These batteries are still in the early stages of utility applications.

Iron-chromium battery: only one demo project was completed for this battery, which is still in the research and development stage. Due to the separation of power and energy, iron-chromium batteries are less dangerous than integrated cell storage systems (such LA, NaS, and Li-ion). Due to the low toxicity of the used iron and chromium species, they are also environmentally safe. The above discussion is summarized in Table 15.2.

Table 15.2 Performance comparison of various battery technologies are given above.

Battery storage type	Discharge duration	Size	Nominal voltage	Number of cycles	Number of projects ≤ 1 MW
Lithium-ion	10 min – 9h 21 min	1 kW – 48 MW	3.7 V	~5,000	430
Sodium-sulfur	6h – 8h	400 kW – 50 MW	2.1 V	500–10,000	14
Lead-acid	50s – 9h 36 mins	2 kW – 36 MW	2.0 V	~1,500	72
Vanadium redox flow	16 min – 20h	5 kW – 20 MW	1.6 V	~10,000	48
Nickel-cadmium	5 min – 15 min	3 MW – 27 MW	1.2 V	~500–3,000	0
Nickel metal hydride	15 min	300 kW	1.2 V	~600–1,200	1
Sodium-nickel-chloride	42 min – 5h	20 kW – 5 MW	2.6 V	~1,000	24
Zinc-bromine	2h – 6h	3 kW – 25 MW	1.8 V	> 2,000	32
Zinc-air	2h – 48h	250 kW – 10 MW	1.65 V	~5,000	0
Iron-chromium	4h	250 kW	1.18 V	> 5,000	0

15.3 Conclusion

15.3.1 Findings

According to this study, battery storage can be employed in conjunction with FCSs to help meet an FCS's peak energy needs. Batteries serve as energy buffer storage; extra capacity is sold during peak hours, and recharge takes place during off-peak hours. Since Li-ion batteries are the most suitable for charging-station applications, they are the ones that are most frequently advised for these projects, followed by LA and VRFB. These batteries are reasonably mature, scalable, and have operational and safety solutions already in place. They also exhibit the best safety, cost, and technical performance for small power distribution applications. This proposal does not take into account the findings from context; instead, it is based on a review of the literature and a comparative examination of various battery technologies. Depending on whether the FCS needs an extra Level 2 charger, an electric upgrade, or extended utility service, the construction cost of an FCS typically ranges from $58,000 to $171,000. Even in the absence of network fees, the FCS network exchanges less energy from battery units, according to the results of the DES model. At least two-thirds of the FCSs need batteries, which suggests that every FCS might require a battery to function. The most expensive batteries are lithium-ion, whereas zinc-air batteries are the least expensive. The interchange of energy between battery units would be more advantageous for discrete microgrids of FCSs because to the high cost of installing a new distribution line.

15.3.2 Contribution

Information on EV charging, in particular fast charging, and the function of battery technology to enable its power supply are provided by this study. A general overview of EV types, charging types, and charging standards is gained from the literature research, which is helpful in the remaining parts of this study. The best possible battery technologies to be employed in conjunction with an FCS are suggested through a comparison analysis of the performance characteristics of current battery kinds. The installation procedure for the FCS and battery storage system is covered in the context that follows. It offers understanding of the obligations, factors, precautions, and requirements owners and contractors must follow for successful installation of such systems. Cost estimates for several fast charger configurations are provided.

15.3.3 Limitations

Data restrictions, breadth, and assumptions are some of the study's drawbacks. Cost estimates for the charging stations and batteries utilized in this study were based on literature sources and data supplied by vendors. Although these numbers might not perfectly correspond to current industry norms, they do provide potential pre-construction phase estimates. More accurate cost estimates for the various installation phases of such projects may be easily obtained from charging station network operators/suppliers as well as battery manufacturers/vendors once the scope and specifications of such projects are determined. Additionally, the technical aspects of employing batteries with FCSs are not taken into account in this study. Numerous presumptions form the basis of the simulation model.

15.3.4 Future Work

Future research can compare the costs of FCSs in-depth between network providers or vendors. A site inspection of a battery-powered FCS installation and note of the associated costs would be a more practical approach. When using a condensed network of rapid chargers, where energy exchange from battery storage may be more practical and cost-effective, the simulation model should be used. The operational expenses for the battery and FCS can also be assessed using an Life Cycle Cost (LCC) analysis.

References

1. U.S. Department of Energy. (2016, June). *Global Energy Storage Database.*
2. Institute for Energy Research. (2013).
3. McDonald, Z. (2016, February 04). *A Simple Guide to DC Fast Charging.*
4. Bashiri, M., N.B. (2016). *Optimized plan of charging stations for management of demands: An emerging need of hybrid electric vehicle.* Future Technologies Conference (FTC). IEEE.
5. Zakeri, B., S. S. (2014). Electrical energy storage systems*: A comparative life cycle cost analysis. Renew. Sustain. Energy*, **569-596**.
6. Schroeder, T. (2012). The economics of fast charging infrastructure for EVs. *Energy Policy,* 136-144.
7. Plug Share. (2017, February). EV Charging Station Map.
8. Sbordonea, D., I. B. (2014*). EV fast charging stations and energy storage technologies: A real implementation in the smart micro grid paradigm. Electric Power Systems Research*, 96-108.
9. Energy, C. (2015). *About EVs Types & Comparison.* Retrieved November 2016.
10. Corporation, E. (2016). *Types of EVs.* Retrieved October 2016.
11. Herron, D. (2014). *DC Fast Charging Standards.* Retrieved from Green Transportation.
12. Tesla. (2017). Superchargers. Retrieved from Tesla.
13. International Electrotechnical Commission. (2011). Electrical Energy Storage: white paper.

16

Photovoltaic Technology and Environmental Impact

Kavinkumar Ravikumar and Milind Shrinivas Dangate*

Chemistry Division, School of Advanced Sciences, Vellore Institute of Technology, Chennai, India

Abstract

By 2050, it is anticipated that solar energy would generate around 25% of the world's electricity, necessitating massive manufacturing. Photovoltaic (PV) is anticipated to be a clean technology that can lower the amount of carbon used in the production of electricity. The environmental effect of production must be diminished in order to optimise environmental gains and reduce carbon dioxide emissions. This research will help in examining alternative methods and determining the environmental effects of mature silicon (silicon) and nascent organic (organic) photovoltaic life cycles. C_{60} is frequently utilized as an acceptor material for organic photovoltaics. The C_{60} cleaning procedures now in use a lot of energy and hazardous chemicals. Therefore, it is preferable to alter the current C_{60} purification procedure before beginning the large-scale manufacture of organic photovoltaics (OPVs) in order to reduce any potential environmental, financial, and chemical risks linked to the manufacturing procedure. Using a life cycle assessment (LCA), environmental hotspots in the cleaning procedure were found. In addition to LCA, we employed analytical chemistry, green chemistry, and toxicological assessment to find more environmentally friendly options. In comparison to current C_{60} cleaning methods, alternative C_{60} cleaning decreased environmental impact (59%), expense (85%), and chemical risks (42%). Si PV makes up 95% of the PV industry now and is anticipated to continue dominating the market until 2040 (>50%).

Keywords: Photovoltaics (PV), organic photovoltaics (OPV), silicon, life-cycle assessment (LCA), toxicity

**Corresponding author*: milind.shrinivas@vit.ac.in

O.V. Gnana Swathika, K. Karthikeyan and Sanjeevikumar Padmanaban (eds.) Smart Grids as Cyber Physical Systems Volume 1: Artificial Intelligence, Cybersecurity, and Clean Energy for Next Generation Smart Grids, (227–246) © 2024 Scrivener Publishing LLC

16.1 Motivation

In a few decades, photovoltaic (PV) is anticipated to be a green substitute for fossil fuels. As a clean technology, PV is predicted to be able to completely eradicate the greenhouse gas emissions caused by fossil fuels. However, the sustainability of PV is threatened by related environment pollution from current non-green PV production techniques. The political and social effects of providing the resources required for PV manufacturing and the potential use of illicit materials are additional difficulties that might damage the PV industry's reputation. Therefore, it is crucial to assess both the socio-environmental effects of supplying PV materials as well as the sustainability (*i.e.*, environmental, financial, and chemical dangers) of the PV production process.

Before mass production, the organic photovoltaic (OPV) manufacturing method has to be evaluated for sustainability and modified. Due to its distinctive benefits, OPV has recently attracted interest. Organic photovoltaic (OPV) can absorb the whole spectrum of indoor light, can be made over a greater area using visually appealing and flexible materials, and is less expensive, lighter, and simpler to make than standard PV [1, 2]. Because of these characteristics, OPV is more appealing for use in interior design and has a wider range of uses, including screens for electrical gadgets and building windows [3, 4]. Low efficiency is the fundamental barrier to the commercialization of OPV, yet there has been a notable increase in OPV efficiency recorded in recent years. From 5% in 2005 to 12% and 18% in 2015 and 2020, respectively, OPV efficiency has grown and is predicted to continue to do so in the future [5, 6]. Only a few research have looked at the environmental effects of synthesizing the necessary components for OPV; the majority of earlier studies concentrated on increasing OPV efficiency and mechanical durability. Even though fullerene (C_{60}) is the essential component for producing OPVs (Figure 16.1), producing it involves using a lot of energy and toxic chemicals [7]. To guarantee that the OPV production line is sustainable before large-scale production, it is crucial to assess the carbon footprint, cost, and chemical risks of synthesizing fullerenes.

Manufacturing of Si PV is seriously threatened by the material scarcity. In 2020, Si PV accounted for more than 95% of the PV market, and it is anticipated that this percentage would stay over 50% until 2040 [8]. Quantifying the ingredients required to enable Si PV manufacturing over the coming years is critical. The majority of earlier research estimated the required PV material using theoretical models and historical PV data.

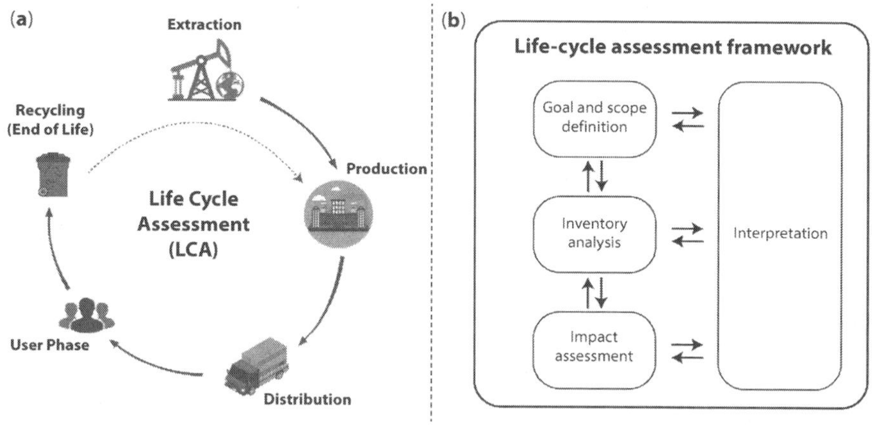

Figure 16.1 (a) Life cycle stages of a product or a process and (b) LCA framework (adapted from) [9].

They disregarded advancements in PV technology like efficiency, market share, wafer thickness, and glass thickness, which have a strong impact on the quantity of materials needed. Projecting Si PV materials in light of Si PV industry advancements is therefore crucial. Additionally, the political and social assessment of distributing the raw materials (such as quartz) required for the production of Si PVs was largely disregarded in the literature. Therefore, it is important to thoroughly research the availability and quality of worldwide resources for building materials that may have an impact on Si PV manufacturing.

16.2 Background

16.2.1 Life-Cycle Assessment (LCA)

Life-cycle assessment (LCA) is a methodical way of assessing a product's or a process's complete life cycle, from raw material extraction to manufacture, distribution, usage phase, and end of life stage (Figure 16.1a) [9]. Four primary phases make up a general LCA framework: defining the purpose and scope, inventory analysis, impact assessment, and interpretation (Figure 16.1b). For the environmental assessment of PV production, LCA is frequently employed. Any stage of the PV manufacturing process may use LCA to assess related environmental consequences, such as carbon footprint (global warming potential - GWP), cumulative energy demand

(CED), water footprint (WD), and E-factor, in order to change procedures for less negative environmental effects.

16.2.2 Green Chemistry Principles

The fundamentals of green chemistry were first presented in 1998 (Table 16.1) [10]. Since then, it has been frequently utilized to plan chemical synthesis from the beginning stages in order to prevent or lessen the negative effects in the downstream. A relatively recent idea in the production of nanoparticles for OPV is "green chemistry." In order to produce the nanomaterials required for OPV production, safer, more energy-efficient, and less toxic synthesis techniques can be identified using green chemistry principles [11].

16.2.3 Toxicity Assessment

An effective approach for assessing the ecological risks of substances is toxicity analysis [12]. Through theoretical models and tests, the toxicity of compounds may be measured. For instance, theoretical models like quantitative structure-activity relationships can be used to predict the toxicity of specific molecules, such as organic compounds (QSARs). On the other hand, toxicity tests should be used to quantify the toxicity of other compounds such nanocarbons. The Toxicity Estimation Software Tool (TEST), created by the U.S. Environmental Protection Agency (EPA) utilizing QSAR techniques, enables users to forecast chemical toxicity based on molecular structures. Additionally, the EPA developed EPA600/8-87/011 standard methods for calculating substance hazard through acute and long-term toxicity tests.

The LC^{50}, or the lowest concentration corresponding to a 50% mortality rate, is the most often used indicator in animal toxicity studies [13]. Early-stage chemical and compound toxicity can be measured using LC^{50} [14]. The EPA advises using Daphnia Magna, a planktonic invertebrate with a length of 0.5 to 5 mm, for toxicity evaluation since it is a trustworthy indicator of freshwater ecosystems at the cellular and animal scales [13]. They can be easily cultured, are toxicant-sensitive, and big enough to be seen under a light microscope [15].

Table 16.1 Twelve principles of green chemistry.

No.	Principle	Description
1	Prevention	Give waste avoidance more importance than cleaning up and addressing garbage that has already been produced. Prepare in advance to reduce waste at every turn.
2	Atom economy (AE)	By optimizing the number of atoms from all chemicals that are integrated into the finished product, waste may be reduced at the molecular level. To assess the effectiveness of a reaction, use atom economy.
3	Less hazardous chemical syntheses	Create safest possible chemical reactions and synthetic pathways. Take into account the risks associated with all materials handled throughout the reaction, including waste.
4	Designing safer chemicals	Directly reduce toxicity by molecular design. Throughout the design process, predict" and assess elements including physical characteristics, toxicity, and environmental destiny.
5	Safer solvents and auxiliaries	Choose the solvent that is the safest possible fit for the given process. Reduce the total amount of solvents and auxiliary materials used because they are a major source of waste.
6	Design for energy efficiency	Should select the low energy consumption of chemical route. Neglect the hot and cold effects at the same time we should consider the pressure and vacuum conditions.
7	Use of renewable sources/ feedstocks	Pick the chemical method requiring the least amount of energy. As well as the starting material (chemical) and waste products from reaction during the synthesis must be renewable i.e., plant-based sources.

(Continued)

Table 16.1 Twelve principles of green chemistry. (*Continued*)

No.	Principle	Description
8	Reduce derivatives	Reduce your reliance on transient derivatives like protective groups. To cut down on the number of formations of the intermediates, resources needed, and biproducts cause carcinogenic, avoid derivatives.
9	Catalysis	Select the effective catalysts to speed up reactions, cut down on waste, improve selectivity, and conserve energy.
10	Design for degradation	Create chemicals that break down and are simple to dispose of. Make sure that neither the chemicals nor the byproducts of their decomposition are hazardous, bio accumulative, or persistent in the environment.
11	Real-time analysis for pollution prevention	To avoid the development and release of any potentially dangerous and polluting compounds, monitor chemical reactions as they take place.
12	Inherently safer chemistry for accident prevention	To reduce the possibility of chemical mishaps, such as leaks, explosives, and combustion, substances and the form of a material utilized in a chemical process should be chosen.

16.2.4 Analytical Chemistry and Material Characterization Techniques for Environmental Assessment

Quantifying and studying molecules in compounds often involve the use of analytical chemistry tools such high-performance liquid chromatography (HPLC), gas chromatography (GC), UV-VIS spectroscopy, and mass spectroscopy. These tools may also be used to monitor the manufacturing processes, target chemicals, and byproducts, as well as their quality and quantity, for the goal of evaluating their impact on the environment [16]. Additionally, to characterize nanomaterials like fullerene (C_{60}) and determine their size, shape, and distribution so that their environmental effects can be studied, material characterization techniques like high-resolution transmission electron microscopy (HRTEM), scanning electron

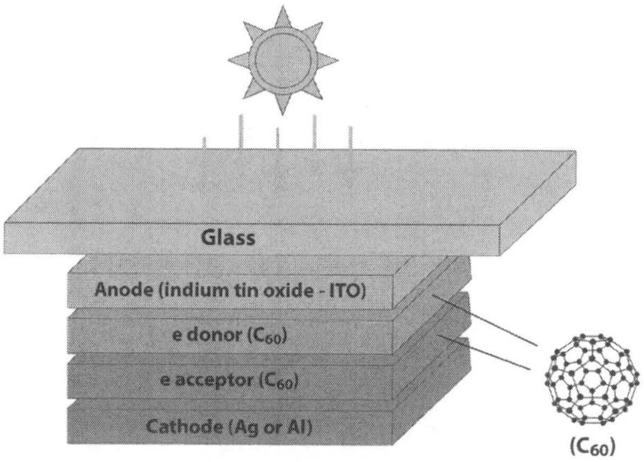

Figure 16.2 A typical structure of organic photovoltaic (OPV).

microscopy (SEM), optical microscopes, and the Zeta Potential Analyzer are used.

16.2.5 Organic Photovoltaic (OPV)

One organic layer, a bilayer (e acceptor and e donor), and bulk heterostructures (an interconnected network of electron acceptor and electron donor) organic cells can all be used to describe OPV. An active organic layer with two different types of materials (e acceptor & e donor) placed between two thin-film electrodes makes up a typical OPV structure (Figure 16.2) [17]. Compared to bilayer organic solar cells, single organic layer solar cells have a substantially lower efficiency. In order to create a commercial product with an efficiency that is equivalent to Si PV and other forms of PV technology, the low interfacial area between the donor and acceptor materials in the bilayer must be overcome. Another construction where components are mixed in a solvent before being deposited on the glass layer is the bulk heterojunction.

This method improves the interfacial area and results in an improved efficiency by forming an interpenetrating network of electron donors and electron acceptors. The most current technique that can produce OPV on a big scale is layer-by-layer processing [18].

16.2.6 Silicon Photovoltaic (Si PV)

The process of making Si PV begins with the extraction of silica sand, followed by the creation of metallurgical-grade silicon (MG-Si), the purification of polycrystalline silicon, crystallization, wafering, cell processing,

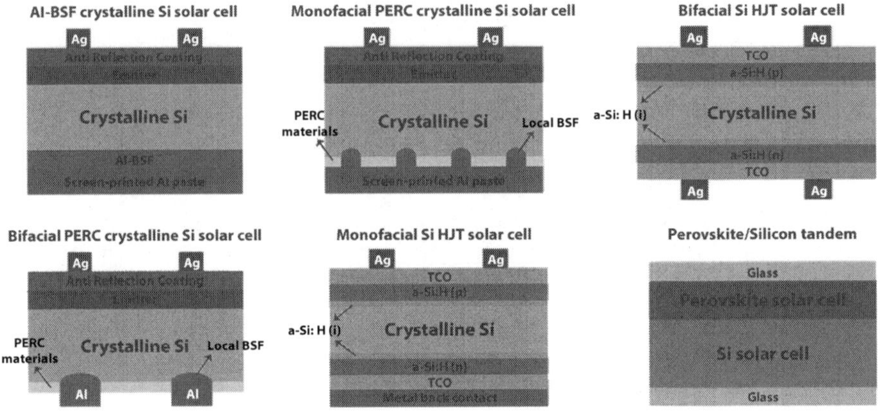

Figure 16.3 Silicon photovoltaic (Si PV) sub-technology structures.

and module fabrication. The purity of quartz deposits determines the purity of silica sand. By eliminating impurities from silica sand in an arc furnace at a high temperature, MG-Si purity is increased to more than 99% (about 1,800°C). To generate solar-grade silicon (SG-Si), which must be 99.9999999% pure in order to make wafers, further purification is required to remove contaminants like boron and phosphorus. Si PV wafers now have a typical thickness of 170 m for crystalline Si PV (c-Si PV) and 180 m for polycrystalline Si PV (mc-Si PV), however this thickness is predicted to decrease to 50 m by 2030. From 243 cm^2 (156 mm x 156 mm) to 441 cm^2 (210 mm x 210 mm), the wafer sizes shift. A typical Si PV module is 1.6 m^2 and has 60 distinct cells with a surface area of 243 cm^2. The present efficiencies for c-Si PV and mc-Si PV are 19.5% and 18.0%, respectively, and are predicted to rise to 21.5% and 20% in 2031. Figure 16.3 depicts the common Si PV sub-technology architectures.

16.3 Photovoltaic Technology and Environmental Impact

16.3.1 Environmental Impact Assessment

Using life-cycle analysis, we evaluated the environmental impacts of producing 1 kg of 99% pure C_{60} (LCA). Life-cycle analysis (LCA) was performed using SigmaPro 9.1.0.7 to model C_{60} purification processes and analyses inventory data, including Eco invent 3.3, US-EI from DATA SMART, published research, and data obtained during testing to determine associated carbon footprints. We used the TRACI 2.1 V1.05 approach, the Cumulative Energy Demand V1.11 method, the (water scarcity) V1.00 method for WD analysis, and the E-Factor computation to analyze the GWP [19].

16.3.2 Cost and Chemical Hazard Analysis

In the earlier study [20], the expenses and the usage of the hazardous chemicals were discussed. In conclusion, NFPA 704 standard scores, comprising health danger, combustible hazard, reactivity hazard, and special hazard, were used to underpin the study of chemical hazards. The target substrate's production rate and the life cycle cost (LCC) of the necessary resources were used to establish the price evaluation. These results were compared into a single cost scaling factor for analysis after being normalized using the numbers from the baseline process.

16.3.3 Toxicity Assessment

To calculate the toxicity of crude oil solvents, this study employed the Toxicity Estimation Software Tool (TEST, V5.1). To enable users to anticipate chemical toxicity based on molecular structures, the U.S. EPA developed TEST employing QSARs techniques. Based on the toxic effect (48 h) results for the Daphnia Magna, we used TEST to compare the toxicity of the crude oil solvents. Plant-based oils' toxicity was discovered by earlier research.

16.3.4 Green Chemistry Principles

Based on the work of Jessop *et al.*, green chemistry concepts were outlined in Table 16.1 [10]. In order to decrease the quantity of solvents needed (principle #5) and minimize toxicity (principles #12 and 4), this study discovered substitute solvents with high C_{60} solubility. These solvents also had decreased toxicity. In accordance with principle #7, which is about using sustainable feedstock, we also took into account plant-based solvents.

16.3.5 Experiments

By analyzing the financial, chemical, and environmental risks associated with current purification techniques, the baseline procedure was determined. The baseline procedure was dissolving the fullerene mixture, which included C_{60}, C_{70}, and higher fullerenes, in 5 ml of TMB using an ultrasonication for five minutes at room temperature [21]. In order to prevent a vortex, the solution was constantly stirred at a low speed for 10 min before adding 103 µl of DBU and 4 µl of DI water to begin the complexation process. To allow DBU and C_{70} and higher to create a solid complex, the fullerenes form complexation was maintained for 5 h under the nitrogen environment. The combination comprising C_{70} and higher fullerenes was separated from the solution containing C_{60} using a 0.2-µ filter. Using 0.2 M acetic acid, the residual DBU was eliminated from the method. In 24 h, the crystallization procedure was performed to extract the C_{60} particles from the solution using isopropanol. The remaining liquids were eventually extracted by drying the C_{60} particles. To change the baseline procedure, we adopted an iterative methodology (Figure 16.4) [20]. We were able to

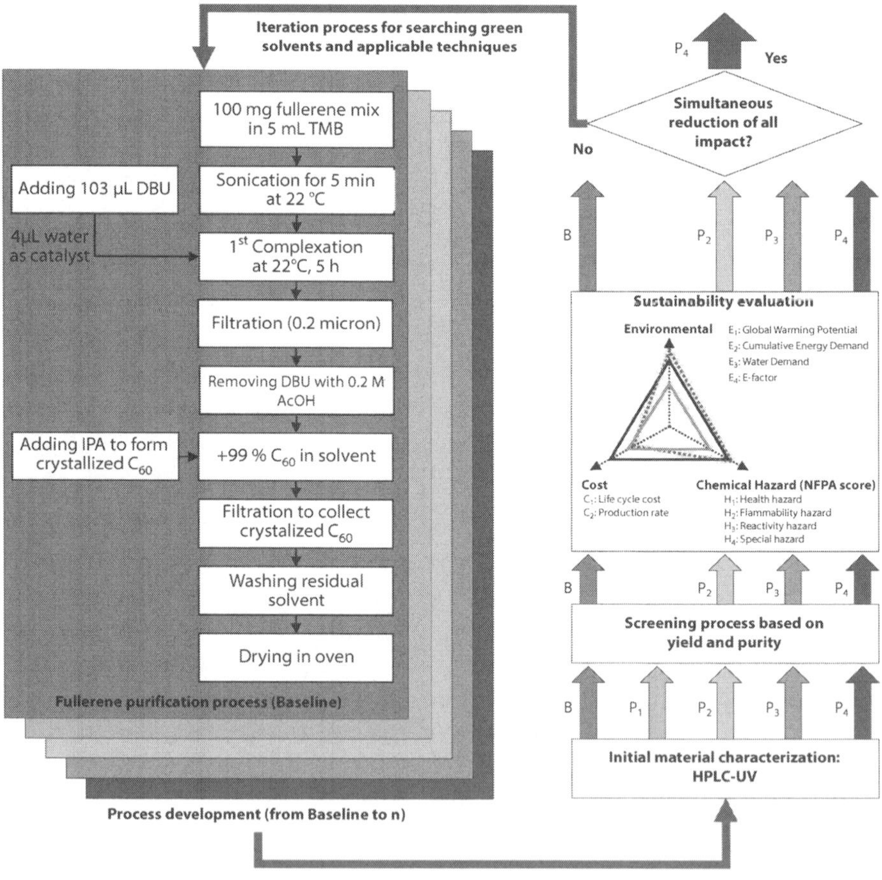

Figure 16.4 The experimental approach to identify alternative C_{60} purification methods for the baseline process.

assess purifying yields and implement in-process changes thanks to the iterative methodology.

The iterative method began with the baseline, as illustrated in Figure 16.4, and proceeded through initial material characterization, screening methods based on C_{60} purity and yield, and sustainability evaluation of viable alternative approaches with a similar purity and yield to the baseline. Based on cost, chemical danger, and environmental consequences, the optimum alternative purifying technique was determined. C_{60} purity was assessed using high-performance liquid chromatography (HPLC-UV). The mobile phase was hexane/2-propanol (70/30), the column was YMC-Pack ODS-ATM (5 μm, 120Å, 150 × 6 mm I.D.), the flow rate was 0.7 ml/min, the wavelength was 350 nm (0.08 AUFS), and the injection volume was 4 L.

16.4 Results

16.4.1 Existing C_{60} Purification Methods

DBU crystalline structure was used as the initial procedure. Together with colleagues, we evaluated the ecological, chemical, and economic aspects of the purification techniques now in use. The system boundaries and material flow of the commercially available C_{60} purification procedures, such as chelation, crystallization, and crystallization, are shown in Figure 16.4.

16.4.2 Baseline Evaluation

The environmental assessment (CED) findings of the complex formation method used to create one kilogram of C_{60} are shown in Figure 16.5. With the aid of CED, we could monitor the amount of energy used during the C_{60}

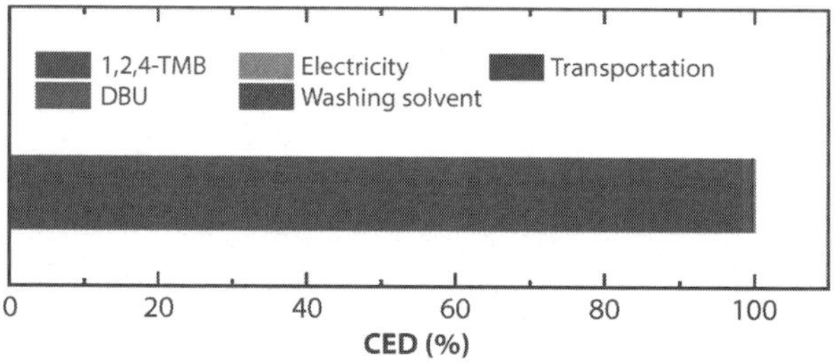

Figure 16.5 Environmental evaluation of the baseline purification process (complexation with DBU).

purification procedure and pinpoint any related environmental hotspots. Due to the quantity of solvents used and the embodied energy of TMB, it was responsible for more than 95% of the total CED of the baseline purification process. It was found to be the ecological issue and needed to be changed. The system boundaries and method specifics are shown in Figure 16.5.

16.4.3 Alternative Solvents for TMB

TMB is a very toxic solvent that has to be replaced with one that is less risky or damaging. Table 16.1: #3: Use substances with no or little toxicity; #4: design safer chemicals with lower environmental toxicity; #5: design to use a smaller number of solvents; #7: use renewable feedstocks; and #12: safer solvents for accident prevention. These green chemistry guidelines were used to determine alternative solvents using the C_{60} solubility and toxicity score. The toxicity ratings and C_{60} solubility of the available solvents are displayed in Figure 16.6. Solvents with similar C_{60} solubility to TMB and lower TMB toxicity ratings are shown in the purple zone. Two kinds of potential replacements were selected: petroleum-based and plant-based solvents. Linseed oil and olive oil were selected because they are nontoxic, more soluble than TMB, and come from renewable sources. Xylene and toluene were chosen because they have a similar C_{60} solubility to TMB and are less dangerous than TMB. TMB exhibited a standardized toxicity level

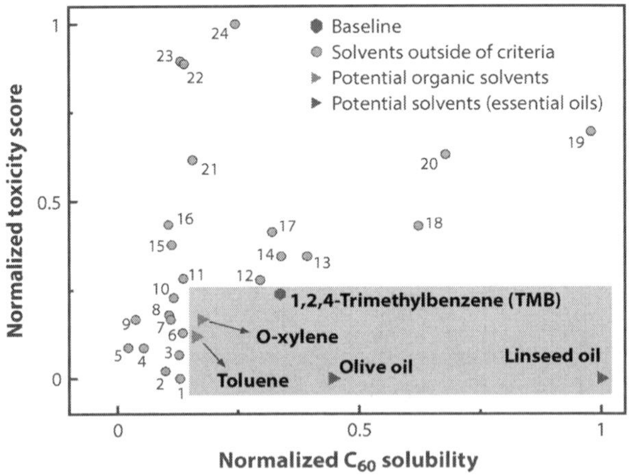

Figure 16.6 The toxicity and C_{60} solubility of solvents to identify replacements for TMB. The purple area highlights potential alternative solvents. Table 16.1 presents the detailed information of solvents.

of 0.24, whereas xylene, toluene, linseed oil, and olive oil had values of 0.17, 0.12, 0.17, 0.12, and 0.38, respectively. TMB scored 0.34 on the normalized solubility scale, compared to 0.18 for xylene, 0.16 for toluene, 1.0 for linseed oil, and 0.44 for olive oil.

16.4.4 Purification Experiments

16.4.4.1 Modified Baseline Process

The modified baseline has no effects which affect nature, low cost and eco-friendly of the baseline. The baseline purification yield was approximately 44%. The low yield is due to filtration in which a significant portion of the C_{60} was lost. We used centrifuges to increase production yields to 76% and reduce environmental, cost, and chemical hazard impacts. The modified baseline had approximately 43% less chemical hazards and environmental impact and was 22% cheaper than the baseline. Experiments were conducted to mitigate environmental, cost, and chemical hazards of the modified baseline. Figure 16.4 shows the material flow of the modified baseline.

16.4.4.2 Alternative Plant-Based Oil Solvents (P1 to P6)

Flaxseed oil and olive oil were selected as plant-based alternatives to TMB due to their low toxicity, similar C_{60} solubility, and being derived from renewable raw materials. Various methods have been proposed for dissolving fullerenes in linseed oil and olive oil. In P1, fullerenes were dissolved in olive oil and stirred for 72 h at room temperature under a nitrogen atmosphere, as described by Cataldo *et al.* proposed. Purification yields for C_{60} were less than 5%. This is because the mixed fullerene particles were not completely dissolved in the olive oil at the start of the experiment and complex formation probably occurred only at the surface. Therefore, C_{70} and higher fullerenes did not form tight complexes with DBU during complexation. In P2, the fullerene mixture was placed in olive oil at 75°C and the solution was stirred under a nitrogen atmosphere by him for 2 h as described by Cataldo *et al.* [22]. Production yields were still less than 5% because the oil cracked at high temperatures and reacted with C_{60}. P3 was based on the suggestion that he used a sonication bath for 15 min at 50°C to dissolve fullerenes in olive oil [22, 23]. Production yield increased to 26%, much higher than P1 and P2, but still lower than the modified starting production yield. In P4, we replaced the sonication bath with a sonication probe to improve the amount of C_{60} dissolved in olive oil prior to complex formation. 100 mg of the fullerene mixture was dissolved in

4 ml of olive oil using an ultrasonic probe for 3 min and the container was placed in a water bath to keep it near room temperature. Production yield increased to 44%. P5 and P6 should identify the optimal production yield for the refining process using linseed oil instead of TMB.

16.4.4.3 Alternative Petroleum-Based Solvents (P7 to P8)

The C_{60} purifying method using toluene and xylene had low environmental impacts, cost, and chemical hazard than the modified baseline. Toluene and xylene were chosen as petroleum-based alternatives to TMB due to their higher C_{60} solubility and lower toxicity ratings compared to TMB (Figure 16.4). 100 mg of fullerene mixture was dissolved in 7.5 ml xylene in P7 and 8 ml toluene in P8. The rest of the procedure was similar to the modified baseline. The production yield of C_{60} was 70% for P7 and 77% for P8. Figure 16.4 shows the material flow for P7 and P8. Figure 16.4 shows the HPLC analysis of baseline, modified baseline, P4, P6, P7, and P8.

16.4.5 Environmental, Cost, and Chemical Hazard Evaluation of Potential Methods

16.4.5.1 Chemical Hazard Assessment

Chemical hazard impacts were quantified for potential alternative solvents as well as alternative purification methods. Table 16.1 presents the detailed calculation of solvents chemical hazard impacts. The chemical hazard score was 1.3 for TMB, 2.0 for toluene and xylene, 0.33 for linseed oil, and 0.67 for olive oil. The chemical hazard score of alternative purification methods was calculated and normalized to the baseline. The modified baseline chemical hazard score was 57. The purification procedure utilizing olive oil had the highest hazards and risk score is 187. (P3). Due to the high boiling point (447 K) [23] of olive oil, which makes it an energy-intensive process to regenerate, all of the olive oil utilized in the procedure was taken into account when determining the chemical hazards. The linseed oil method's chemical hazard rating was 34. (P6). It was almost six times lower than the chemical hazard score of P3 since less amount of solvent was required due to the higher C_{60} solubility of linseed oil. The chemical hazard score was even lower for the alternative methods using petroleum-based solvents mostly because toluene and xylene can be regenerated; therefore, a smaller number of solvents were considered for chemical hazard evaluation. The lowest chemical hazard score was 13 for the purification process using toluene instead of TMB.

16.4.5.2 Cost Assessment

The C_{60} cleaning process using toluene (P8) had the lowest cost impact. Life cycle costs were calculated based on the raw materials and energy required to manufacture the solvents and materials for C_{60} cleaning. The process using xylene and toluene as alternative solvents had less cost impact due to lower production costs from crude sources.

16.4.6 Environmental Impact Assessment

The C_{60} cleaning process with toluene (P8) had the lowest environmental impact (GWP, IED, WD, and E-factor). Table 16.1 provides the LCA details for baseline, modified baseline, and alternative (P3–8) purification methods. The modified baseline has approximately 57% less environmental impact than the baseline due to higher purification yields than the baseline, resulting in less material used to produce 1 kg of C_{60}. became. C_{60} cleaning with olive oil (P3 and P4) gave the best environmental impact ratings. It was almost five times higher than baseline and ten times higher than modified baseline. Olive oil production requires large amounts of energy, water and substrates, increasing the associated environmental impacts. His C_{60} washes with linseed oil (P5 and P6) had a lower environmental impact than washes with olive oil, but a higher environmental impact than the modified baseline. C_{60} cleaning with xylene (P7) had a lower environmental impact than cleaning with olive oil, linseed oil, baseline, and modified baseline. C_{60} cleaning with toluene (P8) had the lowest environmental impact compared to baseline, modified baseline, and other alternative methods. Toluene can be recycled and reused in the process, reducing the material consumption required for C_{60} purification. The environmental impact rating for C_{60} cleaning with toluene was approximately 60% lower than the baseline and 28% lower than the modified baseline.

16.4.7 Overall Evaluation

The C_{60} purifying process using toluene is recognized as an environmentally friendly alternative to C_{60} cleaning. Figure 16.7 summarizes the evaluation of C_{60} cleaning baselines, modified baselines, and alternative methods (P3–8) based on cost and chemical hazard impacts. Although linseed oil and olive oil are non-toxic, plant-based solvents with high C_{60} solubility, life cycle analyzes show that the overall impact of their use is less than that of petroleum-based solvents (such as toluene and xylene). was shown to be

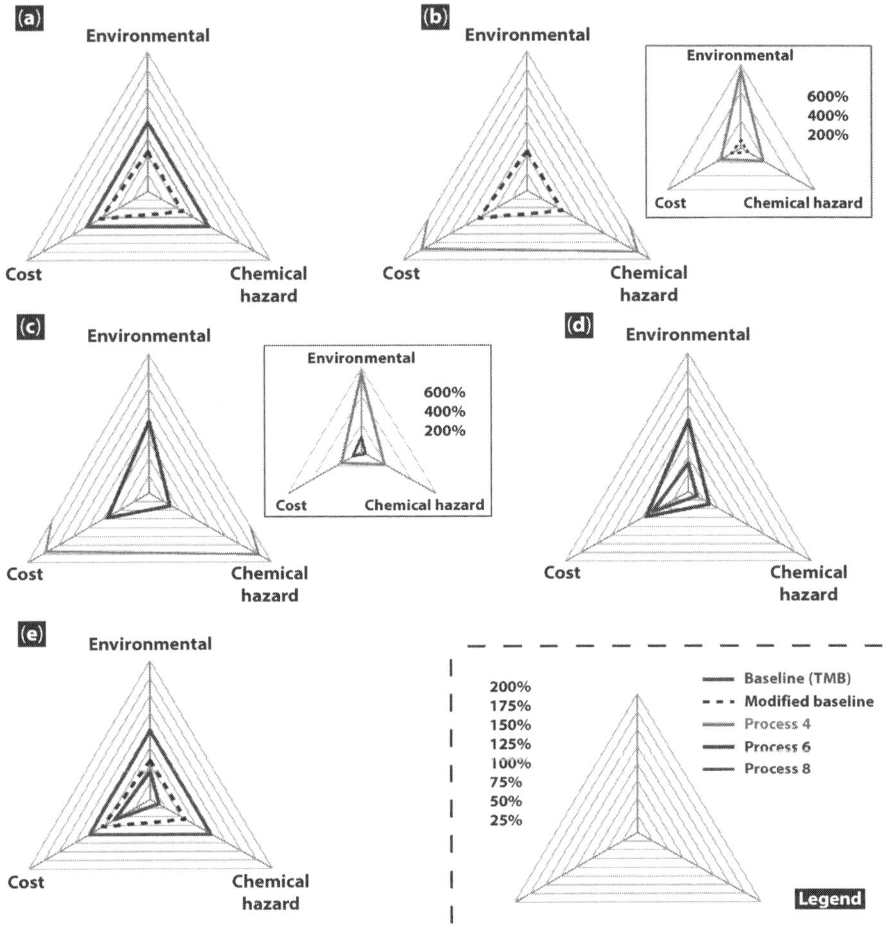

Figure 16.7 The evaluation of baseline, modified baseline, and alternative methods for C_{60} purification based on cost and chemical hazard impacts.

higher than this is mainly due to the large amounts of energy, water and substrates required to produce olive and linseed oils.

16.5 Conclusion

Before large-scale production, identifying a green process for C_{60} purification will reduce the environmental impacts of emerging C_{60}-containing products, such as organic transparent solar cells. Existing purification processes require a large amount of toxic solvents resulting in high cost and chemical hazard impacts. The complexation is best existing purification process; however, more than 95% of the environmental impact (CED) is due to using toxic TMB. Olive oil, linseed oil, xylene, and toluene were identified as potential alternative solvents for TMB since they are less toxic and have a comparable C_{60} solubility to TMB. An iterative approach was used to employ LCA and green chemistry to determine if using potential alternative solvents could result in a greener purification process.

Identifying a green solvent is highly dependent on the production process of the target product. Olive oil and linseed oil are nontoxic and have a higher C_{60} solubility than TMB, xylene, and toluene. However, the plant-based oil production process requires more energy, water, and materials than xylene and toluene in the upstream and the purification stage. Therefore, considering the life cycle of producing potential alternative solvents from raw materials and the production stage of the target product is vital for green solvent selection. It is also essential to study the chemical hazard of using potential alternative solvents. Sometimes selected solvents require extra materials to have a similar capability as the baseline solvent has. For example, olive oil is a nontoxic solvent, but a large amount of heptane, compared to what requires in the baseline, is necessary to separate C_{60} from C_{60}-containing solution results in a higher chemical hazard score.

References

1. Y. Cui, H. Yao, T. Zhang, L. Hong, B. Gao, K. Xian, J. Qin, J. Hou, "Organic Photovoltaic Cells for Indoor Application with over 20% Efficiency", *Adv. Matter.,* **2019**, 31, 1–7.
2. Gambhir, P. Sandwell, J. Nelson, The future costs of OPV – A bottom-up model of material and manufacturing costs with uncertainty analysis, *Sol. Energy Mater.* Sol. Cells., **2016**, 156, 49–58.
3. K.S. Chen, J.F. Salinas, H.L. Yip, L. Huo, J. Hou, A.K.Y. Jen, Semi-transparent polymer solar cells with 6% PCE, 25% average visible transmittance and a color rendering index close to 100 for power generating window applications, *Energy Environ. Sci.*, **2012**, 5, 9551–9557.
4. F. Liu, Z. Zhou, C. Zhang, J. Zhang, Q. Hu, T. Vergote, F. Liu, T.P. Russell, X. Zhu, Efficient Semitransparent Solar Cells with High NIR Responsiveness Enabled by a Small- Bandgap Electron Acceptor, *Adv. Mater.*, **2017**, 29.
5. H. Yao, J. Wang, Y. Xu, S. Zhang, J. Hou, Recent Progress in Chlorinated Organic Photovoltaic Materials, Acc. *Chem. Res.,* **2022** 53, 822–832.

6. C.J.M. Emmott, J.A. Röhr, M. Campoy-Quiles, T. Kirchartz, A. Urbina, N.J. Ekins-Daukes, J. Nelson, Organic photovoltaic greenhouses: A unique application for semi-transparent PV, *Energy Environ. Sci.*, **2015**, 8, 1317–1328.
7. Anctil, C.W. Babbitt, R.P. Raffaelle, B.J. Landi, Material and energy intensity of fullerene production, *Environ. Sci. Technol.*, **2011**, 45, 2353–2359.
8. Elshkaki, T.E. Graedel, Dynamic analysis of the global metals flows and stocks in electricity generation technologies, *J. Clean. Prod.*, **2013**, 59, 260–273.
9. W. Klopffer, Life cycle assessment, *Environ. Sci. Pollut. Res.*, **1997**, 4, 223–228.
10. P.G. Jessop, S. Trakhtenberg, J. Warner, The Twelve Principles of Green Chemistry, in: Innov. *Ind. Eng. Chem., American Chemical Society*, **2008**, 401–436.
11. H. Duan, D. Wang, Y. Li, Green chemistry for nanoparticle synthesis, *Chem. Soc. Rev.*, **2015**, 44, 5778–5792.
12. J. Beyer, K. Petersen, Y. Song, A. Ruus, M. Grung, T. Bakke, K.E. Tollefsen, Environmental risk assessment of combined effects in aquatic ecotoxicology: A discussion paper, *Mar. Environ. Res.*, **2014**, 96, 81–91.
13. L. Guilhermino, T. Diamantino, M. Carolina Silva, A.M.V.M. Soares, Acute toxicity test with Daphnia magna: An alternative to mammals in the prescreening of chemical toxicity?, *Ecotoxicol. Environ. Saf.*, **2000** 46, 357–362.
14. E.E. Kenaga, Test organisms and methods useful for early assessment of acute toxicity of chemicals, *Environ. Sci. Technol.*, **1978**, 12, 1322–1329.
15. M.L. Dini, J. O'Donnell, S.R. Carpenter, M.M. Elser, J.J. Elser, A.M. Bergquist, Daphnia size structure, vertical migration, and phosphorus redistribution, *Hydrobiologia.*, **1987**, 150, 185–191.
16. S. Kayasth, K. Swain, Role of analytical chemistry in environmental monitoring, *J. Radioanal. Nucl. Chem.*, **2004**, 262, 35–42.
17. M.A. Fusella, Y.L. Lin, B.P. Rand, Organic photovoltaics (OPVs): Device physics, in: Handb. *Org. Mater. Electron. Photonic Devices, 2nd ed., Elsevier*, **2019**, 665–693.
18. M.D.M. Faure, B.H. Lessard, Layer-by-layer fabrication of organic photovoltaic devices: Material selection and processing conditions, *J. Mater. Chem.* C., **2001**, 914–40.
19. P. Anastas, N. Eghbali, Green Chemistry: Principles and Practice, *Chem. Soc. Rev.*, **2010**, 39, 301–312.
20. E. Lee, C.J. Andrews, A. Anctil, An Iterative Approach to Evaluate and Guide Fine Chemical Processes: An Example from Chloroaluminum Phthalocyanine for Photovoltaic Applications, *ACS Sustain. Chem. Eng.*, **2018** 6, 8230–8237.
21. K. Nagata, E. Dejima, Y. Kikuchi, M. Hashiguchi, Efficient and Scalable Method for [60] Fullerene Separation from a Fullerene Mixture: Selective Complexation of Fullerenes with DBU in the Presence of Water, *Org. Process Res. Dev.*, **2005**, 9, 660–662.

22. K.N. Semenov, N.A. Charykov, V.I. Namazbaev, N.I. Alekseyev, E.G. Gruzinskaya, V.N. Postnov, O.A. Krokhina, Temperature dependence of solubility of light fullerenes in some essential oils, Fullerenes *Nanotub. Carbon Nanostructures.*, **2011**, 19 225–236.
23. A. Demirbas, Relationships derived from physical properties of vegetable oil and biodiesel fuels, *J. Fuel.*, **2008**, 87, 1743–1748.

17

Transparent Photovoltaics and Environmental Impact

Kavinkumar Ravikumar and Milind Shrinivas Dangate*

Chemistry Division, School of Advanced Sciences, Vellore Institute of Technology, Chennai, India

Abstract

The near-infrared (NIR) and ultraviolet (UV) portions of the light spectrum can be collected using transparent organic photovoltaics (TPVs) that can be used in new applications such as windows. Transparent organic solar cells not only generate electricity, but also absorb it in the NIR range, reducing the need for cooling energy in buildings in the summer. A new technology called organic photovoltaics (OPV) was developed primarily to replace fossil fuels and reduce greenhouse gas emissions. Quantifying the environmental impact of OPV technology is critical as commercialization approaches to prevent new problems from arising. Life cycle assessment (LCA) is commonly used to evaluate energy technologies and identify environmental issues, but the lack of inventory data can make this process challenging for technology development. This chapter analyzes the energy savings of organic TPVs in multiple urban window and skylight applications, evaluates the impact of organic TPVs on urban heat island effects, and uses LCA and geochemical approaches. This approach was used to identify 'hotspots' or processes with the greatest impact on chloroaluminum phthalocyanine (ClAlPc) production. Compared to the existing process, the improved process has been shown to reduce expenses by 9%, environmental effects by 3%, and chemical hazards by 23%. Using a ClAlPc-based device in a window application, the TPV impact was investigated during the usage phase. Heating, ventilation, and air conditioning (HVAC) energy savings have increased building energy efficiency by up to 20%. The NIR absorbed by a window TPV is more effective at higher temperatures, so efficiency gains are climate dependent. In urban environments, applying TPV to windows can reduce the energy consumption of buildings.

**Corresponding author*: milind.shrinivas@vit.ac.in

O.V. Gnana Swathika, K. Karthikeyan and Sanjeevikumar Padmanaban (eds.) Smart Grids as Cyber Physical Systems Volume 1: Artificial Intelligence, Cybersecurity, and Clean Energy for Next Generation Smart Grids, (247–274) © 2024 Scrivener Publishing LLC

Net energy savings from TPV applications in urban areas are estimated to be up to 2 GJ per month higher than in rural areas.

Keywords: Transparent organic photovoltaics, electricity, organic photovoltaic (OPV), chloroaluminum phthalocyanine (ClAlPc), emerging technologies

17.1 Introduction

Future energy solutions should not create new environmental, social, and cost problems. Electricity generation from natural resources, such as charcoal, natural gas, and fuel oils, is not desirable for a long-term goal. These non-renewable electricity resources cause more background impacts and it may increase the production of greenhouse gases (GHGs) [1]. To solve that issue, we should increase the production of renewable energy resources. The most effective and efficient renewable energy source is solar energy because it helps to produce over hundred thousand terawatts of potential energy from sunlight [2]. In recent research, the photovoltaic (PV) technology has improved and there is increase in the interest for alternative PV sources such as organic photovoltaics (OPV). In the way for durability, low weight, flexibility and low cost, the OPV will aid improving the world economy [3]

These days, countries like the U.S. generate the electricity from non-renewable resources like natural gas and coal. Energy conversion from coal increase air pollution due to the emission of toxic gases like mercury, sulfur dioxide (SO_2), nitrogen oxides (NO), and particulate matter (PM). The next five decades are very crucial to replace conventional energy sources with renewable sources in order to reduce pollution and background impacts. According to the U.S. Energy Information Administration (EIA), the net coal-fired electricity generating capacity was reduced by around 60 GW from 2011 to 2016 as observed by the government of the United States Environmental Protection Agency's Mercury, Sulfur and Air polluted compounds. Figure 17.1 shows that the main source of electric energy generation from renewable energy will reach 1650 billion kWH and it will reach the second largest source of electricity by 2050.

The amount of renewable energy produced from renewable resources like wind, geothermal, and hydroelectric are expected to remain similar when compared with PV, while PV would increase until 2050, as shown in Figure 17.2. Increased awareness in building integrated photovoltaic (BIPV) application due to the effective efficiency of economic and environmental benefits. In order to decrease the usage of land for PV arrangement is an important economic benefit of BIPV in major cities, and BIPV could also manage the micro-climate improvement.

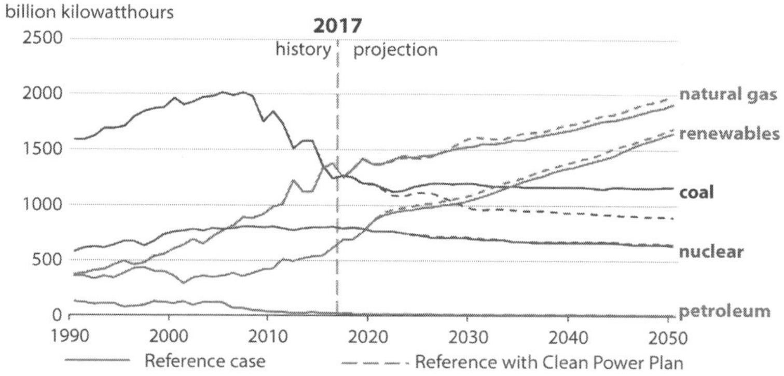

Figure 17.1 Report of generation of electric energy from renewable and non-renewable sources (adapted from Annual Energy Outlook 2018, EIA, 2018).

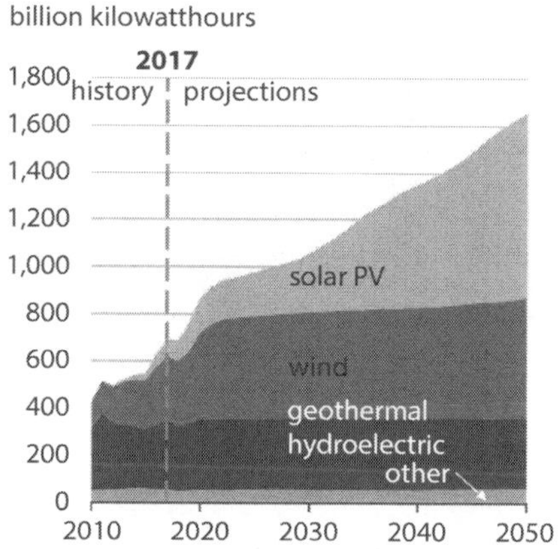

Figure 17.2 Renewable electricity generation, including end-use generation (adapted from Annual Energy Outlook 2018, EIA, 2018).

The building sector contributed to over 50% of total energy use in high-income cities (Berlin, Bologna, Greater London, New York, Seoul, Singapore, and Tokyo) in 2008. The energy use in the building sector was connected with building conservation and operation. Buildings are intricate systems, and their energy consumption is affected by local climate, occupancy profiles, and consumer preferences. Figure 17.3 shows the primary energy usage for electricity generation and future world electricity demand for each sector. A key solution for urban sustainable energy is to increase low consumption of carbon and increasing the efficiency of the

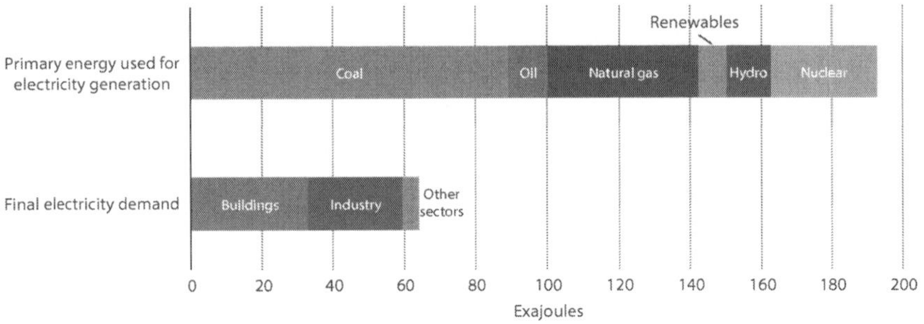

Figure 17.3 Energy use for power generation and total electricity demand in 2010 (adapted from Transition to Sustainable Buildings, IEA, 2013).

buildings, production of electricity in a city where land is limited, which can mostly be attained using BIPV.

It is hard to find but these windows are responsible for the energy consumption in buildings in Organisation for Economic Co-operation and Development (OECD) countries. It is important to grow the low thermal transmittance and highly performing windows and, climate-appropriate solar heat gain coefficient (SHGC) to reduce energy consumption from fossil fuels.

Several glazed windows (triple or quadruple) are produced in the way of using the low electric coating, with effective inert gases like krypton and xenon to improve the high-performance windows. The cost of this kind of windows is high because of the use of inert gases, and it might require more time and additional work to reduce the cost to provide the energy consumption of windows.

Transparent OPV (TPV) offers a prospect to produce electricity, at the same time give a chance to improve building energy performance. The absorption from TPV can be tuned from ultraviolet (UV, <435 nm) to near infrared (NIR, >670 nm) range of the solar spectrum. Transparent OPV (TPV) in window applications could control the inward flow of temperature which is produced from solar and decrease the consumption of cooling energy while the absorbed solar energy is used to produce electricity while performing the original function of windows because of its transparency [10]. These are advantages of OPV applications but the potential impacts from OPV raw material manufacturing are not understood well.

A previous work has used life cycle assessment (LCA) to estimate the impact associated with fullerenes synthesis [11], a common acceptor material in OPV as well as various minor molecules and polymer organic solar cells based on increasing energy demand (CED). The energy requirement for OPV manufacturing was 50% lower than for manufacturing conventional inorganic photovoltaics [3]. While the potential energy benefit of OPV compared to inorganic solar cells has been established, the impact of TPV has not been understood.

One restriction of LCA in its application to developing technology is it uses new compounds which have low inventory information. An approach to decrease the health and background impacts of chemicals and help develop alternative chemical process is green chemistry. The framework of green chemistry includes all the levels of the life cycles in chemistry when it shows that the reaction undergoes the 12 principles of green chemistry [12]. Applying these to the new chemical is relatively simple when related to LCA because green chemistry metric is designed to use stoichiometry or mass balance to evaluate environmental performance [13].

17.1.1 Motivation

The kind of energy developed today will definitely be desirable in the future. In the future, solar cells which are made up of organic materials play a vital role in the world economy, so it is important to do research on the production, development, and usage of the organic solar cells. Life cycle assessment (LCA) studies about all stages of working on the raw material obtainment (cradle) to the ultimate return of the material, which is decomposition (grave).

Infrastructure renewable energy sources requires PV technologies and suitable material from environmental sources [14]. In the time of the production of OPV, the nature of the chemicals which are used in the process of the production should be checked that they does not affect the environment, because prevention is better than solution. Still, there is no establishment of effective ways to find the stability of the production of OPV materials, at least knowing how to increase the usage of OPV is a major thing in manufacturing.

Transparent OPVs have the potential to be affordable by using low price raw materials that can be integrated into the infrastructure of the built environment, thereby decreasing setting up costs and balance-of-systems (BOS) [10]. Nowadays, PV's performance in the environment is calculated from the BOS, module, and production of electricity, and it also helps analyze the traditional function and application of PV. From the perspective of energy efficiency, the trade-off between PV efficiency and energy inputs, such as cumulative energy demand (CED), has been studied and modeled as metrics such as energy return on energy invested (EROI) and energy payback time (EPBT) [16]. However, TPV generates electricity and contributes to generating energy demand through NIR absorption. Near infrared (NIR) uptake by TPV is not well evaluated. However, switching from a conventional double pane window to a semi-transparent photovoltaic window made of amorphous silicon has been found to reduce the building's energy demand by up to 15% [17]. To make a clear comparison with existing renewable energy applications, the advantage of TPV in window application should be integrated with existing methods of PV estimation. Combining energy savings and electricity generation is necessary to evaluate the benefit of TPV in comparison to other renewable energy options.

The application of TPV in window application will help to fulfill this goal of renewable energy in urban areas. An additional impact of energy use in urban areas is the loss of heat from anthropogenic activities and solar radiation, resulting in the urban heat island effect (UHI). Large-scale deployment of BIPV in urban areas, such as rooftop or façade application, has been studied for microclimate impact, and studies conclude that PV application could reduce local temperature because solar energy is converted to electricity [18, 19]. In the same application, studies have shown that TPV in urban areas may not exacerbate UHI effects. Impact is required.

17.2 Background

17.2.1 Green Chemistry

We know that green chemistry has the 12 principles of chemistry (summarized in Table 17.1) and measures [12] which decrease the impacts of the hazardous nature of the chemicals. Using green chemistry as an integrated set of principles or design criteria means that you must (1) take into account all phases of the chemical life cycle, (2) minimize the inherent hazardous chemicals through structure, and (3) consider green chemistry when making decisions [12].

Due to the large quantity of waste produced by the use of solvents, the E-factor (Equation 17.1) is used to define fine chemical processes (a chemical category of OPV materials) as an illustration of green chemistry metrics and applications [20]. A pharmaceutical business similarly indicated that the majority of the resources squandered came from the use of solvent as a reaction medium or work-up. Up to 35–50% of the materials used in a pharmaceutical process were contributed by the usage of solvents [21]. This problem serves as the driving force behind the hunt for green solvents that lessen the negative effects of fine chemical and drug manufacture

Table 17.1 Twelve principles of green chemistry and description.

S. no.	Principle	Description
1	Prevention	Give waste avoidance more importance than cleaning up and addressing garbage that has already been produced. Prepare in advance to reduce waste at every turn.
2	Atom economy (AE)	By optimizing the number of atoms from all chemicals that are integrated into the finished product, waste may be reduced at the molecular level. To assess the effectiveness of a reaction, use atom economy.
3	Less hazardous chemical syntheses	Create safest possible chemical reactions and synthetic pathways. Take into account the risks associated with all materials handled throughout the reaction, including waste.
4	Designing safer chemicals	Directly reduce toxicity by molecular design. Throughout the design process, predict and assess elements including physical characteristics, toxicity, and environmental destiny.
5	Safer solvents and auxiliaries	Select the safest solvent you can for the particular procedure. Reduce the overall number of solvents and auxiliary materials utilized because they contribute significantly to waste production.
6	Design for energy efficiency	Choose the least energy-intensive chemical route. Avoid heating and cooling, as well as pressurized and vacuum conditions (*i.e.*, ambient temperature and pressure are optimal).
7	Use of renewable feedstocks	Pick the chemical method requiring the least amount of energy. Avoid pressured and vacuum situations, as well as heating and cooling (*i.e.*, ambient temperature and pressure are optimal). Instead of other, equal compounds originating from petrochemical sources, use chemicals that are derived from renewable (*i.e.*, plant-based) sources.

(*Continued*)

Table 17.1 Twelve principles of green chemistry and description. (*Continued*)

S. no.	Principle	Description
8	Reduce derivatives	Reduce your reliance on transient derivatives like protective groups. To cut down on the number of reaction steps, resources needed, and waste produced, avoid derivatives.
9	Catalysis	Use catalytic reagents in reactions rather than stoichiometric ones. Select catalysts to speed up reactions, cut down on waste, improve selectivity, and conserve energy.
10	Design for degradation	Create chemicals that break down and are simple to dispose of. Make sure that neither the chemicals nor the byproducts of their decomposition are hazardous, bio accumulative, or persistent in the environment.
11	Real-time analysis for pollution prevention	To avoid the development and release of any potentially dangerous and polluting compounds, monitor chemical reactions as they take place.
12	Inherently safer chemistry for accident prevention	Choose and create safer chemical processes that automatically reduce the likelihood of accidents. Know the potential hazards and evaluate them in advance.

on the environment [22], and the majority of pharmaceutical firms have embraced a green chemistry strategy.

$$E - Factor = \frac{\text{Mass of waste}}{\text{Mass of product}} \tag{17.1}$$

Green chemistry applications in the field of OPV have received relatively little attention, despite the fact that several studies have documented efforts to embrace green chemistry methodologies in the drug industry. A prior paper covers methods for creating interlinked polymers using environmentally friendly chemistry for long-term OPV [14]. This essay critiques latest work on sustainable power for frequently failing to take into account embodied energy and environmental effects. Further recommendations include using catalysts, generating ecologically friendly byproducts, avoiding protected groups, consuming less energy, and producing minimal trash [14].

17.2.2 Life Cycle Assessments

The term "life cycle" refers to the duration of a product and involves the raw materials extraction, their processing and production installation, usage of the product, and disposal, as shown in Figure 17.4 (a) [23]. Goal formulation and scope, cost estimation, impact assessment, and interpretation are all parts of the systematic process known as LCA, which is illustrated in Figure 17.4 (b) [23, 24].

Any product, process, or service's potential environmental effects may be evaluated using LCA. The quality and correctness of the study will depend on how precisely and thoroughly the information were gathered during the life cycle inventory (LCI). A relationship between a product or method and its possible effects is found through the life cycle impact assessment (LCIA). Finally, interpretation assesses the outcomes of LCI and LCIA and properly conveys the outcomes [24].

17.2.3 Transparent Organic Photovoltaic Applications

Excitonic materials that selectively capture electricity from UV and NIR light have recently been used in TPV technologies. Organic small molecules and organic salts are components used to create TPV [25]. Near infrared-selective TPV for small molecule TPV devices was established in 2011 using C60 fullerene (C60) as the acceptor and chloroaluminum

Figure 17.4 (a) Cradle-to-grave life cycle assessment and (b) LCA framework according to ISO standards 14040 and 14044 [23].

phthalocyanine (ClAlPc) as the donor among two transparent conducting electrodes (indium tin oxide, ITO).

The shown device has an average visual transmittance (AVT) of 65% and a power conversion efficiency (PCE) is 1.3 ± 0.1% [25]. Theoretically, a single junction device may achieve a PCE of TPV of 20.6% with 100% AVT, and three to four junction devices can achieve a PCE of roughly 30% and 100% AVT [10].

Material quality affects how well tiny compound OPV functions. According to a study, the PCE of copper phthalocyanine (CuPc) solar cells varies depending on the quality of the material from 0.26 ± 0.01% to 1.4 ± 0.1% [26]. The thermodynamic efficiency of an OPV device, which is a measurement of PV's environmental performance, depends heavily on the material's purity. For instance, the embodied energy is doubled when C60 is purified to electronic grade purity [11].

Refined chemicals are sophisticated, singular, and high-grade compounds used in the drug, agricultural, animal anatomy, dye, and electrical industries. They are manufactured in small amounts (up to 1000 MT/y) and expensive (>$10 per kg). Green chemistry, that is well recognized for its 12 principles, is one approach that may be utilized to lessen the environmental effect of fine chemical manufacture [12]. They serve as broad recommendations for decreasing influence of the chemical synthesis on the environment, such as lowering secondary products and derivative levels, boosting energy effectiveness, and promoting the use of safer chemicals. In order to decrease the inherent risk of chemical products via design, evaluate all phases of the cycle of life sciences, use green chemistry as a coherent set of design principles, and use green chemistry as a design criterion [12]. Life cycle assessment (LCA), a thorough technique for evaluating effects associated with the entirety of a life cycle assessment studies about all stages of working on the raw material obtainment (cradle) to the ultimate return of the material, which is decomposition (grave). The accuracy of the mass and energy flow data gathered to create the life cycle assessment determines the quality of the LCA.

Rarely does the fine chemical business publish in-depth information through patents or journal article articles in order to stay competitive. Since fine chemicals are manufactured in multifunctional factories with common facilities and equipment used for several chemicals, it is challenging to determine the exact energy consumption specifically related to one molecule [28]. In addition to academics, the drug and chemical industries have developed a number of sustainability techniques that direct their product development and incorporate LCA and green chemistry. To do a thorough LCA or quantify company-specific effects, current industrial application approaches call for considerable data collecting.

On the other hand, lab scale approaches frequently employ simplified effect assessment that concentrates on the production stage of the process without taking into account the increase in the usage of the chemicals as would LCA. From the information we have, there is no approach shown that can be used to direct chemical research at the laboratory usage.

Life-cycle Assessment (LCA) is a technique which is helpful for evaluating chemical and drug synthesis processes, provided that the list of chemicals and methods, which is already present in the journals, that is not the event for the majority of novel compounds. Additionally, it is frequently lacking the characterization parameters required to compute the problems. The evaluation will overestimate the impact if certain compounds are not characterized in the LCA database. In order to collect the data from daily based life, concepts of green chemistry could be utilized to detect difficulties with novel compounds related to toxicity and material efficiency. This research suggests a novel evaluation technique that may be used by both business and academics to quickly screen possible environmental, health, and financial implications for fine compounds with few LCA data. The main theme of this green chemistry is combining all the uses of the LCA and it may lead to pollution-less environment and low impacts that cause the impacts in the environment. The suggested technique is utilized with tiny molecules found in OPVs [15]. The majority of the health and background impacts of OPV materials are unknown, and nowadays, research focuses on OPVs that lower expenses and improve the proficiency of solar panels [7]. The combination of acceptor and donor, like fullerenes, as an electron acceptor and metal (M-Pc) phthalocyanine as a donor, is necessary for the fundamental structure of an OPV [8].

Because it exclusively absorbs in the NIR spectrum, ClAlPc has special relevance for OPVs. It enables transparency in the solar cell, which is crucial for uses like electricity-producing windows [10, 25]. Even though metal phthalocyanines are frequent, often-used small molecules in the fabric sector, the high transparency in the quality is necessary for solar applications [26]. to make it a good product for society. At the moment, the cost of high quality (98%) ClAlPc is \$157/g, whereas the cost of lesser purity (85%) is \$55/g. The energy-intensive purifying procedure required for electronic applications is to blame for the price discrepancy [11].

The suggested approach is repeated and adheres to the phases in Figure 17.5 (a). The beginning of production of ClAlPc will serve as the starting point, under the circumstances detailed in a published patent, is first step in the process. To create the life cycle inventory, the direct energy consumption from the raw materials and yield during the synthesis are measured. The material is evaluated using UV-Vis and HPLC mass spectroscopy in

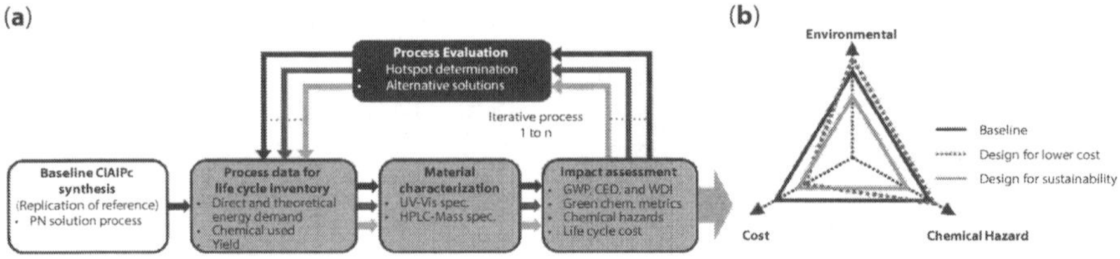

Figure 17.5 (a) Proposed repeated methodology and (b) sustainability classes.

order to determine if it satisfies the purity criteria for OPV applications. The life cycle inventory is used to do impact assessment, which combines LCA and green chemistry impact factors. Based on the three evaluation criteria depicted in Figure 17.5, the procedure is assessed (b).

The chemical or process that contributes the most to one or more of the effect groups is known as the "hotspot." Green chemistry concepts are used to create an alternative, which is then contrasted with the standard. Until a substitute is identified that concurrently lowers the ecological, financial, and chemical hazards, the repeated approach is used. As shown in Figure 17.5(b), the objective is to prepare to stabilize the infrastructure for our environment, thus, there must be a reduction in all effect categories and no trade-offs in the alternatives. For instance, planning for cheaper costs could unintentionally result in increased chemical risk and environmental effects.

17.3 Method

17.3.1 Life Cycle Assessment for Fine Chemicals Process

To fulfill the purity requirements of the OPV application, the research covers every step from the extraction of raw materials to the manufacturing of 1 kg of ClAlPc. SigmaPro (Pre-Consultants, Netherlands) is used for this project along with the already collected data from (Eco invent 3.3), (US-EI from DATASMART), the previously papers, and the data collected in the experimental part of the synthesis. The use of LCA software makes it easier to compile and analyze the inventory data. Utilizing default parameters and reactions based on the research paper released recommendations, inventory data for novel compounds is approximated. In order to evaluate the effects of material synthesis and quickly identify possible issues, it is intended to combine life cycle evaluation methods for current compounds with green chemistry metrics.

Since none of the 12 green chemistry principles were intended to be standalone objectives, they are all regarded as being present [12]. The information from the life cycle evaluation is supplemented by green chemistry measures rather than being duplicated. The simultaneous reduction of the environmental, financial, and social consequences is required for sustainable chemical process design. The research and development of the sector only considers the sub-category of the stages and this has a major role in screening the production of the synthesis. The tools and sub-category used for the technique are listed in Table 17.2. Since LCA is frequently used to evaluate environmental implications, certain sub-categories are chosen for existing chemicals and are supplemented with green chemical metrics. The only part of the social effect that can be reduced throughout the study and improvement is the inherent dangers that the chemicals pose to employees. Aside from exposure, the harmful evaluations of chemicals should be taken into account for risk categorization; there are so many risk factors but the impacts to the environment depend on the chemical, which is the third level risk factor.

Due to the selection of LCA techniques, only two of the principles are utilized as sub-criteria in Table 17.2. Out of twelve principles that does not obey the four rules, including #11: *real-time analysis for pollution prevention* and #7: *use renewable feedstocks* since we do not want to change the chemical. The remaining five principles out of twelve such as *#6: design for energy efficiency* is used to guide alternative synthesis rather than in the evaluation stage (Table 17.2).

17.3.2 Selection of Impact Category for Sustainability Assessment

17.3.2.1 Environmental Impact

Both the CED (E2) and the global warming potential (GWP) (E1) are widely used environmental impact indicators with a proven method. The previously indicated new materials inventory criteria are utilized in the absence of real-time data, although they are only applicable to CED and GWP.

(E3), WDI (The water demand indicator) is an LCA measure for usage of the water that takes into account water consumption from utility input and chemical processes that is not taken into account in the E-factor. Despite the recent addition of water use to the E-factor [4], the assessment technique still adheres to the E-original factor's definition. To create an alternative ClAlPc method, the reaction which obey the principles of green chemistry is shown in Table 17.2.

Table 17.2 Outline of green chemistry tools and LCA effects and the alter method for the synthesis.

		Green chemistry		Life cycle assessment	
		Principle (#)	**Metrics or tool**	**Impact category**	**Method**
Sustainability methodology index	Environmental	**E4:** waste prevention (1)	E-factor: amount of waste generated per kg of product	**E1:** global warming potential (GWP)	Tool for reduction and assessment of chemicals and other background impacts (TRACI) 2.1- U.S. 2008 normalization
				E2: cumulative energy demand (CED)	CED 1.09
				E3: water demand indicator (WDI)	Water accounting and vulnerability evaluation (WAVE)
	Chemical hazard	**H1–H4**: inherently safer chemistry for accident prevention (12)	NFPA 704 standard (0-4 score): **H1:** health issue **H2:** flammability hazard **H3:** reactivity hazard **H4:** special hazard		
	Cost			**C1:** cost of life cycle	Price of natural resources as raw material inputs
				C2: reaction rate	Reaction capacity based on vessel size and time per batch

(*Continued*)

Table 17.2 Outline of green chemistry tools and LCA effects and the alter method for the synthesis. (*Continued*)

	Green chemistry		**Life cycle assessment**	
	Principle (#)	**Metrics or tool**	**Impact category**	**Method**
Guide alternative synthesis	Atom economy (2)	Atom economy: formula weight (FW) of atoms utilized/ FW of all reactants in percentage	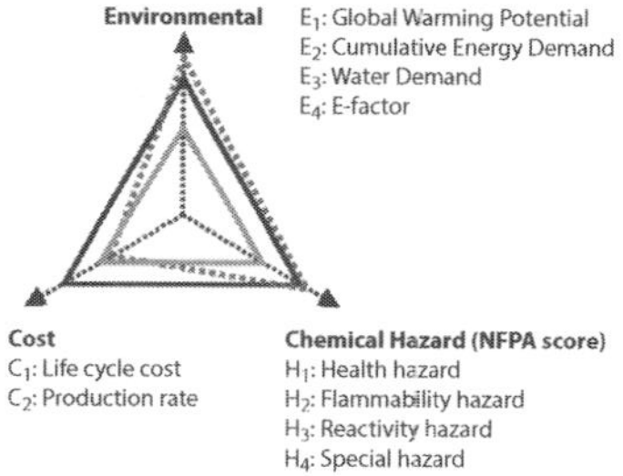	
	Less hazardous chemical synthesis (3)	Globally harmonized system		
	Safer solvents and auxiliaries (5)	GSK solvent selection guide		
	Design for energy efficiency (6)	Ambient temperature and pressure process		
	Reduce derivatives (8)	Atom economy: FW of atoms utilized/ FW of all reactants in percentage		
	Catalysis (9)	M-Pc synthesis with organic base		
Not applicable	Designing safer chemicals (4)	Preferred molecular design to reduce chemical toxicity		
	Use of renewable feedstock (7)	Raw material or feedstock should be renewable		
	Design for degradation (10)	Designing small molecules for biodegradability		
	Real-time analysis for pollution prevention (11)	Real-time, in-process monitoring and control prior to the formation of hazardous substances		

We know that *#1: waste prevention* is the first principle of green chemistry, it says to reduce the bi-products which is hazardous to the environment. The E-Factor (E4), a weight of molecule that used in the reaction process, and the lost material is used to compute it. The difference between the number of starting materials which show the impacts of the mass of the product is known as process mass intensity (PMI). Already, the water demand indicator clearly shows the consumption of the water, E-factor is utilized as an effect category rather than PMI, which covers water use [2].

17.3.2.2 Chemical Hazard

The social effect is calculated using chemical hazard. The persons who worked on the field who have health problems and their safety are regarded by the World Business Council for Sustainable Development (WBCSD) as necessary subjects in social life cycle metrics for chemical goods. The agreement model, USEtox, created by UNEP/SETAC and based on the probable toxicity consequences and chemical emission, is the suggested approach for LCA of human health impact. Since these chemicals employed in ClAlPc procedures are uncharacterized and their toxicity and emissions are unknown, this technology cannot be used. Quantitative structure-activity relationship (QSAR) models are used by EPI suite and ECOSAR respectively, to calculate the pollutants in the ecological substances.

Since chemical characteristics are inappropriate for inorganic compounds and also an organometallics like aluminum chloride and ClAlPc employed in our alternative procedures, their usage is restricted to organic chemicals [5, 6]. The NFPA 704 criteria is constructed on the assumption that there would be an instantaneous chemical contact due to an explosion, accident, or other crisis. It offers details on the quality, physical and chemical nature of the substances and unique risks of hazardous things that are evaluated by very short-term and show a severe impact and employ a grading scale of 0 to 4, here compared with other 0 is a low risk. A categorization and labelling system, the GHS (globally harmonized system) also addresses extra topics including chronic toxicity and environmental danger.

According to the decision tree technique, occupational exposure limits (OELs) or GHS words are used to calculate the health hazard score. A chemical's health hazard rating is 4 out of 10 if it lacks an OEL value and its yearly output is less than 1000 tons. Since their annual output is less than the 1000 tons/y threshold in the ClAlPc example, the majority of chemicals utilized in the processes would earn this score for health hazards. The NFPA 704 standard was chosen as a result to represent the real health threat.

The weight of chemicals required to generate 1 kg of ClAlPc is multiplied by the risks ratings for each chemical, and the results are then normalized using the preliminary score. The reaction rate of (H3) and dangerous and supernatural (H4) is equal to the weight of one-sixth. Because the corresponding chemicals generally have an oxidizing nature, it denotes both the reaction rate and toxic nature of those products. Health and flame nature are taken in the ratio of one third.

17.3.2.3 Cost

Expenses is defined by two factors: C1 (production rate) and C2 (life cycle cost); some have used PMI as an indication that green chemistry is not applicable here, and as a tool with which one evaluates the cost of the process [21]. The usage of high-cost chemicals, particularly for compounds having purity, is not effectively reflected by the statistic since PMI is generally based on the weight of starting materials, solvents, and the cost of the chemicals.

17.3.3 Material Synthesis and Data Collection

ClAlPc's standard procedure makes use of a heated mantle and is based on a patent. P1 through P5, which corresponds to the sequence in which alternative processes are generated, in a different procedure (Table 17.3). Using a mantle or a microwave, the synthesis of ClAlPc from phthalonitrile and phthalic anhydride is carried out (CEM Corporation Discover SP).

The molar ratio of phthalonitrile or phthalic anhydride to metal ions is fixed to ratio of one fourth, in the M-Pc synthesis. Aluminum chloride serves as the metal ion source. In the reaction medium, phthalonitrile or phthalic anhydride are combined with the metal alkaline compounds. All synthesis requires nonvolatile solvents because to the high reaction temperature of 230°C. Polar solvents are necessary for microwave synthesis since the heating process is dependent on rotation. 1-chloronaphthalene (1-ClNP), 2,4-dichloroanisole (2,4-DCA), and diethylene glycol (DEG) are used as reaction media. Auxiliaries such as urea, ammonium chloride (NH_4Cl), and 1,8-Diazabicyclo [5.4.0] undec-7-ene (DBU) are added for P2–P5. During this synthesis of ClAlPc, direct energy requirements, including usage of the chemicals (starting materials) and electrical requirements from the vacuum pump and reactor, are tracked. The theoretical energy required to preliminary method is determined by a previously described scheme and that is transferred to vapor in order to more accurately reflect industrial processes [9].

Table 17.3 Reaction condition and chemical use for ClAlPc processes.

Process	Energy source	Precursor		Reaction media		Auxiliaries		Rxn. time
		Type	Use (g)	Type	Use (g)	Type	Use (g)	
Preliminary	Heating mantle	PN: $AlCl_3$	0.9:0.25	1-ClNP	5.38 g	.	.	6 h
i.	Microwave	PN: $AlCl_3$	0.9:0.25	1-ClNP	5.38 g	.	.	45 min
ii.	Microwave	PA: $AlCl_3$	0.9:0.25	1-ClNP	2.72 g	Urea	1.08 g	45 min
iii.	Microwave	PN: $AlCl_3$	0.9:0.25	2,4-DCA	3.39 g	Urea	0.27 g	45 min
iv.	Microwave	PN: $AlCl_3$	0.9:0.25	2,4-DCA	3.39 g	NH_4Cl	0.27 g	45 min
v.	Microwave	PN: $AlCl_3$	0.9:0.25	DEG	3.33 g	DBU	1.24 g	45 min

In sonication, the reference (Tokyo chemical Industry, 98%) material and ClAlPc samples are soluble in ethanol and are placed for 15 min in sonication process in order to characterize the substance. Using a 22-m syringe filter, undissolved phthalocyanine and contaminants are eliminated. Shimadzu UV-Vis 2600 Spectrometer is used to measure the extracts UV-vis spectrum. ClAlPc sample mass spectra are analyzed using a Waters Xevo G2-XS UPLC-MS. By electrospray ionizing extracts and using a column chromatography method with 95% acetonitrile and 5% H_2O as solvent us for enhancing agent, mass spectra are produced.

17.4 Results and Discussion

17.4.1 Methodology Assessment

ClAlPc is synthesized using the repeated approach, which is depicted in Figure 17.5 (a), and alternative procedures that were created are compiled in Table 17.2. The repeated procedure is shown in Figure 17.6 (a), along with examples for the preliminary, P1, P4, and P5. The effect evaluation of each suggested alternative in comparison to the preliminary or the prior best chemical synthesis method is shown in Figure 17.6 (b)–(e). The alternate procedures from the Preliminary to P5 are outlined in the sections that follow.

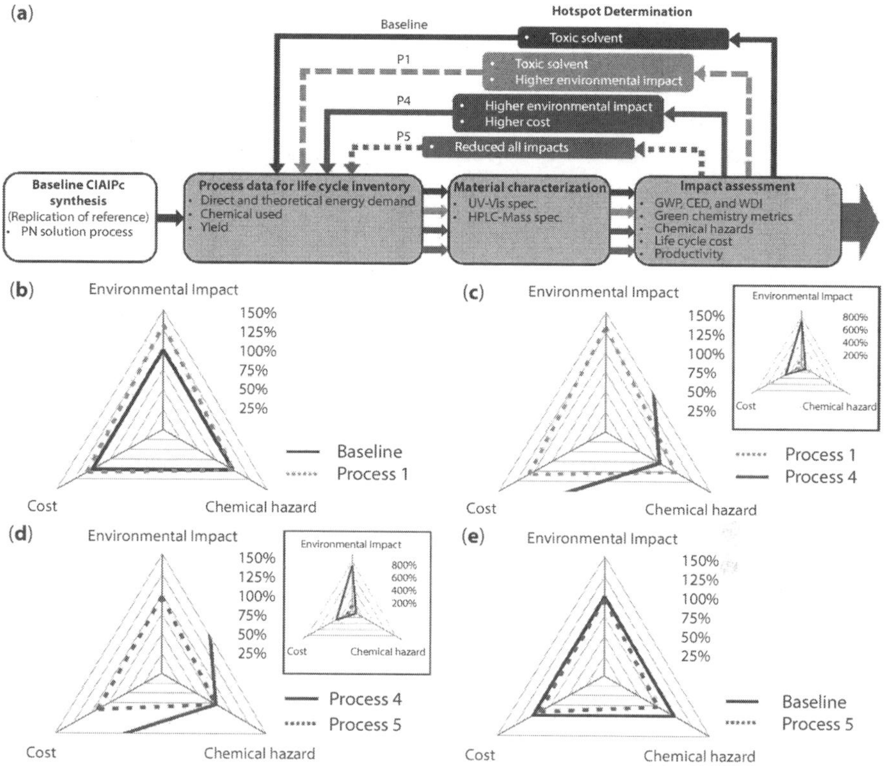

Figure 17.6 (a) Outline of problems described by repeated evaluation, and (b)–(e) estimation of alternative ClAlPc methods based on expenses, environmental effects and chemical impacts.

Preliminary to Process 1. The preliminary method, which uses solution processing and has the highest yield (84%), is a well-established approach for the synthesis of ClAlPc. However, high yield does not always imply a sustainable method. The time of the reaction occur (6 h), which affects the rate of the product formation, and also shows its reflection on the toxic medium of the reaction are the two key issues or "hotspots" from the preliminary procedure. Because of the formation of the partial chlorination of ClAlPc, it is regarded as an impurity, the quality standards for OPV application are also not reached.

To find these problems, we use the microwave instead of mantle, another suitable solvent is measured, which is based on research articles for other metal phthalocyanine [27]. P1 maintains all other variables as-is but substitutes a microwave reaction for a mantle. Although the reaction time has been cut from 6 h to 40 min, this is still longer than the 5–10 min described for other metal phthalocyanine, which is an improvement over the preliminary [24].

Figure 17.6 illustrates how the microwave response has a negative influence on both the environment and the economy (b). Even when a reduced reaction volume is taken into account, the production rate is increased by 27% by the microwave method (3.5 L for microwave reactor vs 20 L of pilot scale reactor). Like that, microwave method uses more power and produces less chemical yield (73% vs. 84%) than the preliminary, increasing the environmental effect by 31%. The cost impact is essentially unchanged.

Preliminary to Process 2. The second alternative pathway (P2) takes into account alternate precursors for ClAlPc since P1 results in an increase in environmental effect but no decrease in cost. Because PA is inexpensive, the dye industry uses it to make CuPc. Because PA's bulk chemical is three times less expensive than PN, the cost of ClAlPc might be drastically reduced. Various ClAlPc methods are outlined in the Porphyrin Handbook along with reaction conditions and product formation rate, but none of them employ PA as a precursor. Regrettably, the influence for all three categories increases and the P2 response yield is only 25%. Our group's prior research has demonstrated that the production of contaminants during the ClAlPc synthesis from PA as a result of urea breakdown lowers the efficiency of photovoltaic devices. Since using PA decreases material purity while raising environmental and financial consequences, it does not appear to be a good method for the synthesis of ClAlPc. P2 is thus removed from Figure 17.7, and subsequent repeated aspect phase is based on Principle #1.

Process 1 to Process 3. The preliminary method change Principle #1 attempted to decrease the problem of preliminary method, where a slow reaction time occurs, doing so increased the amount of electricity used. Using the green chemistry concept, the second major problem, which one shows more impacts of 1-ClNP, is taken into consideration for Principle 3. *#3: less hazardous chemical synthesis.*

Because it persists in aquatic environments and is recognized to be hazardous to aquatic life, chlorinated naphthalene was banned from industry usage in Japan [28]. We clearly know that the fifth principle of green chemistry #5: *safer solvents,* so we can change the solvents depending on their efficiency and effectiveness when it is used as a reaction medium.

It is challenging to replace 1-ClNP because of the high reaction temperature. We know that anisole has a boiling point of 155.5°C, and it would be an excellent substitute solvent according to the CHEM21, which is a guide to select the solvent. A high crystallinity chemical is reportedly produced when anisole is used in place of phenoxy derivatives during the synthesis of ClAlPc [6]. Since 2,4-dichloroanisole (2,4- DCA) has a boiling point of

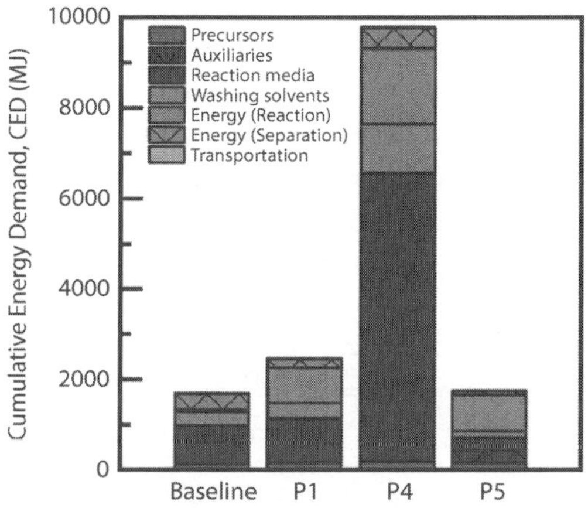

Figure 17.7 Increasing energy demand for preliminary, 1-chloronaphthalene (P1), 2,4-dichloroanisole (P4), and diethylene glycol (P5) ClAlPc processes.

285°C, we choose this as an alternative to anisole. To avoid chlorinating the phthalocyanine during the preliminary and P1 synthesis, an additional amount of solvent was utilized, although this constrains the amount of reaction per batch and therefore the production pace. While decreasing the amount of 2,4-DCA, 30 w/w percent of urea is applied to reduce the formation of partial chlorination of ClAlPc.

Chemical danger is 23% lower in P3 than in P1, but cost and environmental implications are higher (by 893% and 352%, respectively). Due to 2,4-DCA's upstream process, elevated CED and GWP are linked to an increase in environmental effect. Large quantities of organic solvents like CH_3COOH (Acetic acid) are required for the chlorination of anisole, which accounts for 40% of 2,4-DCA's effects on CED and GWP. The value of LCA in our technique is shown for P3 since it shows the effects of the upstream phases, which green chemistry alone does not take into account. Filtration is a problem caused by the 2,4-DCA synthesis. After the reaction was finished, the crude ClAlPc solution became viscous, and the separation requires more organic and inorganic solvents than the preliminary technique, which raises the water and E-factor demand.

Due to the price of 2,4-DCA and additional cleaning agents, raw material costs 637% more than the preliminary. Because of a greater reaction rate and less usage of the chemicals to form a byproduct, which balance out expensive prize of raw materials, the overall price of P3 is 352% cheaper than the preliminary. The material's HPLC-MS characterization reveals the same impurity as that in P2. To address the purity issue, P4 focuses on a different ammonium source.

Process 3 to Process 4. Because carbamides like urea decomposes between 200 and 250°C, that compound is used as a reagent in a reaction mixture because it obeys the P2 and P3 at the same time it forms an insoluble product and byproducts and it is hard to filter. Because of its high boiling point (338°C), ammonium chloride serves as a substitute supply of ammonium for P4. After reaction, the salt is dissolved in water and does not decompose, allowing for removal during filtering. P4 increases yield by 10% over P3 and eliminates contaminant generation from urea breakdown. Figure 17.7 illustrates the ecological and financial implications of P4, which are lesser than those of P3 but still significantly greater above the preliminary (733 and 284%, respectively) (c). As of now, moving from the preliminary to P4 may reduce the usage of the hazardous chemical and shows a good result in the low cost and reduced background impacts. As a result, the method shows how focusing on a particular issue has the capacity to lead to the development of brand-new problems along with alternative syntheses. Since the usage of 2,4-DCA in P3, the issue with the hazardous solvent has not been resolved, and P4 transfers a toxic chemical to some background impacts and financial complications. Therefore, the next stage is to lessen the effects of 2,4-DCA synthesis on the environment and financial costs.

Process 4 to Process 5. The green solvent list maintained by GSK is used to locate 2,4-DCA substitutes. Despite the fact that ethylene glycol and diethylene glycol both show the same effects so ethylene glycol's boiling point is lower, but this reaction requires very high boiling point.

Direct power usage from microwave is a huge amount but the GWP and CED related with the basic processes of DBU and DEG are lesser than those of 1-ClNP. In Principle 5, a general cleaning agent is a methanol, a cleaner agent (principle #5)—reduces the E-factor and WDI, lowering the total environmental effect. Diethylene glycol (DEG) is less expensive than 2,4-DCA and 1-ClNP because it is a byproduct of ethylene glycol, which is manufactured in huge quantities worldwide. Because of this, it is mandatory for all compounds with a construction higher than 1000 tons/y, to use chemicals produced in big quantities which has the benefit of having had their toxicity evaluated. When compared to P4, using DEG as a reaction medium has less negative financial and environmental effects, but the chemical danger is the same, as shown in Figure 17.6 (d).

17.4.2 Overall Evaluation

As illustrated in Figure 17.7, the preliminary process is adjusted to P5 using this repeated approach to lessen its negative effects on the environment, costs, and chemical hazards (e). Alternative procedures were effectively directed by the hotspots for each step. In P2, #2 principle: AE is very helpful in calculating the screening test of the atoms used in the reactions, but when used on its own, it can result in synthesis methods give a greater negative impact in the environment due to the poor yield, that shows more consumption of energy and the raw material used in the process. It is known that using a microwave reactor speeds up the process of phthalocyanine production [27], is highly complicated because of the extreme energy consumption.

Energy is required for chemical synthesis, and the environmental effects of energy use can make up over half, and perhaps up to 80%, of global warming potential. The CED impact of some of the processes taken into consideration is compared in Figure 17.8. When taking into account the

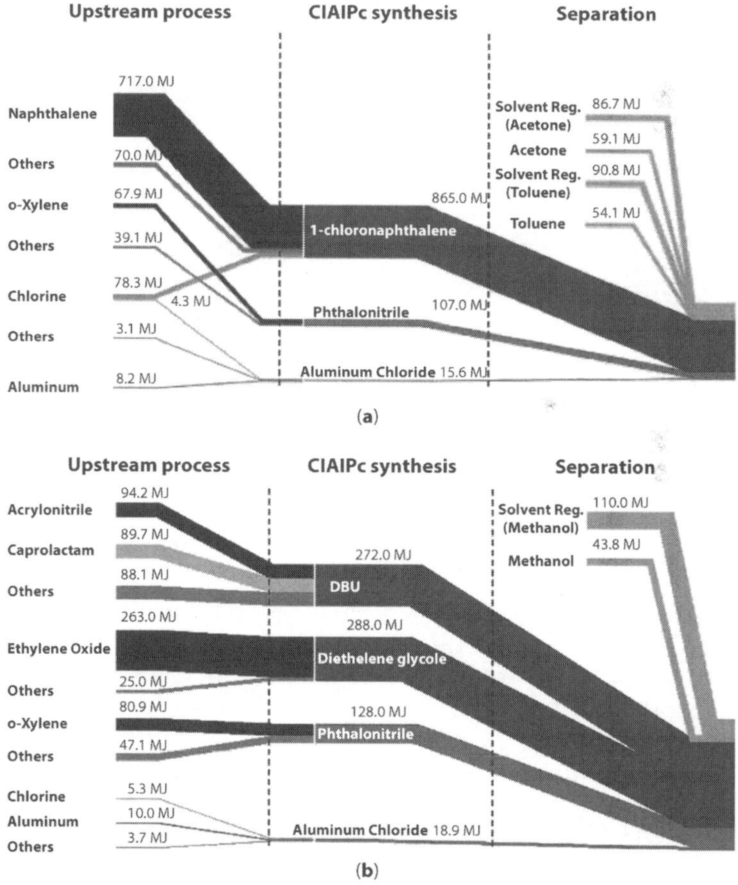

Figure 17.8 Sankey diagram of the increasing energy flow for ClAlPc synthesis from (a) the preliminary and (b) diethylene glycol and DBU (P5) excluding transport and electricity.

chemical synthesis and separation, P1 and P5 are the two where the power consumption from the microwave has the most effect. The solvent used in the process of upstream for P4 is more significant than the direct electric consumption use, as was previously noted.

To highlight the impact of chemicals, the energy transportation and electricity usage (shown in Figure 17.7 orange) are not included in the figure. Figure 17.8 shows that the 1-ClNP upstream pathway has a greater influence on CED than the combination of DEG and DBU (865 MJ compared to 560 MJ).

In order to assess and direct the production of fine chemicals, this effort created a novel technique that combines LCA with green chemistry metrics. The construction of life cycle inventories for the four processes designed for ClAlPc based on a microwave reactor is a supplementary contribution. The final step is the development of a new method of DBU (P5) which generate ClAlPc and simultaneously decrease the global warming and other environmental issues by 3%, the cost by 9%, and the chemical hazard by 23% when related to the preliminary process. This method also produces high-quality material suitable for OPV applications.

17.5 Conclusions

Organic photovoltaic (OPV) material process is improved by using the repeated evaluation method developed in this work, which is based on green chemistry and life cycle approach. The material produced from the new process reduces the overall impact of OPV device, especially TPV made of ClAlPc. The application for window reduces the energy demand of building and waste heat from an HVAC system, and shows potential for urban climate improvement. These benefits would be considered as net environmental benefit of TPV application and could be quantified with a projection of overall impacts through LCA.

Synthesis pathways of ClAlPc are established by applying green chemistry principles and metrics. The work identifies a greener pathway for ClAlPc synthesis by starting from phthalonitrile as the precursor. The AE of phthalonitrile pathway is about 30% higher than that of phthalic anhydride pathway. The phthalic anhydride pathway has a theoretical yield of 60%, however, because of byproduct generation during intermediate step, the actual yield lowers to 25%. The application of microwave reactor to the phthalonitrile pathway reduces reaction time from 6 h to 40 min, which suggests better energy efficiency and yield.

The performance of OPV devices from ClAlPc synthesized from different processes was evaluated. Two precursors of ClAlPc, phthalonitrile and phthalic anhydride, are studied to evaluate the tradeoff between material purity, device efficiency, and material cost. The results show that UV-vis and mass spectra of ClAlPc from phthalonitrile and phthalic anhydride processes are in agreement with the reference material. The best device from phthalonitrile sample shows similar energy conversion efficiency with the reference device although the fabrication yield is low. For phthalic anhydride samples, all devices work, but have low energy conversion efficiency due to impurities from urea decomposition.

A new methodology is developed by combining LCA and green chemistry metrics to evaluate OPV material processes. The hotspot of preliminary process of ClAlPc was identified as toxic solvent use and high energy demand. Applying the repeated evaluation method could solve issues related to each intermediate process regarding environmental, cost, and chemical hazards. Finally, a new process using diethylene glycol under microwave reactor is developed to produce ClAlPc that decreases the cost by 9%, the effect of global warming and other environmental issues by 3%, and chemicals hazard by 23% simultaneously when related to other preliminary method, during the production of high efficiency and pure raw materials which can be used in the OPV applications.

The application of TPV made of ClAlPc is assessed to evaluate the energy benefit of power generating window. The energy benefit is a key aspect of TPV application, and it will be used for use phase input data in LCA for net environmental benefit of TPV application for windows over its lifetime. The building energy demand simulation of the medium office and the small office identifies energy benefit in locations representing different climates and background impacts from the electricity grid. The reduction of cooling energy demand in all study location and type of applications results in annual energy saving while the natural gas demand in Detroit shows contrary trends by application type. The window application saves natural gas while skylight application consumes more natural gas due to heat loss from the skylight. However, overall net energy benefit is notable for both applications in terms of energy saving and PV electricity generation. The energy savings from HVAC by TPV application in windows is seen under cold climate condition as well, even though the cold climate condition, such as Detroit, requires higher heating energy demand.

The application considers not only location but also urban impact as well. Transparent OPV application for a window in an urban area could result in additional benefits in terms of saving land use and reducing waste heat contributing to UHI effect. From the research data, the energy saving in the urban is 2 GJ when compared to the rural area per month while we use the TPV. The overall result implies that TPV window in the urban area reduces energy demand and has additional potential for deployment of renewable energy devices and mitigation of UHI effect, leading towards urban sustainability.

Transparent photovoltaic (TPV) application in windows was studied and provided motivating research questions for future study. Regarding cooling energy demand, the absorption profile of TPV plays an essential role since NIR absorption reduces energy demand. Regarding the TPV absorption, the next step in this subject is to study different types of TPV that harvest electricity from the different countries. The requirement of thermal transmittance of a building is recommended by climate conditions, to save energy in both cooling and heating, and thus, there could be optimal configurations of TPV specific to climate conditions.

Waste heat from HVAC could be reduced by TPV application since TPV application aids to reduce energy demand from HVAC used in the office buildings. HVAC systems vary depending on the purpose and type of building. Therefore, the benefit of TPV application from the reduction of waste heat from HVAC will differ. Various types of buildings exist in an urban area, and application of TPV to each building will result in benefits of differing magnitude and temporal patterns. Another study for TPV application will be to study TPV in windows of all the different Department of Energy (DOE) commercial reference buildings to assess the patterns of energy saving and reduction of waste heat from HVAC.

References

1. Keoleian, G. A.; Lewis, G. M. Modeling the life cycle energy and environmental performance of amorphous silicon BIPV roofing in the US. *Renew. Energy* **2003**, *28* (2), 271–293.
2. Cho, A. Energy's Tricky Tradeoffs. *Science (80).* **2010**, *329* (5993), 786–787.
3. Anctil, A.; Babbitt, C. W.; Raffaelle, R. P.; Landi, B. J. Cumulative energy demand for small molecule and polymer photovoltaics. *Prog. Photovoltaics Res. Appl.* **2013**, *21* (7), 1541–1554.
4. Sheldon, R. A. The E factor 25 years on: the rise of green chemistry and sustainability. *Green Chem.* **2017**, *19* (1), 18–43.
5. Toropova, A. P.; Toropov, A. a.; Benfenati, E.; Gini, G. QSAR modelling toxicity toward rats of inorganic substances by means of CORAL. *Cent. Eur. J. Chem.* **2010**, *9* (1), 75–85.

6. Melnikov, F.; Kostal, J.; Voutchkova-Kostal, A.; Zimmerman, J. B.; T. Anastas, P. Assessment of predictive models for estimating the acute aquatic toxicity of organic chemicals. *Green Chem.* **2016**, *18* (16), 4432–4445.
7. Zimmermann, Y.-S.; Schäffer, A.; Hugi, C.; Fent, K.; Corvini, P. F.-X.; Lenz, M. Organic photovoltaics: potential fate and effects in the environment. *Environ. Int.* **2012**, *49*, 128–140.
8. Williams, G.; Sutty, S.; Klenkler, R.; Aziz, H. Renewed interest in metal phthalocyanine donors for small molecule organic solar cells. *Sol. Energy Mater. Sol. Cells* **2014**, *124*, 217–226.
9. Jiménez-González, C.; Kim, S.; Overcash, M. R. Methodology for developing gate-to-gate Life cycle inventory information. *Int. J. Life Cycle Assess.* **2000**, *5* (3), 153–159.
10. Traverse, C. J.; Pandey, R.; Barr, M. C.; Lunt, R. R. Emergence of highly transparent photovoltaics for distributed applications. *Nat. Energy* **2017**, 2 (11), 849–860.
11. Anctil, A.; Babbitt, C. W.; Raffaelle, R. P.; Landi, B. J. Material and energy intensity of fullerene production. *Environ. Sci. Technol.* **2011**, 45, 2353–2359.
12. Anastas, P.; Eghbali, N. Green Chemistry: Principles and Practice. *Chem. Soc. Rev.* **2010**, 39 (1), 301–312.
13. Calvo-Flores, F. G. Sustainable chemistry metrics. *ChemSusChem* **2009**, 2, 905–919.
14. Burke, D. J.; Lipomi, D. J. Green chemistry for organic solar cells. *Energy Environ. Sci.***2013**, 6, 2053.
15. Asatkar, A. K.; Bedi, A.; Zade, S. S. Metallo-organic Conjugated Systems for Organic Electronics. *Isr. J. Chem.* **2014**, 54 (5–6), 467–495
16. Bhandari, K. P.; Collier, J. M.; Ellingson, R. J.; Apul, D. S. Energy payback time (EPBT) and energy return on energy invested (EROI) of solar photovoltaic systems: A systematic review and meta-analysis. *Renew. Sustain. Energy Rev.* **2015**, 47, 133–141.
17. Chae, Y. T.; Kim, J.; Park, H.; Shin, B. Building energy performance evaluation of building integrated photovoltaic (BIPV) window with semi-transparent solar cells. *Appl. Energy* **2014**, 129, 217–227.
18. Wang, Y.; Tian, W.; Zhu, L.; Ren, J.; Liu, Y.; Zhang, J.; Yuan, B. Interactions Between Building Integrated Photovoltaics and Microclimate in Urban Environments. *Sol. Energy*, **2005**, 499–504.
19. Taha, H. The potential for air-temperature impact from large-scale deployment of solar photovoltaic arrays in urban areas. *Sol. Energy* **2013**, 91, 358–367.
20. Sheldon, R. a. Fundamentals of green chemistry: efficiency in reaction design. *Chem. Soc. Rev.* **2012**, 41 (4), 1437.
21. Cespi, D.; Beach, E. S.; Swarr, T. E.; Passarini, F.; Vassura, I.; Dunn, P. J.; Anastas, P. T. Life cycle inventory improvement in the pharmaceutical sector: assessment of the sustainability combining PMI and LCA tools. *Green Chem.* **2015**, 17 (6), 3390–3400.

22. Curzons, A. D.; Constable, D. C.; Cunningham, V. L. Solvent selection guide: a guide to the integration of environmental, health and safety criteria into the selection of solvents. *Clean Prod. Process.* **1999**, 1 (2), 82–90.
23. Anctil, A.; Fthenakis, V. Life Cycle Assessment of Organic Photovoltaics. In Third Generation Photovoltaics; *InTech*, **2012**.
24. Curran, M. A. Life cycle assessment: principle and practice; Cincinnati, OH, **2006**.
25. Lunt, R. R.; Bulovic, V. Transparent, near-infrared organic photovoltaic solar cells for window and energy-scavenging applications. *Appl. Phys. Lett.* **2011**, 98 (11), 3–5.
26. Salzman, R. F.; Xue, J.; Rand, B. P.; Alexander, A.; Thompson, M. E.; Forrest, S. R. The effects of copper phthalocyanine purity on organic solar cell performance. *Org. Electron. physics, Mater. Appl.* **2005**, 6 (5–6), 242–246.
27. Shaabani, A.; Maleki-Moghaddam, R.; Maleki, A.; Rezayan, A. H. Microwave assisted synthesis of metal-free phthalocyanine and metallophthalocyanines. *Dye. Pigment.* **2007**, 74 (2), 279–282.
28. Wernet, G.; Conradt, S.; Isenring, H. P.; Jiménez-González, C.; Hungerbühler, K. Life cycle assessment of fine chemical production: A case study of pharmaceutical synthesis. *Int. J. Life Cycle Assess.* **2010**, 15 (3), 294–303.

18

Design of Greedy Approach-Based Vulnerability Detection Framework for Smart Grid Systems

Manas Kumar Yogi*, Sai Pathrudu Lanka Vatapatra, Navya Sri Medapati, Devi Krishna Arumilli and Vaishnavi Lingam

CSE Department, Pragati Engineering College (Autonomous), Surampalem, A.P., India

Abstract

The smart grid is a digital actual framework that incorporates equipment, programming, and actual parts fittingly coordinated, collaborating, and interrelating to detect the fluctuating condition of the actual world. Smart grid provides on-demand electricity to customers from centralized and distributed generation stations using information and communication technologies. Power companies can deliver reliable power at reduced cost and control power demand. In this chapter, we formulate a novel vulnerability detection technique and propose a universal robust vulnerability detection algorithm in a smart grid system that generates a program model based on various attack situations and sets rules to perform model checking. Attack detection issues are posed as dynamic learning problems for various online attack scenarios. In this approach, algorithms will determine whether measurements of the components involved in a smart grid system are secure or compromised. A framework for attack detection is offered to take advantage of any prior information about the system that may be accessible and get over limitations brought on by the sparse nature of the problem in the suggested technique. We plan to use a variant of the greedy approach to identify the degree of the node vulnerability, which helps operators of the smart grid ecosystem to take profitable measures and detect the vulnerability in a minimum amount of time and effort. The main motivation of using the greedy approach variants is that it is provably effective to optimize the overall cost of vulnerability detection in a smart grid.

**Corresponding author*: manas.yogi@gmail.com

O.V. Gnana Swathika, K. Karthikeyan and Sanjeevikumar Padmanaban (eds.) Smart Grids as Cyber Physical Systems Volume 1: Artificial Intelligence, Cybersecurity, and Clean Energy for Next Generation Smart Grids, (275–294) © 2024 Scrivener Publishing LLC

The simulation results demonstrate the effectiveness of our proposed approach for determining the vulnerabilities in smart grid systems with significant efficiency.

Keywords: Smart grid, greedy, feasible, optimal, vulnerability

18.1 Introduction

The human need for electricity is sufficient for decades by using conventional, conceptually straightforward methodologies with the National Grid Model Approach. The smart grid is currently replacing the conventional electric power grid. It is a fusion of information and communication technology along with the usual electric-powered strength grid (ICT) [1]. These systems engage with many distribution devices with more sophisticated sensors to improve efficiency. Smart grid is more efficient and intelligent when related to the usual powered grid model. It is a digital technology for transmission line sensing as well as two-way communications between the utility and its customers. Electric networks and cutting-edge digital communication networks are combined by smart grid technology. This makes it possible to offer a better power distribution that is simple to manage. Among the fundamental technologies of the smart grid.

Energy suppliers would not be able to incorporate renewable energy sources into current grid infrastructures without smart grids. Making the most of the available energy resources, conventional or renewable, requires the efficiency enhancements that smart electrical networks provide. An eco-friendly and energy-efficient grid, such as a smart grid, helps to ease the transition to a low-carbon economy. From power generation, transmission, storage, and distribution to electricity consumption, a smart grid connects every element of the electrical system at different stages. An integrated data and energy grid is produced by plugging each device in and becoming online. Due to its two-way interactivity, a smart grid may instantly redirect electricity in the case of an outage or equipment failure.

Detecting and isolating an outage, for instance, can stop a widespread and protracted blackout.

18.1.1 Characteristics of Smart Grid

Some of the frequent traits assigned to smart grids shown in Figure 18.1 are as follows:

Self-healing: online comparison and pre-manipulation of device failures, computerized faults, and restoration.

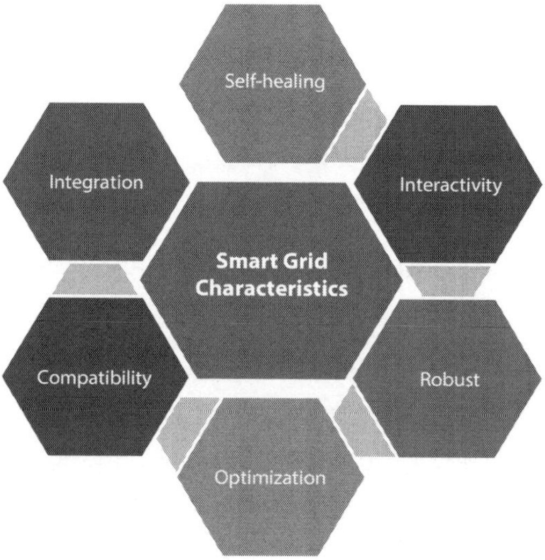

Figure 18.1 Essential features of a smart grid.

Interactivity: the capability to execute interactions between the distribution community and the client to get the energy waft records and two-way interaction to go with the flow of capital.

Robust: better response time in instances of herbal human disasters, protection, and records security.

Optimization: improvement of asset utilization, a discount on working and protection costs, and a discount on losses.

Compatibility: large generators, choice of power sources, allotted generation, etc.

Integration: improvement of processes, integration of information, and standardization/administration refinements.

18.1.2 The Functionality of Smart Grid

This more recent grid type must be monitored and controlled for its numerous dispersed energy sources to operate consistently and effectively. This is achievable, as was already mentioned, if the smart grid is viewed as a typical Internet-of-Things (IoT) application. A centralized control point with the ability to carry out analysis and control operations can receive data that is gathered in real-time from line sensors, users, and generators [2]. This enables managing distribution, resolving outages, and balancing power demands.

- The ability of the smart grid to self-heal thanks to control systems' ability to identify small issues and make fixes without human intervention is another benefit.
- Integrated communications offer real-time control, information, and data exchange to maximize system dependability, resource usage, and security.
- The main duties of sensing and measuring include maintaining equipment health, stopping energy theft, assessing grid stability and congestion, and assisting control strategy.
- Distributed power flow controls clamp onto active transmission lines.
- In intelligent power-generating systems, which can start, stop, and run efficiently at a chosen load, independently of the others, energy supply and demand are matched. They are hence suitable for generating base load and peaking power.
- Power system automation offers quick diagnosis and targeted fixes for smart grid faults or disturbances.

18.1.3 Advantages of Smart Grid

Some of the advantages or benefits of the smart grid are as follows:

- It minimizes electricity losses (transmission, distribution, etc.)
- It lowers the fee for electricity, meter readings, test and measurement (T&M) operations, maintenance, and different costs.
- Automatic functioning based on shifting load conditions decreases equipment failures. Demand response lessens the strain placed on the smart grid system's assets during peak hours, which lowers the likelihood of failure.
- It lessens prolonged outages and the expense of the subsequent restoration.
- It decreases the need for oil and widespread blackouts. As a result, the smart grid gives individuals security by supplying constant power.
- The smart grid can accommodate rising consumer demand without adding to the infrastructure.

18.1.4 Threats and Weaknesses in Smart Grid

The traditional electricity network becomes more complicated and prone to various forms of assaults because of the smart grid network's introduction of improvements and better capabilities. The smart grid is more vulnerable to cyberattacks because of its size and improved communication capabilities. The smart grid system contains flaws that might allow the unauthorized access that will not ensure the confidentiality and integrity of data that is being transferred which will further disable the service.

Some of the threats and weaknesses of smart grid are as follows:

- Security for the customer: smart meters capture enormous amounts of data on their own and send it to the utility, the customer, and the service providers.
- Additional intelligent technologies: the management of network demand and power supply is done by a variety of intelligent devices that are part of a smart grid.
- More stakeholders: having many stakeholders could result in insider attacks, a potentially deadly type of attack.

Attackers could be novice programmers, skilled hackers, terrorists, workers, rival businesses, or clients such as:

- Attackers without malevolent intent who visualize the system's operation and security functionalities as a puzzle to be solved.
- Consumers with vindictiveness against other consumers may find ways to turn off the power supply in their homes.
- The smart grid is seen as a desirable target by terrorists because it has a significant impact on millions of people and makes their cause more widely known.

18.2 Background Work

Vulnerability detection in smart grid systems as shown in Figure 18.2 is an important task as it helps to identify potential security weaknesses and potential threats to the system as indicated in Table 18.1. There are several approaches that have been used in the past 10 years to detect vulnerabilities in smart grids [4]. Some of these approaches are mentioned below.

18.2.1 Vulnerability Detection Approaches in Smart Grid

Network analysis: this approach involves analyzing the structure and connections within a smart grid network to identify vulnerabilities.

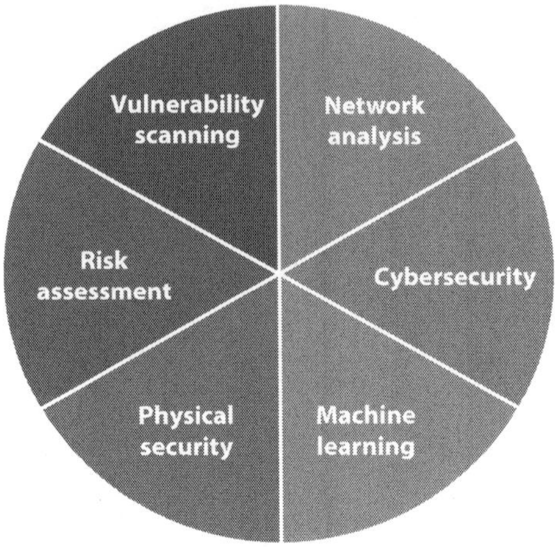

Figure 18.2 Challenges in a smart grid.

This can be done through techniques such as network topology analysis, traffic analysis, and intrusion detection.

Over the past 10 years, there have been a number of methods developed for analyzing smart grid systems using network analysis techniques. These methods are designed to help utilities and other stakeholders better understand the behavior and performance of smart grids and identify potential areas for improvement.

One common approach is to use network analysis tools to model and analyze the power flow in a smart grid system. This can help identify bottlenecks and other issues that may be impacting the system's performance. Other methods include the use of network analysis tools to identify and evaluate the impacts of different types of distributed energy resources (DERs) on the grid and to optimize the integration of these assets into the grid [5].

The use of network analysis techniques to detect and lessen the effects of cyberattacks on smart grid systems has been another area of concentration in recent years.. This includes the use of tools to monitor network traffic and identify potential vulnerabilities, as well as the development of algorithms and other techniques for detecting and responding to cyberattacks.

Overall, the use of network analysis techniques has become increasingly important in recent years as the complexity and connectivity of smart grid systems continue to grow. These methods are helping utilities and other stakeholders to better understand and manage the performance of these systems, and to identify and address potential issues before they become major problems.

Machine learning: machine learning techniques, such as supervised and unsupervised learning, have been used to identify patterns in smart grid data that may indicate vulnerability.

Since smart grid innovation is so confounded, there are numerous ways for assailants to obstruct the framework. Thus, a totally guarded technique is expected to address every possible danger and shortcomings a smart grid might confront.

It has been shown that machine learning procedures are an especially powerful device for foiling disavowal of administration assaults. Machine learning can perceive attacks, go to precautionary lengths to fix existing issues, and ready security experts.

Machine learning methods should be ceaselessly worked on for the electric business to push toward smart grids since their utilization can consistently incorporate every one of the parts utilized. In this manner, quality, proficiency, and congruity are guaranteed in smart electric frameworks, as well as dependability [7]. Our leaders and governments must guard against this fragility as we go in a world where infrastructure disasters like those seen in Texas become the norm by taking a more thoughtful approach to increasingly flexible and sustainable smarter energy systems.

Physical security: physical security measures, such as access controls and security cameras, can be used to detect and prevent unauthorized access to smart grid infrastructure.

Physical security measures are an important part of protecting smart grid systems from various types of threats, including physical attacks, vandalism, and natural disasters. These measures can include a range of different technologies and practices, such as:

Perimeter security: this includes measures such as fencing, gates, and security personnel to prevent unauthorized access to smart grid facilities.

Access control: this includes technologies such as security badges, biometric scanners, and other measures to ensure that only authorized personnel are able to enter smart grid facilities.

Surveillance: this can include the use of security cameras and other monitoring technologies to detect and deter potential threats.

Asset protection: this includes measures such as hardened enclosures and other physical protections to protect critical smart grid assets from damage.

Physical security training: providing training to smart grid personnel on physical security best practices can help to ensure that they are aware of potential threats and know how to respond to them.

Generally speaking, the purpose of physical security measures in smart grid systems is to make sure that these systems are safe from physical dangers and that they can continue to function dependably and efficiently.

Cyber security: there are several cyber security techniques that have been used in smart grid systems over the past decade.

Encryption: encryption is used to protect data transmitted over the smart grid from being intercepted and read by unauthorized parties.

Firewalls: firewalls are used to protect the smart grid network from external threats such as hacking and malware attacks.

Access control: to guarantee that only authorized workers have access to the smart grid system, access control procedures are utilized.

Intrusion detection and prevention systems: these systems keep an eye on the smart grid network for unusual activity and can notify administrators if any risks are found.

Network segmentation: network segmentation is the practice of dividing the smart grid network into smaller, more secure sub-networks to reduce the risk of a successful cyberattack.

Security awareness training: providing training to smart grid personnel on how to identify and prevent cyber threats can help to reduce the risk of a successful attack.

Risk assessment: smart grid system vulnerabilities may be found and ranked using risk assessment techniques like the STRIDE model [9].

Globally, the growing complexity related to traditional electrical furnishings and the rising want for electrical energy among clients is forcing companies to shift toward a digital environment. On the other hand, the rising want for improvements in the strength area is additionally permitting businesses to shift towards automation and more modern applied sciences to respond digitally to the altering electric-powered demand. In this new technology of reliability and availability, handing over promising options to clients is necessitating the want for an enterprise risk assessment solution. In the smart grid manufacturing space, enforcing a strong threat evaluation answer helps groups in the smart grid manufacturing house limit risks, higher comply with the place of work safety, and efficaciously create and preserve a more healthy work environment [10].

Vulnerability scanning: vulnerability scanning tools can be used to scan smart grid systems for known vulnerabilities and provide recommendations for remediation.

Over the past 10 years, there have been a number of advances in the development of methods for detecting vulnerabilities in smart grid systems. These methods are designed to help utilities and other stakeholders identify and address potential vulnerabilities before they can be exploited by attackers.

One common approach is the use of vulnerability assessment tools, which scan smart grid systems for known vulnerabilities and provide reports detailing any issues that are found. To make sure that vulnerabilities are found and fixed right away, these tools may be used on a regular basis.

Another approach that has gained popularity in recent years is the use of penetration testing, in which simulated attacks are launched against smart grid systems to identify vulnerabilities that might not be detected by other means. This can provide valuable insights into the vulnerabilities of a smart grid system and help identify potential weaknesses that need to be addressed.

In addition to these methods, there has also been a focus on the development of algorithms and other tools for detecting anomalies in smart grid systems that may indicate the presence of vulnerability. This includes the use of machine learning and other AI techniques to analyze network traffic and identify patterns that may indicate the presence of a threat [16].

In general, over the last 10 years, there has been a tremendous advancement in the approaches for identifying weaknesses in smart grid systems. In the face of a threat environment that is becoming more sophisticated and dynamic, these techniques aid in ensuring the security and dependability of these systems.

In order to effectively identify and address vulnerabilities in smart grid systems, it is crucial to use a variety of techniques. The methods used for vulnerability identification in smart grid systems will vary based on the specific needs and constraints of the system, the availability of data and resources, and other factors.

Combining different approaches to detect and mitigate vulnerabilities in smart grid systems can help to improve the overall effectiveness of the vulnerability management process. Some potential ways to combine different approaches include:

Network-based and asset-based approaches: analyzing the network traffic and communication patterns in the smart grid, as well as the assets in the system, can help to identify potential vulnerabilities and their causes.

Data-driven and machine learning-based approaches: analyzing data from various sources, such as sensors, control systems, and supervisory control and data acquisition (SCADA) systems, using machine learning algorithms can help to identify patterns and anomalies that may indicate a potential vulnerability [18].

Human-centered and technical approaches: engaging with human experts, such as system operators and security professionals, can help to identify potential vulnerabilities and provide insight into how to mitigate them [21]. Technical approaches, such as network-based and asset-based approaches can then be used to implement the necessary security measures.

18.2.2 Study of Detection and Prevention Techniques of Different Types of Attacks in a Smart Grid

Table 18.1 Literature study of smart grid vulnerabilities.

References	Type of attack	Attack target	Type of technique used for detection	Solution after research
[15]	Social engineering attack (SEA)	SCADA and Advanced Metering Infrastructure (AMI)	——	Two intrusion detection systems were subjected to a security evaluation in a smart grid SCADA environment.
[20]	Denial of service (DoS)	AMI network	——	AMI network effects of DoS
[22]	Denial of service (DoS)	Load frequency controller	——	Effects of DoS on the management of load frequency.
[12]	Distributed denial of service (DDoS)	Communication network	Cloud computing	Cloud-based firewall
[13]	Distributed denial of service (DDoS)	Communication network	Cloud computing	Cloud computing capacity
[8]	False data injection attack (FDIA)	Communication system to taint data on PV generation	Machine learning	Model for an artificial neural network that can detect and foretell cyberattacks
[6]	False data injection attack (FDIA)	SCADA network	Deep machine learning	Conditional deep belief network (CDBN)
[21]	False data injection attack (FDIA)	AMI network	Machine learning	Ensemble-based learning

(*Continued*)

Table 18.1 Literature study of smart grid vulnerabilities. (*Continued*)

References	Type of attack	Attack target	Type of technique used for detection	Solution after research
[6]	False data injection attack (FDIA)	Communication network	Machine learning	Using a multilayer neural network, to examine the effects of FDIA on a smart grid powered by artificial intelligence.
	False data			Blockchain-based secure
[14]	Injection attack (FDIA)	Smart meters	Blockchain	Message transit for smart meters and service providers.
[3]	Social engineering, data manipulation, and session hijacking	SCADA	Cloud computing	Deep belief network
[11]	Social engineering attack (SEA)	Humans at organizations	——	Secure message delivery for smart meters and service providers using blockchain technology.
[17]	Flooding attack	Communication network	Time measurement of flooded packets	Bait-message-based detection
[19]	Data manipulation	AMI and SCADA	Cloud computing	Online/offline searchable encryption method based on Attributes.

18.3 Proposed Mechanism

In this chapter, we consider a variant of the greedy algorithm to find the minimum cost to find the vulnerability in the smart grid. For this, we chose a sample graph SG{V, E} where V is vertices/nodes of smart grid network and E is edges/wires for transmission of messages between nodes in the smart grid. In this, we consider houses as the vertices or nodes and routers

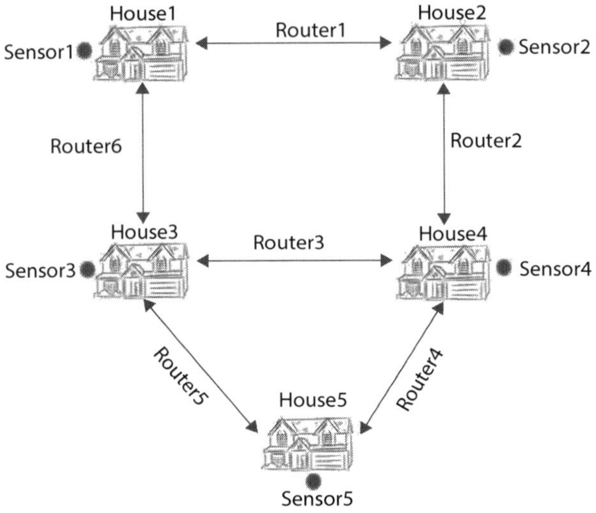

Figure 18.3 Service component of smart grid architecture.

as the edges. For instance, let us consider a network of six houses and six routers connected as shown in Figure 18.3.

Each house consists of sensors that are responsible for the major chance of vulnerability problems. Sensors are electronic devices based on micro-controllers. They measure physical quantities and convert them into signals which are read by an instrument or observer. Smart grid security can be achieved by a comprehensive study of sources of threats and vulnerabilities. To identify the smart grid vulnerabilities, it is not only sufficient to separately analyze the physical components, but also we need to thoroughly look into the interaction between them. Here, there are some parameters on which the cost of the vulnerability of each node depends. Some of them are:

- Time configuration of the network.
- Complexity of the operation to be performed.
- Number of sensors per unit length of transmission
- Number of adjacent nodes in the network.

Let us take a set {S, R, H} where S represents the sensors, R represents the routers, and H represents the Houses. We consider subsets of the above set {S, R, H} such as {{},{S},{R},{H},{S,R},{S,H},{R,H},{S,R,H}}.

Now in these subsets, let us consider the single-element subset for our convenience.

STEP-1: Summarize cost(VD(n1))+cost(VD(n2))+... +cost(-VD(nx)) [integration].
STEP-2: Apply the greedy principle and minimize the cost obtained in step 1.
STEP-3: Again, repeat step 2 until we get the minimum cost.

We have to consider the efficient parameter among the four identified parameters. For this, we have to apply a greedy algorithm considering the four parameters as K1, K2, K3, and K4 as shown in Figure 18.4.

STEP 1.1: Select minimum among K1, K2, K3, K4.

Say K1'= cost(VD(n1))+cost(VD(n2))+ ...+cost(VD(nx))

Cost(VD(n1))=K1,cost(VD(n2))=K2,cost(VD(n3))=K3, cost(VD(n4))=K4

Consider K1' such that K1'= K1+K2+K3+K4

Let us assign ranks for each parameter according to their effect on the cost of vulnerability.

1. Time configuration of the network.
2. Complexity of the operation to be performed.
3. Number of adjacent nodes in the network.
4. Transmission length per unit of power transfer

STEP 1.2: First, let us consider the cost of vulnerability using time configuration as K1'.

STEP 1.2.1: Select minimum among K1, K2, K3,K4 .

Say K1'=K1+K2+K3+K4

Where K1, K2, K3, and K4 represent the minimum costs for vulnerability detection.

STEP-1.2.2: Summarize K1'=cost(VD(n1))+cost(VD(n2))+ ... +cost(-VD(nx)) [integration].

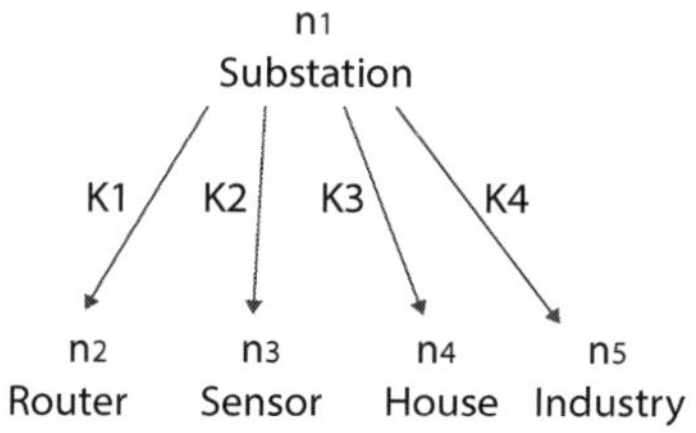

Figure 18.4 Components of a substation in a smart grid.

STEP-1.2.3: Apply the greedy principle and minimize the cost obtained in step-1.2.2.

STEP-1.2.4: Again, repeat step 1.2.3 until we get the minimum cost.

STEP 1.3: Now, let us consider the cost of vulnerability using the complexity of operation to be performed as K2'.

STEP 1.3.1: Select minimum among K1, K2, K3,K4 .
Say K2'=min{K1,K2,K3,K4}.
Where K1, K2, K3, and K4 represent the minimum costs for vulnerability detection.

STEP-1.3.2: Summarize K2'=cost(VD(n1))+cost(VD(n2))+ ... +cost(-VD(nx)) [integration].

STEP-1.3.3: Apply the greedy principle and minimize the cost obtained in step-1.3.2. STEP-1.3.4: Again, repeat step 1.3.3 until we get the minimum cost.

STEP 1.4: Then, let us consider the cost of vulnerability using the number of adjacent nodes in the network a K3'.

STEP 1.4.1: Select minimum among K1, K2, K3, K4 .
Say K3'=min{K1,K2,K3,K4}.
Where K1, K2, K3, and K4 represent the minimum costs for vulnerability detection.

STEP-1.4.2: Summarize K3'=cost(VD(n1))+cost(VD(n2))+ ... +cost(VD(nx)) [integration].

STEP-1.4.3: Apply the greedy principle and minimize the cost obtained in step-1.4.2. STEP-1.4.4: Again, repeat step 1.4.3 until we get the minimum cost.

STEP 1.5: Lastly, let us consider the cost of vulnerability using Length per unit as K4'.

STEP 1.5.1: Select minimum among K1, K2, K3, K4.
Say K4'=min{K1,K2,K3,K4}.
Where K1, K2, K3, and K4 represent the minimum costs for vulnerability detection.

STEP-1.5.2: Summarize K4'=cost(VD(n1))+cost(VD(n2))+ ... +cost(VD(nx)) [integration].

STEP-1.5.3: Apply the greedy principle and minimize the cost obtained in step-1.5.2.

STEP-1.5.4: Again, repeat step 1.5.3 until we get the minimum cost.

```
Logic involved in proposed mechanism: while (cost!=min cost)
do{
cost= cost (VD(n1))+cost(VD(n2))+... +cost(VD(nx)) greedy();
}
Algorithm Greedy(c,n)
{
Mincost=0;
for i=0 to n do
{
x:=select(c);
if feasible(Mincost,x)
{ Mincost:=union(Mincost,x);}

return Mincost;
}
}
```

Where c is an array of costs n is the total no. of nodes.
Feasible() checks for a possible solution.
Union() checks for an optimal solution.
Select() selects any cost from the array of costs.
Mincost is the minimum cost among an array of costs

18.4 Experimental Results

We considered the N-BaIoT dataset for detection of IoT-based attacks in a smart grid from Kaggle. This dataset was constructed from nearly nine commercial IoT devices data with size up to 2 GB. This dataset can be used to identify weakness in IoT ecosystem and classification can be done with user defined parameters. Figure 18.5 shows the comparative analysis of the current popular methods like anomaly based vulnerability detection, rule-based and cloud-based vulnerability detection methods. Our proposed variant of greedy algorithm outperforms the other methods as shown in the plot with respect to time of configuration of the smart grid. It can be

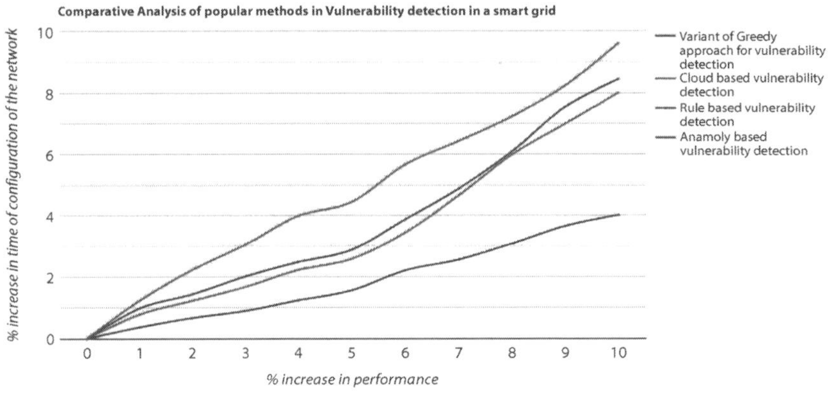

Figure 18.5 Comparative analysis of proposed mechanism with current methods.

observed that our proposed method has a percentage increase in performance of 10% with only 4% increase in time of configuration whereas other methods need over 7% increase in time of configuration to reach 10% increase in performance. Hence, we can say that in terms of overall time of configuration, our proposed method is better than its counterparts.

In Figure 18.6, it can be observed that in terms of complexity of operation in a smart grid, when we compare our proposed method with other popular methods for vulnerability detection, then our proposed method keeps a stable rate of 7% increase in terms of percentage increase in complexity of operation to achieve a 10% increase in performance. Due to the inherent complex nature of a smart grid, it is difficult for any designer to obtain such high level of performance within limit of 5% growth in complexity of operation in a smart grid.

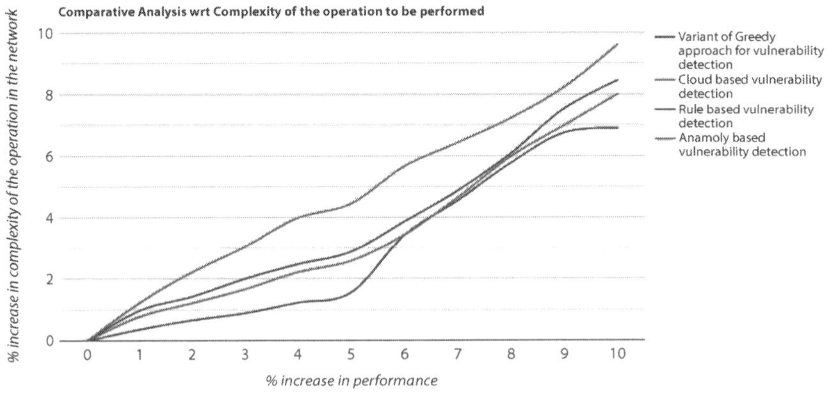

Figure 18.6 Comparative analysis with respect to complexity of operation in smart grid.

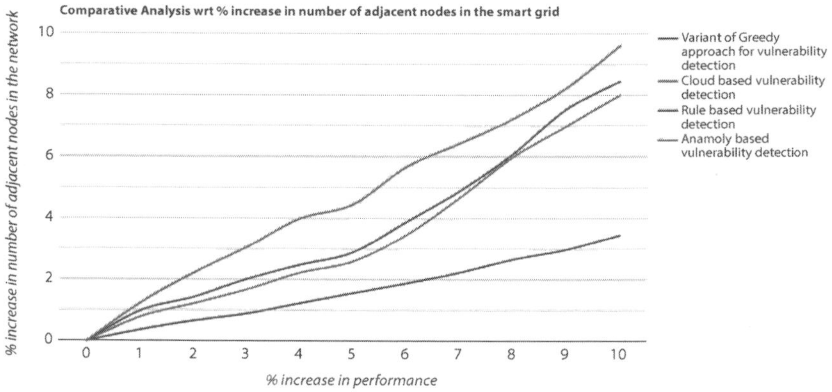

Figure 18.7 Comparative analysis with respect to percentage increase in number of adjacent nodes in smart grid.

In Figure 18.7, it can be observed that in case of increase in the count of adjacent nodes in the smart grid, our method outperforms its counterparts by rate of nearly 5% to obtain the same increase in performance. The anomaly-based vulnerability detection method takes the highest rate of increase in adjacent nodes in a network to achieve a 10% increase in performance but the proposed method is so efficient that it takes only a 3.8% increase in number of adjacent nodes for obtaining a 10% increase in performance.

18.5 Future Directions

A smart grid is characterized by high speed communication network. The main challenge in detecting a weakness takes few assumptions as observed in our proposed approach which seek latency. The cost of threat detection or vulnerability detection increases for a large scale smart grid due to this latency. In the future, we plan to reduce the scope of such assumptions which leads to increase in the cost of vulnerability detection in a smart grid. Another challenge for vulnerability detection in a smart grid is that the factor of cohesiveness among the smart grid components is at a high level so the cost of threat detection at component level is not discrete but rather it is linked with cost of vulnerability detection at another component or sometimes at another level of operational also. So, the cost of vulnerability detection is cascading in nature. This cascading nature has to be restricted in future work. This is possible only when security and privacy concerns are addressed at design level of a smart grid rather than be implemented at operational level of the smart grid architecture.

18.6 Conclusion

This chapter aims at delivering an efficient mechanism for minimizing the cost of threat detection in a smart grid. The variant of greedy approach deployed is better than other popular methods currently used for threat detection in a smart system. It is proved as per the experimental results and performance wise also, the proposed technique outperforms its counterparts to an appreciable level. As state-of-art smart grid techniques are being developed at a rapid rate, the weakness aspects of a smart grid system must also be worked upon as the various stakeholders of the smart grid depend heavily on their usage with great level of concern for security and privacy. Our research work can be regarded as a sincere attempt towards this direction such that with minimum cost and effort, the vulnerabilities in smart grid architecture can be uncovered and make the smart grid more robust for future usage.

References

1. Kurt, M.N.; Yilmaz, Y.; Wang, X. Real-Time Detection of Hybrid and Stealthy Cyber-Attacks in Smart Grid. *IEEE Trans. Inf. Forensics Secur.*, 2019.
2. The Essential Role of Cyber Security in the Smart Grid. https://electricenergyonline.com/energy/magazine/312/article/The-Essential-Role-of-Cyber-Security-in-the-Smart-Grid-.htm.
3. Chen, L.; Liu, J.; Ha, W. "Cloud service security evaluation of smart grid using deepbelief network." *Int. J. Sens. Netw.*, 2020.
4. Gunduz, M.Z.; Das, R. Cyber-security on smart grid: Threats and potential solutions. *Comput. Netw.*, 2020.
5. Rajendran, G.; Sathyabalu, H.V.; Sachi, M.; Devarajan, V. Cyber Security in Smart Grid: Challenges and Solutions. In *Proceedings of the 2019 2nd International Conference on Powerand Embedded Drive Control (ICPEDC)*, Chennai, India, August 2019.
6. He, Y.; Mendis, G.J.; Wei, J. "Real-Time Detection of False Data Injection Attacks in Smart Grid: A Deep Learning-Based Intelligent Mechanism." *IEEE Trans. Smart Grid*, 2017.
7. V. J. Forte, "Smart Grid at National Grid," *2010 Innovative Smart Grid Technologies (ISGT)*, Gothenburg, 2010.
8. Riggs, H.; Tufail, S.; Khan, M.; Parvez, I.; Sarwat, A.I. "Detection of False Data Injection of PVProduction." In *Proceedings of the 2021 IEEE Green Technologies Conference (GreenTech)*, Denver, CO, USA, April 2021.

9. Flake H, Structural Comparison of Executable Objects, In *Proceedings of the IEEE Conference on Detection of Intrusions and Malware & Vulnerability Assessment (DIMVA)*, 2004.
10. M. Wagner, M. Kuba and A. Oeder, "Smart grid cyber security:A German perspective," *2012 International Conference on Smart Grid Technology, Economics and Policies (SG- TEP)*, Nuremberg, 2012.
11. Green, B.; Prince, D.; Busby, J.; Hutchison, D. "The Impact of Social Engineering on Industrial Control System Security." In *Proceedings of the First ACM Workshop on Cyber-Physical Systems-Security and/or PrivaCy, CPS-SPC '15*, Denver, CO, USA, 16 October 2015; Association for Computing Machinery: New York, NY, USA, 2015
12. Diovu, R.C.; Agee, J.T. "A cloud-based openflow firewall for mitigation against DDoS attacks in smart grid AMI networks."
13. Califano, A.; Dincelli, E.; Goel, " Using features of cloud computing to defend smart grid against DDoS attacks".
14. Zhang, H.; Wang, J.; Ding, Y. "Blockchain-based decentralized and secure keyless signature scheme for smart grid." *Energy*, 2019.
15. Singh, V.K.; Ebrahem, H.; Govindarasu, "Security Evaluation of Two Intrusion Detection Systems in Smart Grid SCADA Environment". In *Proceedings of the 2018 North American Power Symposium (NAPS).*
16. Costin A and Zaddach J, "Embedded Devices Security and Firmware Reverse Engineering, Black Hat", 2013.
17. Zhang, F.; Mahler, M.; Li, Q. "Flooding attacks against secure time-critical communications in the power grid." In *Proceedings of the 2017 IEEE International Conference on Smart Grid Communications (SmartGridComm)*, Dresden, Germany, October 2017.
18. YIN Ming, ZHANG Gongxuan, "Buffer overflow detection method based on source code analysis ", *Journal of Jiangsu University (Natural Science Edition)*, 2016.
19. Alohali, B.; Merabti, M.; Kifayat, K. "A cloud of things (cot) based security for home areanetwork (han) in the smart grid." In *Proceedings of the 2014 IEEE Eighth International Conference on Next Generation Mobile Apps, Services and Technologies*, Oxford, UK,September 2014.
20. Yi, P.; Zhu, T.; Zhang, Q.; Wu, Y.; Li, "A denial of service attack in advanced meter in infrastructure network". In *Proceedings of the 2014 IEEE International Conference on Communications (ICC).*
21. Ashrafuzzaman, M.; Das, S.; Chakhchoukh, Y.; Shiva, S.; Sheldon, F.T. "Detecting stealthy false data injection attacks in the smart grid using ensemble-based machine learning." *Comput. Secur.*, 2020.
22. Liu, S.; Liu, X.P.; El Saddik, "Denial-of-Service (dos) attacks on load frequency controlin smart grids". In *Proceedings of the 2013 IEEE PES Innovative Smart Grid Technologies Conference (ISGT).*

19

Smart Grid Technology Development and Intellectual Property Law Protection: Opportunities and Challenges

E. Prema* **and S. Suganya**

VIT School of Law, VIT Chennai, Tamil Nadu, India

Abstract

Integrating renewable energy into the large grid requires an innovative and flexible grid management system. Modern machine learning techniques help intelligent systems adapt quickly and accurately to massive input data. Moreover, computers can outperform even the most imaginative humans thanks to their superior problem-solving abilities. Technologies have opened up many possibilities in the energy sector as it evolves toward a more sustainable and efficient power generation model. It is everyone's goal to leave future generations with a world that is both habitable and environmentally friendly. Also, countries' need for energy has prompted their pursuit of renewable and environmentally benign options as fossil fuel reserves run low. Scientists are working tirelessly and consistently to develop new ideas to address the energy dilemma. In the meantime, the government implements several similar plans on a national scale to improve the country's economy and social infrastructure. Solutions developed by robust and adaptive computing algorithms may outperform human intelligence. The question is whether and how to use intellectual property law to provide the right incentives. This chapter provides an initial discussion of the primary operations of the smart grid and the consequent need for machine learning. The following section examines how leading patent offices are now responding to the difficulties posed by smart grids incorporating artificial intelligence. Ultimately, it will be shown that technologies that generate new and exclusive intellectual property are less promising than real problems that consider the possibility of market entry.

Keywords: Artificial intelligence, intellectual property law, machine learning, patent law, smart grid

**Corresponding author*: Prema.e@vit.ac.in

O.V. Gnana Swathika, K. Karthikeyan and Sanjeevikumar Padmanaban (eds.) Smart Grids as Cyber Physical Systems Volume 1: Artificial Intelligence, Cybersecurity, and Clean Energy for Next Generation Smart Grids, (295–314) © 2024 Scrivener Publishing LLC

19.1 Introduction

Under the smart grid concept, the traditional power grid will be transformed from an electromechanical management system into an electronic network. According to the smart grid system report of the US Department of Energy, the smart grid system consists of information management, control technology, digital sensing, communication technology and field devices to coordinate various power operations. These smart grid technologies change traditional grid planning and operations issues in at least three key ways: they can now monitor or measure processes, send data back to operations centres, and often react automatically to any changes; and systems; they can process, analyze and assist operators in accessing and using data generated by digital technologies used throughout the grid.

The term "smart grid" refers to the communication between the many components of an energy system, from generation to distribution to storage to consumption. The successful integration of renewable energy is a vital component of the new structure and functionality of this integrated data and energy network part. The use of renewable energy sources increases the risk of grid instability and power fluctuations, as they often have periodic conditions such as sunlight or wind. Smart network operators can learn more about and adjust feeds, increasing the proportion of renewable energy. Smart grid shows that exemplary implementation requires fast processing of largc amounts of data, thus requiring intelligent systems and substantial computing power (Halgamuge & Guruge, 2022).

However, a significant focus of smart grid efforts is increasing the share of renewable energy. Since the system must deploy generators to meet demand as production changes, forecasting future renewable energy generation is critical. While it might be feasible to manually create complex forecasting models for huge solar parks, doing so for a distributed generation across millions of homes across the grid is a challenging task. This problem can be solved by implementing a computer-based method (Bassiliades & Chalkiadakis, 2018).

Modern power grids' composition and structure are becoming more complex. As the power grid continues to generate a large amount of high-dimensional and multi-type data, higher requirements are put forward for the power grid. The traditional modelling, optimization and control methods have many limitations. Artificial intelligence (AI) is expected to play an essential role in energy, transportation, healthcare, security applications, and more. Artificial intelligence has become one of the fastest-growing areas of technology. Most problems in power systems revolve around prediction and optimization. Artificial intelligence can provide solutions to assess energy consumption, grid balancing, and energy production.

The use of AI in the electrical field has grown significantly. Artificial intelligence is an applied method of self-learning and computing. It can combine computers' powerful data processing capabilities and integrate human vision, perception, cognition, communication and other abilities. After summarizing the evolution of smart grid and AI, this chapter will examine some applications of AI in smart grid, including the use of AI in the following situations: power system stability control, power system fault detection, power load forecasting, power generation forecasting, power grid security protection.

19.2 Machine Learning and Smart Grid

First, it is vital to demonstrate the fundamental machine learning systems to clarify the specific effects machine learning may have on energy systems which communicate legally.

For software to be capable of handling complicated and abstract problems, it must learn how to build its answers using machine learning. The software is said to have a "learning ability". Artificial intelligence has been an enormous success, and there are high hopes for more advancements, even a "fourth industrial revolution."

Artificial intelligence has yet to have a clear concept, legally or technically. The standard definition used in academia is "the science and engineering of constructing intelligent machines", first mentioned at a computer science conference in 1956. Because of its emphasis on the development of AI, the definition needs to be validated in the scientific community. Address new legal issues. A lack of specificity complicates the discussion of the law. Numerous attempts have been used to define or characterize AI in the legal literature, often with less than satisfactory results. By telling the difference between the two, you can better understand the technological advances that make a regular computer programme or algorithm into AI (Bassiliades & Chalkiadakis, 2018).

Algorithms are logical systems that technical tools can carry out. They have a set order of operations that, when given a specific input, provide a predetermined output, similar to the "if-then" operator. While an AI may take the shape of an algorithm, the term AI encompasses much more than that. If you want to think of AI as something, picture it as the means by which an algorithm may be constructed, not by a human programmer but by an artificial one. So, AI can create and improve algorithms by testing and improving them repeatedly. Artificial intelligence does not need to outsmart humans in terms of intelligence; all it has to do is harness the computational power and storage capacities of technology, which are already much superior to any human intellect (He, 2011).

A big challenge is making non-humans appear to be able to learn or adapt. Artificial intelligence is partly inspired by the belief that it is easy to replicate certain aspects of nature. To build intelligent machines, it is necessary first to understand what makes humans so perceptive. Tracing intelligence back to its biological origins in the human brain is at the heart of one approach. One of the best-known strategies in machine learning is the use of artificial neural networks, sometimes called "neural network methods." Similar to how neurons in the brain connect and communicate, neural networks take a connection-oriented approach. Edges connect an artificial neural network's nodes (neurons).

In contrast, the genetic breeding model mimics the evolutionary process of improving skills and fitness throughout multiple generations by drawing inspiration from the development and growth of natural organisms. The result, the so-called "target algorithm," is independently created by the machine learning environment, and even the developers need to learn precisely how the problem is solved. However, individual lines or clusters of code can still be understood by humans. The fact that AI-generated algorithms are often built in object code rather than source code and require large amounts of specific training data is also highly relevant to the legal evaluation (He, 2011).

19.3 Smart Grid Definition

Authors and government agencies have defined the "smart grid" in various ways. The widespread implementation of intelligent communication monitoring and management systems has resulted in a "smart grid," an enhanced energy grid network allowing two-way information and power exchange between providers and consumers.

To enhance demand-side management and energy efficiency, as well as build self-healing, dependable grid protection against deliberate sabotage and natural calamities, the smart grid's early components revolved around the notion of "advanced metering infrastructure" (AMI) (Brown & Suryanarayanan, 2009).

As new challenges and opportunities arose, however, the power sector, research institutions, and governments rethought and broadened the original scope of the smart grid. The US Energy Independence and Security Act of 2007 told the National Institute of Standards and Technology (NIST) to study and expand a framework to make smart grids and devices work together. In its paper, NIST lists the expected benefits and needs of a smart grid, even though a clear and complete definition has yet to be given.

- Increasing the capacity and efficiency of existing electrical power networks;
- increasing their resilience to disruptions; enabling predictive maintenance and self-healing responses to system disturbances;
- facilitating the extended deployment of renewable energy resources;
- accommodating distributed power sources;
- automating maintenance.

While the smart grid described in this Special Issue may sound like something out of a science fiction novel, it is, in fact, a fundamental global transformation that will see hundreds of billions of dollars equivalents spent within the next decade on existing technologies that will provide intelligent management of the electric grid over the coming decades. Some parts of the proposed smart grid system may be too pricey, though, and will have to be put on hold until either more affordable technologies are developed, or the social advantages of implementing them are deemed sufficient to justify the costs. The ideal smart grid is a concept, and it must be cost-justified at each stage before implementation, followed by testing and verification before widespread deployment (Fjelland, 2020).

A clean, safe, secure, reliable, resilient, efficient, and sustainable electric system that integrates electricity generation, transmission, substations, distribution, and consumption through the use of information, two-way, cyber-secure communication technologies, and computational intelligence is what is meant by the term "smart grid." This concept encompasses the whole breadth of the energy system, from the initial stages of energy production to the last stages of energy consumption. Suppose the intelligent grid is characterized as smart metering, for example. In that case, it covers the consumption and, to some extent, the distribution elements of this definition but only covers part of the spectrum of integration necessary to realize the smart grid (Brown & Suryanarayanan, 2009).

Achieving a fully functional smart grid will be a long and winding road that will take decades. Each element does not need to be implemented simultaneously to be considered part of an intelligent grid. Instead, they can be added in whenever they are ready. Of course, each one will need an explanation of costs and a possible return on investment. However, after the technologies have been tested, the interoperability of open systems will allow each addition to "plug and play" into the smart grid. Assuming it is fully implemented, the smart grid will have several advantages over the current electric power grid.

- System-wide, end-to-end, secure communication
- Cables, connections, terminations, transformers, consumer loads, power quality, etc., will all be monitored in real-time for anomalies.

Large volumes of incoming data will be generated by the features mentioned above, and they must be transformed into an understanding of the current status of the grid. So that information is streamlined, problems are diagnosed immediately, corrective actions are identified and executed dynamically in the field, and feedback loops provide metrics verifying that the work is producing the desired effects, controller technologies will need to automate data and energy management (Halgamuge & Guruge, 2022). Here are some of the features you may expect from such smart grid controllers:

- Self-healing: automatically repairing or removing possibly problematic equipment from service before it breaks; reconfiguring the system to redirect energy supply to keep all consumers connected;
- Flexible: allows distributed generation and energy storage to be connected quickly and safely at any time,
- Predictive: using machine learning, reinforcement learning, weather impact forecasts, and stochastic analysis to offer predictions of the following most probable occurrences so that corrective steps may be taken to reconfigure the system before the following worst events can occur:
- The system must be interactive so that everyone involved in the energy system, from operators to end users, can take part in the best way to handle problems.
- Secure: because the smart grid has two-way communication capability across the entire system, physical and cybersecurity of all critical assets are critical.
- Optimized: knowing the status of every major component in real-time or near real-time and having the control equipment to provide alternative routing paths provides the capability for autonomous optimization of the flow of electricity throughout the system.

19.4 Features of Smart Grid

Even though many countries and organizations define the term "smart grid" differently, its primary meaning remains the same. Power and data can freely move back and forth between any two points on the smart grid. In addition, smart grid performance is excellent and can provide customers with a wide range of additional benefits. Collectively, it is these characteristics that make the smart grid desirable.

- An electrical grid that can repair itself if something goes wrong. Self-healing control systems used in smart grids can automatically detect and improve problems, bringing power back online quickly and reducing outages and their duration, frequency, and extent.
- All renewable energy must be provided to consumers free of charge. The smart grid uses new materials, energy storage technology, and information technology to support the two-way flow of electric energy, has more substantial resource allocation capabilities, and is conducive to large-scale power generation from renewable energy.
- Easy to implement and enforce. A smart grid enables grid companies to allocate resources more efficiently. For example, dynamic review and adjustment can improve the efficiency with which an organization uses its resources by ensuring that its assets are used to their full potential.
- The ability to accept and respond to bidirectional user input and feedback. Users play a more passive role in traditional grids. Thanks to the two-way, real-time connection enabled by the smart grid, users can receive information and provide it on demand through feedback. In a two-way dialogue, changes in how people consume and invest in electricity are mutually beneficial. If grid managers successfully implement demand management strategies, they can benefit from better supply-demand balance and higher system reliability.
- Improve efficiency in the power sector. Maintaining a solid network and technology backbone is critical to developing the power industry. A smart grid is a network of electricity distribution and consumption facilities and related information and communication technologies that ensures the

smooth functioning of energy markets by linking together buyers and sellers of electricity.

19.5 Challenges of Achieving in Smart Grid

Smart grids increase the efficiency with which we use electricity to achieve improved global living standards at a lower cost. No matter how hard we try, we can only get from here to there once the move is affordable. Assume the project will take 20–30 years to fully transform any country into an integrated smart grid. In that case, key performance indicators must be tracked to continuously and automatically score the benefits created by the smart grid.

Establishing an initial baseline for all major components of the current grid is critical to documenting these enhancements, as is ongoing monitoring of the impact of new construction and implementations relative to this baseline. The result of this document is the validation of an adaptive stochastic controller for an intelligent grid that can redistribute load around congestion and manage peak demand, weather, and equipment failures, thereby reducing the need for expensive new power plants and substations. Abroad, the need to build additional generation capacity could be avoided if computers running these adaptive stochastic controllers managed each level of the future smart grid.

Over time, smart grids improve the capacity factor of the power system by facilitating better supply and demand management. In addition, it enables more efficient reuse of existing hardware infrastructure by intelligently integrating state-of-the-art controllers into the system. Suppose intelligent grid utilities thoroughly understand the risks and impacts to consumers of using available resources optimally. In that case, they should be able to reduce peak demand and reduce capital and Operating and Maintenance (O&M) costs (Fjelland, 2020).

As an industry, we are responsible for ensuring that the expected benefits of the smart grid are realised and demonstrated through global tracking measures. Sharing what works should be quick and easy. Several factors threaten the long-term viability of the smart grid. These include: consumer reluctance to invest until they see tangible benefits; improved regulation; government oversight to keep pace with technology and keep citizens connected; cost justification; Innovative grid components need to be cost-effective individually and systematically; and education; Utilities, service companies, and universities need to develop educated consumers and a knowledgeable workforce.

19.6 Legal Challenges Implementing Smart Grid

Several challenges exist in intellectual property law related to machine learning systems' classification and legal protection and their accompanying target algorithms or innovations. Chinese and Western authorities have established rules for machine learning and related technologies. The smart grid is an excellent example of how new software patentability standards can give computer programs a technical identity.

Models and algorithms that are "mathematically characteristic per se" are not patentable unless applied to another technological advancement. This limitation's purpose is to limit overly broad language in patent claims. This can be solved by referring to the technical programme that implements the method.

19.6.1 EU Approach

In 2019, the European Patent Office (EPO) put out rules for patenting systems that use AI and machine learning.

Depending on the context, terms such as "support vector machine," "inference engine," and "neural network" (all of which are types of machine learning) may refer to abstract models or technical methods, as the EPO clearly states. But this approach is helpful because it looks at the words used in the patent application rather than abstract ideas. The EPO, therefore, advises its examiners to pay careful attention to these definitions.

If the goal of your machine learning strategy is to classify data automatically, filing a patent is not feasible due to a lack of complexity or other technical reasons. For example, linear regression analysis defines the automatic operation of printing presses by establishing a link between subjective and objective harmonic analysis data on print quality and establishing regression parameters. With this knowledge, you can use accurate data to predict your personal information. Of course, this approach is unique only if the problem can be solved using these records. However, if determining the features used to classify a dataset is trivial, the application may need to be more patentable (EPO Guidelines for Examination, 2019).

Therefore, innovations emphasising machine learning results have relatively strict protection requirements. The EPO's rules account for the invention's technical nature. Due to the recent addition of a subsection on mathematical methods to the guidelines, the process for determining eligibility for mathematical methods may be similar to that used to determine the patentability of computational models and algorithms that are essential to AI-related inventions (AI). Therefore, if a patent claim only describes an abstract mathematical process without using any technical means to obtain the desired result, the patent claim will be considered invalid.

In general, once the applicant files a patent application, the claimed subject matter must be a technical feature of a device (such as a computer or processing unit) or a technique for applying an algorithm to a technological device. For the EPO, an invention having technical, if it has "further technical impact," applies to the technical field and is designed for technical implementation. The defining characteristic of a technical application is its ability to perform a specified technical function. However, if "managing technical systems" is the primary goal, then mathematical methods lack technical quality. Nevertheless, mathematics enhances the technical features of an invention, even if the claims do not explicitly mention the technical application of mathematical methods. It is done even without a technological advantage.

Algorithms must contribute in ways other than mathematics, such as faster processing speeds, to be helpful in practice. Algorithms, as one of the "technical" elements, need to be directly related to the technical realisation of the invention to be eligible for patent protection. EPO standards give examples of computer programs that describe procedures for adjusting a car's anti-lock braking system, detecting emissions from X-ray equipment, compressing video, repairing damaged digital photographs, or encrypted electronic communications, among other things. Add "further technical effects" as long as they are computer-based.

In smart grids, technical or practical implementations are often feasible. Artificial intelligence will run on complex computer networks and have access to electronic control and measurement equipment, so the execution of AI-generated algorithms goes beyond the "normal" physical interaction between a programme and the computer processing it. According to the EPO, currents flowing around a computer are an average material effect of running a program, but they are insufficient to give a programme a technical identity. The ability to affect technological processes, the inner workings of the computer itself, or its user interface is explicitly considered another technological consequence (EPO Guidelines for Examination, 2019).

Grid control systems are likely already based on computer software that runs algorithms and has a built-in user interface. It's one thing to install a computer program, but quite another to improve it to be smarter than the previous generation of software for the same non-technical purpose. Smart grid systems require new hardware, such as AI-accessible control interfaces and measurement devices, to function correctly. The smart grid initiative is also a useful invention because some of its parts are based on computer networks and are put into action with the help of computer programs. Current grid controls are often unable to accommodate the flexible and fast adjustments required by smart grids.

(a) Distributing energy from renewable sources and collecting data on energy flows and peaks requires smart grid devices to quickly transmit large amounts of data. The technical process of calculating and predicting energy use is the purview of these devices. The ability to predict energy peaks in real-time through real-time data analysis sets this technology apart. This part of the technical process controlling energy distribution must have some "additional technical consequences." A complex network of programmable devices controlled by software facilitates this smart grid functionality. It is possible to create new software with the help of AI (Fu, *et al.*, 2012).

(b) Similarly, smart grids use computer programs to identify attack patterns as an "additional technical consequence." To identify attack trends, AI must capture all available data on potential electrical energy entry points into the system. After processing the data, the AI may learn what was illegally input or output. Likewise, by giving AI access to the controls of the generation and transmission lines, the technical process is under the program's power to prevent such attacks. Moreover, when implemented by computer programs, this innovative feature of automated and adaptive cyber-attack defence is enough to give AI solutions a technical flavour (JianJiao, 2020).

(c) Because of their ability to integrate individual homes, smart grids are often patentable. Every home need smart device to monitor energy usage and provide controls over energy production, storage, and distribution to keep it running smoothly. Smart grid AI relies heavily on the data provided by these inspections and measurements to successfully perform functions such as integrating renewable energy and spotting cyberattack patterns. The technical management system and the establishment of the network between these individual devices in each residence and the grid control centre hold the computer software.

19.6.2 US Patent Law

In January 2019, the US Patent and Trademark Office (USPTO) changed its rules about what can be patented to make it easier to decide if something is patentable. Although the amendments do not explicitly address AI-related innovations, they provide guidance on assessing an invention's patent eligibility (USPTO, 2019). It starts by analyzing patent claims to see if they fall into one of three categories of abstract concepts: mathematical concepts, specific organisational techniques of human activity, or mental processes. If the claim is accepted, the examiner decides whether the abstract concept contained in the claim is incorporated into a usable application (USPTO, 2019). Next, examiners need to consider whether the claims integrate abstract ideas into practical applications, thereby imposing significant constraints on mathematical concepts similar to the EPO criteria. This strategy ensures that applicants can maintain concrete and abstract concepts.

Examples of such constraints imply necessary integration into practical applications. They include enhancements to the operation of computers or other technologies or technical fields, the use or application of specific machines, and the transformation or reduction of particular items into different enhanced states or things. To demonstrate patentability, it is often necessary to apply an abstract concept in another way than its technical application. What the EPO means by "technical means" or "achieving technical properties" is very different (Fu, *et al.*, 2012).

The USPTO uses the same two-part Alice/Mayo test to determine whether an invention is patentable before the patent office acknowledges a patent. The first test is to see if the patent claims anything abstract, such as a law of nature or a natural phenomenon. For example, this judicial loophole is at the heart of the complaint. The next thing to do is determine whether the claim has any elements other than judicial exceptions. The second step, "identifying the inventive concept," is significant for software development. As an alternative to the "special purpose computer" test that suggests software or algorithms are not eligible for patents, this new test requires only a general-purpose computer, which has been included in the claim as a "special purpose" device. This other method is equivalent to the EPO test but more stringent (JianJiao, 2020).

Given that the procedure is a legitimate exception to the general principle, if the algorithm passes the first test, it would be necessary to prove many other things that have nothing to do with the code. Even though US patent law calls a general-purpose computer a "specially designed" or "specific machine," adding a general-purpose computer or a part of a general-purpose computer would make a software patent claim invalid. Putting judicial exceptions into place is one way to go "far beyond judicial reform."

In other words, "the machine must be an essential part of the claimed method and not just an obvious way to get to a faster solution" for the method to work. The phrase "additional solution activities" refers to situations in patent claims where using the machine doesn't significantly affect how the claimed process works. Patents are less likely to cover claims that include what used to be called "additional elements" but are now considered common knowledge. One of these steps could be telling doctors that there is a link between the number of thiopurine metabolites in the blood and how well a specific drug dose works and telling them to use any standard method. "Physicians routinely analyse metabolites as part of their studies to see if there is a link between metabolite levels and the effectiveness and toxicity of thiopurine drugs," which governed false (USPTO, 2019).

While this approach differs from the usual purpose of the Internet hyperlink protocol, which typically directs visitors from a host website to a target website, it is considered suitable for the dynamic construction of dual-source hybrid web pages. Furthermore, a computer browser's back and forward buttons exemplify well-understood, habitual, and routine behavior. A computer browser's back and forward buttons exemplify well-understood, habitual, and routine behavior. Other examples include:

- Performing double calculations.
- Keeping electronic records.
- Storing and retrieving information from memory.
- Receiving or transmitting data over a network.

Another technology-enforcement-focused patent succeeds because it provides substantive limitations, not just limiting judicial exceptions to specific technical circumstances. In this case, the claim is about how detailed equations can be used to automate the operation of a rubber moulding machine. This is both an abstraction and a legal exception. Other factors that determine procedures, such as inserting the rubber into the press, closing the mould, continuously monitoring the temperature inside the mould, and automatically opening the press at the appropriate time, sufficiently limit the judicial exception to the actual process of moulding rubber products. We were unable to distinguish these cases due to their extensive overlap. Therefore, patent claims are assessed as a whole and not individually.

When deciding whether or not to use the judicial exception, there are other things to think about, such as whether a claim reduces or changes something into a new state or object. However, these things are less important regarding smart grid technology.

Artificial intelligence (AI), which can be found in software programs or shown through them, is a vital part of how smart grid technologies work. Therefore, the innovation will pass the first component of the test as a whole. Any smart grid device or process patent claim will focus on an abstract concept. The second part of the analysis is to determine if a smart grid patent application claims to list extra features that go far beyond the judicial exception for abstract ideas.

(a) According to the Alice/Mayo test, adding a general-purpose computer or a component of a general-purpose computer to a software claim would impair patent eligibility, regardless of whether the general-purpose computer is a "specially programmed" or "specially programmed" machine. For example, peak energy prediction using real-time data analysis is currently not possible with today's energy network infrastructure. Therefore, new "smart meters" must be installed and connected to the grid. So, a smart grid often requires more than a "general purpose computer" running some specialised software.

(b) The AI powering the smart grid contributes to the use of the claimed technology and does not merely serve as a prominent tool to help solve problems more quickly. For example, estimating energy consumption, dynamically allocating energy, and identifying attack patterns, would only be possible today with AI (WIPO, 2019). This is simply due to the need for better data collection that the computing power of the smart grid can efficiently process. It is a new way to manage energy networks, not just a faster or more efficient way to use existing methods.

(c) If using machines does not make a big difference in how well the technology works, then the smart grid is not likely to exist. So, for this strategy to work, AI must automatically collect the data that AI needs to learn. Understanding that AI in the smart grid is a dynamic process of changing algorithms and not just a static piece of computer software is critical. Making the smart grid viable first requires using smart metres and more efficient energy access points.

(d) Finally, whether AI can be described as previously recognized, routine, and well-understood behavior must be

> assessed. It would be equally simple to reject the idea in the present case since it had already been denied in cases where the strategy employed deviated from the usual procedure.

While the Alice/Mayo test does not block smart grid patent applications, it is more stringent than the EU's approach. However, to meet USPTO requirements, more attention must be paid to how new software technology compares to industry standards and how important and good smart metres are from a technical standpoint.

19.6.3 China's Approach

As in the EU and the US, algorithms are laws and methods that consist entirely of algorithms or mathematical concepts. All applicable countries agree that the claims do not contain any technical solution since they neither disclose nor claim any specific technical method or technical feature. Such proposed innovations are only patentable if the necessary conditions are met. Examiners of Chinese patent applications often refer to Articles 2.2 and 25 of the Chinese Patent Law, which are interrelated in practice.

Innovations related to AI are certainly no exception. For example, machine learning is a subfield of AI that uses algorithms to analyze data and apply what it learns to new situations. Common learning strategies in this field include supervised and semi-supervised methods. Nevertheless, the learning method claimed in the application applies only to numerical calculations and does not improve hardware performance. In that case, the application cannot be granted a patent for the learning method (Mahmood, *et al.*, 2008).

Similar to EU regulations, there must be a clear relationship to the technical realisation of the software. For example, the CNIPA rejected AI-related applications because they needed more patentability. However, the Patent Re-examination Board held that the content claimed in the patent application was essentially a function of using a mathematical algorithm on the sample set to classify the elements of the set and therefore upheld the ruling of the State Intellectual Property Office (CNIPA, 2014).

In most countries, patents cannot be issued for abstract mathematical models or logical systems. This is done so as not to hoard these concepts. The problem is that monopolies only sometimes foster creativity, according to art. Article 25 of China's patent law on patentability and examination procedures for AI-related innovations stipulates that most AI-related patent applications are rejected. Meanwhile, China said further investment in AI would be "inevitable" if laws were clarified and patent protections improved (CNIPA, 2017).

Accordingly, the rules have been amended to strengthen the intellectual property protection of works produced in recent IT research and to facilitate the development of such businesses. As a result, the State Intellectual Property Office revised its policies, notably re-examining the clauses that lead examiners to reject applications related to AI (Fu, 2014).

This involved two major adjustments: first, it was thought that AI-enhanced business processes might make companies more efficient in their operations. A smart grid project might be a hypothetical use for this. A new provision has been introduced in the rules to address this issue, providing that claims related to business models are patentable if they include both the business process and the technical features. Although the clause does not explicitly address AI innovations, it can still be used for business model inventions such as smart grid programs.

An example of a recently filed patent application is "A Targeted Advertising Method Based on Regional User Habit Big Data." Multiple data processing components make up the system. The system can activate these modules to remind merchants to stock up on upcoming products or to remind customers to buy specific items before they sell out. Reporting business strategies for collecting and analyzing data, such as information about targeted consumers and merchants, can significantly improve transaction efficiency. The State Intellectual Property Office sets patentability based on the claims in the application, which include both rules and technology and produce technical effects. This thinking is relevant to developing smart grids that implement AI-generating algorithms (Mahmood, *et al.*, 2008).

Once AI breakthroughs are isolated from purely algorithmic processes, second chances arise. This makes the former claim to be a hybrid of hardware and computer programme algorithms. Before this, algorithms used in computer programs were not protected by patents because they were considered rules and techniques of mental work. The old rule did not distinguish between computer programs and algorithms, which could lead examiners to believe that computer program inventions are generally not patentable. While algorithms themselves cannot be patented, innovations that incorporate algorithms can. This strategy is expected to influence examiners to grant permission in more cases as AI-related innovations are often implemented (CNIPA, 2014).

As mentioned earlier, one everyday use of AI is to mimic how humans classify and understand information. Image processing and recognition involve one of the earliest instances in China. Algorithms trained by AI can enhance the presentation of medical images, enabling additional diagnostics. Artificial intelligence (AI) teaches algorithms to recognize data patterns to automate complex activities or make predictions. In this case, the multiscale retina algorithm and correction algorithm in the patent application methods for enhancing the presentation of medical images. A claim in a patent application consists of the following four elements:

First, acquire images of the target component or lesion. A combined augmented image sequence is created after these image sequences are upscaled. The next step involves processing the combined image sequences to make them easier to read and increase the chances of an accurate diagnosis. The second stage is where the AI teaching algorithm is most actively used, converting the grayscale values of the image sequence into floating-point types.

The main goal of the algorithm is to use the taught functions and formulas to improve the grayscale images of each image sequence and produce photos with sharper edges. The technical solution suggested in the application makes it easier to see lesions or target parts by improving the way the edge of the image is shown. It also meets the State Intellectual Property Office's standard requirements.

Since they help surgeons distinguish between seemingly identical organs, the statement is technical and meets artistic standards. Article 2.2 of the Chinese Patent Law Therefore, the patent has been approved by the State Intellectual Property Office. The State Intellectual Property Office also authorized another convolutional neural network patent application. Using neural networks in this way can improve gesture detection's reliability while reducing the processing time required for training (CNIPA, 2017).

It is worth remembering that whether or not their claims affect technology is a crucial reason for their patentability. The requirements of the USPTO and EPO are almost met here. However, distinguishing between AI applications with and without technological advantages is challenging across all jurisdictions.

However, this particular aspect differs markedly from the above studies. In CNIPA practice, the connection between algorithms and technical achievements is much weaker. In all the patents mentioned above granted by the State Intellectual Property Office, the impact of the algorithm on the calculation accuracy and speed is the primary determinant of the algorithm output. In the eyes of EU patent law, such progress is not enough; it needs more obvious technical consequences. It argued that even more targets, such as targeted advertising strategies or even the creation or improvement of the user experience, are excluded from technological influence under Chinese patent law. It should emphasize that such an experience should be an objective result of the combination of natural laws and human biological characteristics in the eyes of the examiner and should not base on the subjective feelings of the user (CNIPA, 2014).

19.7 Conclusions

In addition to making it feasible to integrate renewable energy sources into a power grid meaningfully, machine learning can also enable user-friendly pricing and defend the network from recent cyberattacks. This presents an exceptional potential for smart grid efforts. It is not easy to imagine designing such an intelligent energy source without machine learning.

Algorithms are not patentable under either Chinese or Western patent law systems. However, smart grid algorithms can meet technical standards by incorporating AI into a technical system. Despite significant challenges in integrating the differences between AI and pure algorithms, the Chinese approach allows for a greater variety of successful patent filings due to relaxed requirements. The political goal of providing additional protection for AI breakthroughs inspired the legal legacy of China's patent system. Therefore, it is dangerous to accept the concept of China solely based on monetary interests. Chinese policy is driven less by genuine financial research and more by politics. Additionally, the patent application process for slowing algorithms is complex and time-consuming. The result of a machine learning process in the form of an algorithm is still not copyrightable. It does not qualify as a technological invention under the statutes of any jurisdiction with patent laws that we have surveyed. Considering the information, a trade secret seems flawed and inappropriate, given the time and money spent on development. Given the unique nature and economic importance of the machine learning process, it is crucial to have a viable framework under intellectual property law to guarantee the availability of specific smart grid solutions. More protection may be necessary, but this is not likely the only approach. Further research should be done on requiring licences to protect the public's interests.

References

1. Bassiliades, N., & Chalkiadakis, G. (2018). Artificial Intelligence Techniques for the Smart Grid. *Adv. Build. Energy Res., 12*(1), 1–3. https://doi.org/DOI: 10.1080/17512549.2017.1410362.
2. Brown HE, Suryanarayanan S (2009) A survey seeking a definition of a smart distribution system. *North American Power Symposium'09,* pp 1–7.

3. CNIPA Guidelines for Patent Examination (2014).
4. CNIPA Guidelines for Patent Examination (2017).
5. EPO Guidelines for Examination (2019).
6. Fjelland, R. (2020). Why general artificial intelligence will not be realized. *Humanit. Soc. Sci. Commun.*, **7** (10) (2020). https://doi.org/10.1057/s41599-020-0494-4.
7. Fu X (2014), Regenerative medicine research in China: from basic research to clinical practice, *Sci. China Life Sci.*, 57 (1), 155-156.
8. Fu L, Zhao H, Guo S (2012) An analysis on the low-carbon benefits of smart grid of China. *Phys. Procedia,* 24:328–336. https://doi.org/10.1016/j.phpro.2012.02.049.
9. He J (2011) Daxuesheng Zhishichanquan Jiaoyu Tanxi – Yi Shandongsheng Daxuesheng zhishichanquan Yishi Diaocho wei Jichu [Analysis on the education of intellectual property for college students]. *J. Shandong Youth Univ. Polit. Sci.,* 27:27–42.
10. Halgamuge, M. & Guruge, D. (2022). Fair Rewarding Mechanism in Music Industry using Smart Contracts on Public-Permissionless Blockchain. *Multimed. Tools Appl.*, 2022, 81, 1523–1544.
11. JianJiao (2020), Application and prospect of artificial intelligence in smart grid. *IOP Conf. Series: Earth and Environmental Science* 510 (2020) 022012, doi:10.1088/1755-1315/510/2/022012.
12. Mahmood, A., M. Aamir, and M. Anis, Design and Implementation of AMR Smart Grid System, *2008 IEEE Electrical Power & Energy Conference,* 2008.
13. USPTO (2019) Revised patent subject matter eligibility guidance.
14. World Intellectual Property Organization (2019) Technology trends 2019: artificial intelligence. WIPO 85.

20

Architecture for Transactive Energy Management Systems with Different Market Clearing Strategies in Smart Grid

Arun S. L. and Vijayapriya Ramachandran*

School of Electrical Engineering, Vellore Institute of Technology, Vellore, India

Abstract

Peer-to-peer (P2P) energy trading idea has evolved considerably in the last few years, letting end energy users to effectively trade their surplus generation and unmet demand in a more cost-effective manner towards the development of smart grid deployment. This chapter describes the architecture of a local energy market (LEM) with several competitive techniques for P2P energy exchange within the locality of the energy users (prosumers and consumers). Accordingly, relative to the utility costs in a P2P energy market, consumers/prosumers can import/export available demand/generation in LEM at a considerable profit. Also, the concept of shared platform such as transactive energy market operator (TEMO) is also introduced to oversee LEM transaction. The purpose of TEMO is to construct the trading contracts among the P2P participants via fixing each contact cost with respect to the participants' quoted rate and demand. Tested case studies with ten residential P2P users in the considered locality confirm the effectiveness of the recommended TEMO performance.

***Keywords*:** Peer-to-peer, transactive energy, energy management system, smart grid, market clearing, energy market, renewable energy, TEMO architecture

20.1 Introduction

The energy industry has been supported over the previous century by extensive usage of traditional power generators like nuclear, gas, and thermal. However, these conventional power plants produce larger percentage of CO_2,

**Corresponding author*: vijayapriya.r@vit.ac.in

O.V. Gnana Swathika, K. Karthikeyan and Sanjeevikumar Padmanaban (eds.) Smart Grids as Cyber Physical Systems Volume 1: Artificial Intelligence, Cybersecurity, and Clean Energy for Next Generation Smart Grids, (315–336) © 2024 Scrivener Publishing LLC

leading to negative impact on the environment. Besides, the deterioration of conventional power generation necessitates the discovery of new sources of energy to meet the world's growing demand [1]. Nonetheless, non-conventional generations like solar, wind, and mini hydro are gaining prominence due to their enticing qualities. Concurrently, these dispersed energy generations have several drawbacks, such as intermittent power supply and site-limited power generation. On the other hand, to satisfy the demand either completely or partially, consumers have started to implement the small-scale distributed energy resources [2]. Consequently, to reduce the demand on the grid, customers wish to set up more renewable energy resources (RERs) in their residential building. Due to the dynamic nature of demand patterns and RER generation, efficient usage of generation of RER may be impossible, resulting in unproductive RER investment. Hence, users seek alternate methods to enhance their profits and save electricity bills. Considering the classic peer-to-grid (P2G) marketplace, *i.e.*, with the grid at utility-defined price, prosumers independently trade the required net demand (difference between demand and production) [3]. Excess demand will be acquired at the utility retail rate (grid exporting rate) whenever the consumer demand surpasses generation. Likewise, excess energy is also transferred to the grid at the feed-in-tariff (FiT) (grid purchasing rate) if the prosumer's generation exceeds the demand. The FiT strategy is intended to boost RER installation. Renewable energy resource installations has been gradually increased over the course of years, prompting regulatory authorities to substantially lower the FiT price, which may result in longer investment pay-back time. Literature shows that FiT initiatives are gradually reduced in certain parts of the world, for instance, Queensland in Australia [4]. Furthermore, considering the utility operating and steadiness constraints, the energy transferred to the utility must not surpass the limit of the power injection capacity (PIL) in the P2G market [3]. Hence, to produce high returns in the energy sector, prosumers seek innovative commercial frameworks/prototypes.

Transactive energy is one such concept, which allows neighbors to share excess generation or demand for a bigger return than the grid price. To balance the energy and validate the long-term energy trading via transmission networks new opportunities are emerging towards the development of peer-to-peer (P2P) market [5]. This strategy allows the residents (prosumers) to supply energy to the grid in case of excess generation in turn to meet the demand of the other consumer [6].

Consumers will have more alternatives for selecting the suitable lucrative solution to their demands as count of the prosumers expands. Implementation of P2P energy trading with a defined market pricing strategy helps in identifying the most lucrative option. This sort of transaction strategy has multiple benefits:

active utilization of distribution-side resources, which results in reduced total energy consumption; lowers the demand from large power suppliers, that links the demand-side management (DSM) consumption strategies [7], leading to lower overall pricing; promotes the prosumer and consumer monetary growth by letting prosumers to increase the savings from their surplus generation and permitting consumers to pick the best beneficial financial solution for their instant desires. Also, the system's reliability and security can be improved in the absence of a middleman in transactions along with automation [8]. Blockchain is one such technology that provides such a service is the smart contract [9]. The population of prosumers who can generate surplus electricity is projected to surge in the coming years. In terms of economics, an end user with surplus energy supply over demand must benefit by exporting it. Simultaneously, the residential users without RER installation who cannot generate their own electricity strive to purchase the energy at a reduced price. Negotiation in the best interests of two parties, a simple, profitable, and adaptable trading scheme is necessary to meet the needs of both the parties. Utilization of advanced home energy storage with power handling approaches enhances prosumers position in the P2P market. Technologies employed for managing the energy allow for the most efficient energy utilization and distribution to other locality users. In the absence of electricity supply from the PV panels or during a power outage the stored energy from the batteries are utilized.

This chapter discusses a model of transactive energy market operator (TEMO) for simplifying the energy transaction among the neighboring end users through local energy market (LEM). A community of participants involve to trade their available demand or surplus generation to reduce the dependency on the grid and energy expenses. In the proposed technique, the TEMO creates a transaction contract between the participants based on their energy prices and trading power. Different market clearing mechanisms are also suggested to ensure the reliability and profitability of LEM.

20.2 TEMO Architecture

Figure 20.1 depicts the suggested multi-agent-based LEM model and it comprises of several self-governing agents such as players, TEMO, and the utility operator. These players engage, exchange, and collaborate to achieve individual and communal goals. The members of LEM want to reduce their individual energy expenditure, while the community wants to reduce its dependency from the central utility with the provision of increasing self-consumption. Generally, TEMO may be the distribution or the independent element of the system operator [10]. Distinct actor, *i.e.*, the grid

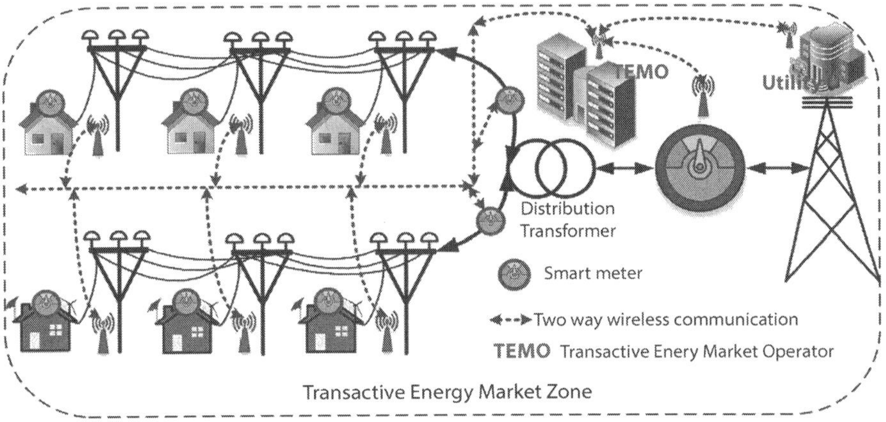

Figure 20.1 Architecture of LEM.

buys and sells energy at specified prices is liable for the system's dependability and security.

Generally, TEMO receives trade details of the participants like quoted demand (available demand/excess generation) and trading price (bid/offer) prior to the commencement of every trading interval. Transactive energy market operator (TEMO) uses a suitable market clearing technique to clear the LEM and develops transaction contracts between participants on basis of the provided data. For an interval, t, the participant, p, predicted demand within the locality is calculated as,

$$ED_p^t = D_{p,NDL}^t + D_{p,DL}^t + D_{p,B}^t \tag{20.1}$$

where $\in \mathcal{P} \triangleq [1,2,\ldots P]$. $\mathcal{P}$ gives the LEM participants set, where P expresses the number of active participants for the given interval, t. Non-deferrable loads (critical appliances like fan, light and kitchen loads) and deferrable loads (flexible loads like well pump and washing machine) demand estimation for trading interval, t, are represented as $D_{p,NDL}^t$ and $D_{p,DL}^t$, respectively. The estimated power exchange from battery is expressed as $D_{p,B}^t$ ($D_{p,B}^t \langle 0 - \textit{during discharging mode}; D_{p,B}^t \rangle\, 0 - \textit{during charging mode}$). The participant's trading power (δ_p^t) may be calculated by taking into account the estimated generation (EG_p^t) from the installed in-house distributed generations as displayed in Equation 20.2.

$$\delta_p^t = ED_p^t - EG_p^t \tag{20.2}$$

Local energy market (LEM) participants would like to sell the available power at a higher rate or fulfil unmet demand at a lower rate compared to the grid electricity buying or selling rate, respectively. Transactive energy market operator (TEMO) acts as a common interface which expects the participants to submit their estimated trading demand and suitable agreement (for excess power capacity)/bid (for unmet demand) to manage the LEM. Concurrently, a suitable market-clearing technique is implemented to compute the energy transactions rate during the given interval. Participants are categorized as importers (having unmet demand) or exporters (having surplus generation) by TEMO depending on their quoted transaction. The participant is considered an importer when a participant's δ_p^t is greater than zero and the participant is considered an exporter when δ_p^t is less than zero. To provision viable and cost-effective energy marketplace construction, TEMO likes to incorporate boundary limits in the LEM. Irrespective of the participant's either being an exporter or importer, reported transaction should be within the specified minimum and maximum limits, as given in Equation 20.3.

$$\begin{aligned} \delta_{exp,min}^t \leq \left|\delta_p^t\right| \leq \delta_{exp,max}^t \qquad & \forall\, p \in \mathcal{P}_{exp}^t \\ \delta_{imp,min}^t \leq \delta_p^t \leq \delta_{\mathrm{imp},max}^t \qquad & \forall\, p \in \mathcal{P}_{imp}^t \end{aligned} \tag{20.3}$$

where, $\delta_{exp,min}^t$ and $\delta_{exp,max}^t$ are the least and highest energy export limit for exporters. Similarly, the least and highest unmet demand limit for importers are expressed as $\delta_{imp,min}^t$ and $\delta_{\mathrm{imp},max}^t$, respectively. Further, the set of exporters and importers for a trading interval, t, are expressed as $\mathcal{P}_{exp}^t$ and $\mathcal{P}_{imp}^t$, respectively. Transactive energy market operator (TEMO) determines net community demand by taking into account of all exporters and importers' reported trading generation and demand as shown in Equation 20.4.

$$\mathcal{D}^t = \mathbb{I}^t - \mathbb{E}^t \tag{20.4}$$

$$\mathbb{I}^t = \sum_{p \in \mathcal{P}_{imp}^t} \delta_p^t \tag{20.5}$$

$$\mathbb{E}^t = \sum_{p \in \mathcal{P}_{exp}^t} |\delta_p^t| \tag{20.6}$$

where $\mathbb{I}^t$ and $\mathbb{E}^t$ are represented as the aggregated unmet demands and surplus generation of all importers and exporters, respectively. Further, the utility is considered to be a LEM by TEMO since it may buy the participants' excess generation and sell the energy to meet the unmet demand of participants at the same time for a predetermined offer. As previously stated, in the P2G paradigm, end users rely exclusively on the utility to trade energy. Under the P2G method, the energy bill of any end user p during a trading interval t may be determined using Equation 20.7.

$$\mathfrak{B}^t_{P2G,n} = \begin{cases} \delta^t_{P2G,p} \cdot C^t_S \cdot dt & \forall p \in \mathcal{P}^t_{imp} \\ \delta^t_{P2G,p} \cdot C^t_B \cdot dt & \forall p \in \mathcal{P}^t_{exp} \end{cases} \tag{20.7}$$

where, C^t_S and C^t_B are expressed as the utility's unit selling price for meeting end users' demand and unit buying price for purchasing energy from prosumers, respectively. Furthermore, LEM members will only profit economically if they fulfil their unmet demand at a lesser energy rate compared to utility and offer the excess energy at a better rate compared to the utility buying rate. The market clearing rate ($\mathcal{M}^t$) within the utility boundary rate for the players in the efficient and cost-effective electricity market is expressed in Equation 20.8.

$$C^t_B < \mathcal{M}^t < C^t_S \tag{20.8}$$

Transactive energy market operator (TEMO) settles the LEM with due consideration to the individuals' reported energy, economic benefit of all community participants and the distribution system operators. Further, active involvement of LEM players may boost the success of the P2P energy market. Adopting good market approaches for LEM may also increase individuals' interest in it. Various market approaches are described in the next section in order to create a profitable energy exchange among participants.

20.3 P2P Energy Market Strategies

Calculating and publishing the common market set rate after collecting demand patterns from the LEM players are main objective of TEMO. To determine the viable internal market rates, in turn to attract market performer to participate in the cost effective marketplace, TEMO uses a number of market clearing processes.

20.3.1 Mid-Pricing Strategy (MPS)

To begin with utility electricity subscribers in a community who wish to participate in LEM is considered. The method described in Equation 20.4 for calculating net community energy demand $(\mathcal{D}^t)$ may be used to calculate the utility dependency of the whole neighborhood. Two sub-strategies will be implemented within the category of MPS based on the calculated $\mathcal{D}^t$.

20.3.1.1 *Higher Community Demand ($\mathcal{D}^t > 0$)*

When a community's total energy demand surpasses its aggregated generation, the community is becoming a grid importer. In this circumstance, the community should depend upon the utility to support the unmet demands. Unlike utilities, LEM allows importers and exporters to exchange their unmet demands/surplus energy generations with neighbors for a better profit. Considering the trading period '*t*', the market clearing price of the exporter using the market strategy is defined in Equation 20.9.

$$\mathcal{M}_{exp}^t = \frac{\mathcal{C}_S^t + \mathcal{C}_B^t}{2} \tag{20.9}$$

The market clearing cost for all LEM importers $(\mathcal{M}_{imp}^t)$ over the trading period '*t*' is computed using the obtained exporter market clearing cost as shown in Equation 20.10.

$$\mathcal{M}_{imp}^t = \frac{(\mathcal{D}^t \times \mathcal{C}_S^t) + (\mathcal{M}_{exp}^t \times \mathbb{E}^t)}{\mathbb{I}^t} \tag{20.10}$$

20.3.1.2 *Higher Community Generation ($\mathcal{D}^t < 0$)*

In case of surpassing the total demand of all importers by the aggregated generation of the community, the community will act as a grid exporter. Under such condition, the market clearing cost of all the importers for the trading duration '*t*' is described as

$$\mathcal{M}_{imp}^t = \frac{\mathcal{C}_S^t + \mathcal{C}_B^t}{2} \tag{20.11}$$

Likewise, the market clearing cost of all the exporters for the trading duration 't' is determined using the obtained importer cost $(\mathcal{M}_{imp}^{t})$ as shown in Equation 20.12.

$$\mathcal{M}_{exp}^{t} = \frac{\left(\mathbb{I}^{t} \times \mathcal{M}_{imp}^{t}\right) + (\mathcal{D}^{t} \times C_{B}^{t})}{\mathbb{E}^{t}} \tag{20.12}$$

The electricity bill under MPS-based P2P energy trading $(\mathfrak{B}_{P2P,p}^{t})$ for any participant, p, during interval, t, is computed, as given in Equation 20.13.

$$\mathfrak{B}_{P2P,p}^{t} = \begin{cases} \delta_{P2P,p}^{t} \cdot \mathcal{M}_{imp}^{t} \cdot dt & \forall p \in \mathcal{P}_{imp}^{t} \\ \delta_{P2P,p}^{t} \cdot \mathcal{M}_{exp}^{t} \cdot dt & \forall p \in \mathcal{P}_{exp}^{t} \end{cases} \tag{20.13}$$

Because the proposed method is simple, market settling processing complexity can be minimized irrespective of large amount of participants. Further, easy predictability of MPS-based LEM increase the participants' interest in market trading. Consequently, irrespective of their level of participation in community net demand, importers earn more profit during higher community generation, exporters make more profit during greater community demand. To overcome this issue and establish profitable transactions for all sorts of participants, the upcoming subsection presents a market clearing approach based on generation-to-demand ratio (GDR).

20.3.2 GDR Pricing Strategy (GDRS)

The ratio of community energy production to demand for the considered trading duration 't' is estimated by taking into account the energy exporters and importers total demand, as given in Equation 20.14.

$$\Psi^{t} = \frac{\mathbb{E}^{t}}{\mathbb{I}^{t}} \tag{20.14}$$

where Ψ^{t} estimated value indicates the state of community. $\Psi^{t} < 1$ indicates as an importer and $\Psi^{t} > 1$ indicates as an exporter As a result, this GDRS market clearing technique includes two different strategies.

20.3.2.1 Higher Community Demand ($\Psi^t < 1$)

When the community behaves as a grid importer, the market clearance cost of community exporters ($\mathcal{M}_{exp}^t$) and importers ($\mathcal{M}_{imp}^t$) for a trading interval, t, is premeditated correspondingly using Equations 20.15 and 20.16.

$$\mathcal{M}_{exp}^t = \frac{C_S^t + C_B^t(1-\Psi^t)}{2} \tag{20.15}$$

$$\mathcal{M}_{imp}^t = \left(\mathcal{M}_{exp}^t \times \Psi^t\right) + \left(C_S^t(1-\Psi^t)\right) \tag{2.16}$$

20.3.2.2 Higher Community Generation ($\Psi^t > 1$)

In case the calculated value of Ψ^t surpasses '1', the community behaves like a grid exporter, and the LEM importers ($\mathcal{M}_{imp}^t$) and exporters ($\mathcal{M}_{exp}^t$) market clearance cost is obtained from Equations 20.17 and 20.18, correspondingly.

$$\mathcal{M}_{imp}^t = \frac{C_S^t - C_B^t\left(1-\dfrac{1}{\Psi^t}\right)}{2} \tag{20.17}$$

$$\mathcal{M}_{exp}^t = \frac{\mathcal{M}_{imp}^t + C_B^t(\Psi^t - 1)}{\Psi^t} \tag{20.18}$$

The GDRS market participant trading power bill may be estimated using Equation 20.13. The suggested market strategy, like MPS, will require less computing time to settle the energy market. Moreover, when compared to P2G scheme, both exporters and importers will have lucrative transactions. However, the current technique is cooperative in nature, the profitability of the every participants is exclusively dependent on the community net energy demand. Importers may not get a large economic gain from LEM in case the community total demand exceeds the generation. Exporters, instead gains a smaller amount in case of negative community's net demand. The following section details the double auction rating approach in order to improve player profit while taking into account the aforementioned restrictions.

20.3.3 Double Auction Strategy (DAS)

Transaction cost and power exchange in P2P energy marketplace is calculated with the aid of double auction strategy (DAS). Further, the TEMO acts as a mediator to monitor discussions between importers and exporters. Since all the surplus generation offers from exporters $(\Phi_p^t, p \in \mathcal{P}_{exp}^t)$ and unmet demand bids from importers $(\Phi_p^t, p \in \mathcal{P}_{imp}^t)$ are reported to TEMO simultaneously, participants may be unaware of other participants' offers and bids. After obtaining the data, TEMO classifies participants as exporters or importers by considering the sign of their reported net demand (δ_p^t). Further, TEMO organizes exporters' offers and importers' bids based on merit. For example, exporters $(\mathcal{P}_{exp}^t = \{E_1, E_2, E_3, E_4, E_5\})$ are placed in ascending order based on reported offers, whereas importers $(\mathcal{P}_{imp}^t = \{I_1, I_2, I_3, I_4, I_5\})$ are organized in descending order based on reported bids, as shown in Figure 20.2. Further, to take part in LEM participants are expected to satisfy the power limit constraints of minimum (δ_{min}^t) and maximum (δ_{max}^t) value, as indicated in Figure 20.2.

The equilibrium point for the proposed DAS-based LEM is the connection point of the quoted bids and offers which enhances the community social welfare, as depicted in Figure 20.2. Successful and unsuccessful bids and offers are placed before and after the market clearing quantity (MCQ). The area encompassing the successful demand and generation are mentioned as market's social welfare. Energy cost at the MCQ is known as the market-clearance cost since it applies to all transactions between exporters and importers. All exporters under DAS-based LEM would sell their surplus generation at a cost over and above or equal to their agreed-upon offer, and all importers

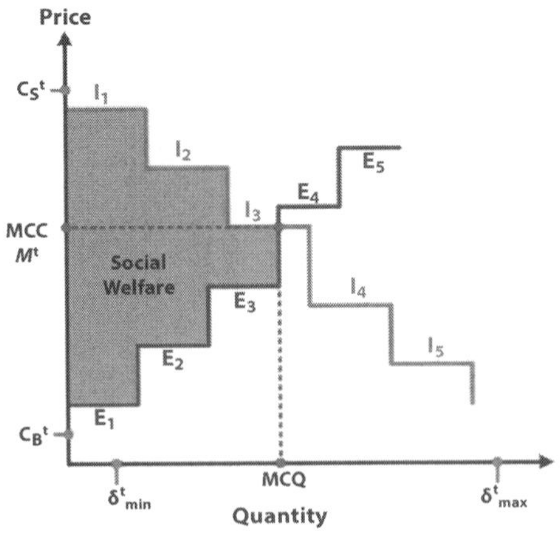

Figure 20.2 DAS-based market settling mechanism.

offer a rate below or equal to their willingness payment. Trading electricity bill of a participant under this strategy can be computed by using Equation 20.19.

$$\mathfrak{B}_{P2P,n}^{t} = \begin{cases} \delta_{P2P,n}^{t} \cdot \mathcal{M}^{t} \cdot dt & \forall p \in \begin{cases} \mathcal{P}_{imp}^{t} \\ \text{Bids before MCQ} \end{cases} \\ \delta_{P2P,n}^{t} \cdot \mathcal{M}_{exp}^{t} \cdot dt & \forall p \in \begin{cases} \mathcal{P}_{exp}^{t} \\ \text{Offers before MCQ} \end{cases} \end{cases} \tag{20.19}$$

Double auction strategy (DAS) may fail to meet the optimal MCQ and MCC in some cases irrespective of the recommended DAS, which significantly increases the trading profit of active participants. For instance, the curve of exporters and importers which does not cross is illustrated in Figure 20.3. Consequently, exporters and importers are not able to establish a trading agreement. In order to address the above limitation, a priority-based auction technique is suggested to enhance active transaction in all circumstances.

20.3.4 Priority-Based Auction Strategy (PAS)

The LEM exporters and importers are prioritized correspondingly based on their submitted offers and bids in the priority-based auction strategy (PAS). The lowest offer in exporters set and highest bid in importer set take

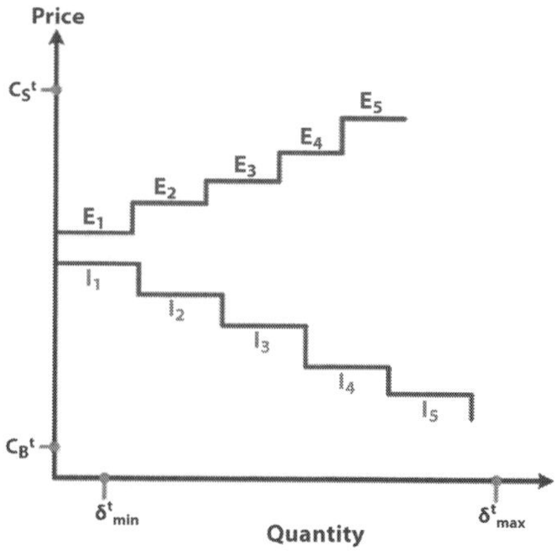

Figure 20.3 The failure scenario of DAS.

high priority. However, the highest offer and lowest bid established by the exporter and importer, respectively, receive the lowest priority. When many participants in each set propose the same offer/bid at the same interval, the player with the higher generation/demand gains priority. Participants that are prioritized have greater opportunity to sell or buy surplus generation or demand through LEM. The prioritization of TEMO for a set of importers ($\mathcal{P}_{imp}^{t} = \{i, j, k\}$) and set of exporters ($\mathcal{P}_{exp}^{t} = \{a, \mathrm{b}, \mathrm{c}, \mathrm{d}\}$) is displayed in Figure 20.4. Further, δ_a^t and θ_a^t represent the exporter *a's* reported offer and surplus generation for the trading interval, t, respectively.

A transaction is described as the interchange of energy among an exporter and importer. Each transaction decreases both the importer's and exporter's prospective trading power. Negotiations between one importer/exporter and others continue up to the existing transactional power hits zero. TEMO utilize suitable framework to determine the transaction price (TP). The present work employs a simple mid-priced strategy. The TP for a deal between exporter, *a*, importer, *I*, and exporter, a, is indicated as $\eta_{i,a}^{t}$, and it may be calculated as shown in Equation 20.20.

$$\eta_{i,a}^{t} = \eta_{i,\epsilon_i}^{t} = \eta_{a,\epsilon_a}^{t} = \frac{\theta_i^t + \theta_a^t}{2} \tag{20.20}$$

where ϵ_i is the total number of transactions completed by importer, *i*. Considering the traded power in each transaction with different exporters ($\delta_{i,\epsilon}^{t}$), the tth trading interval transactions cost of importer '*i*' is estimated totalling all the TP as per Equation 20.21. Similarly, the transaction cost for an exporter '*a*' is determined based on Equation 20.22. ϵ_a is the completion of exporter's 'a' total number of transactions and $\delta_{a,\epsilon}^{t}$ represents the transacted power. Figure 20.5 elaborates on the PAS-based LEM as a flowchart.

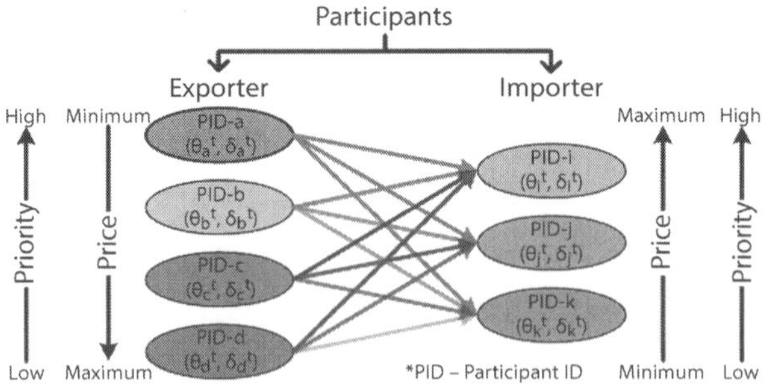

Figure 20.4 LEM participants prioritization using TEMO.

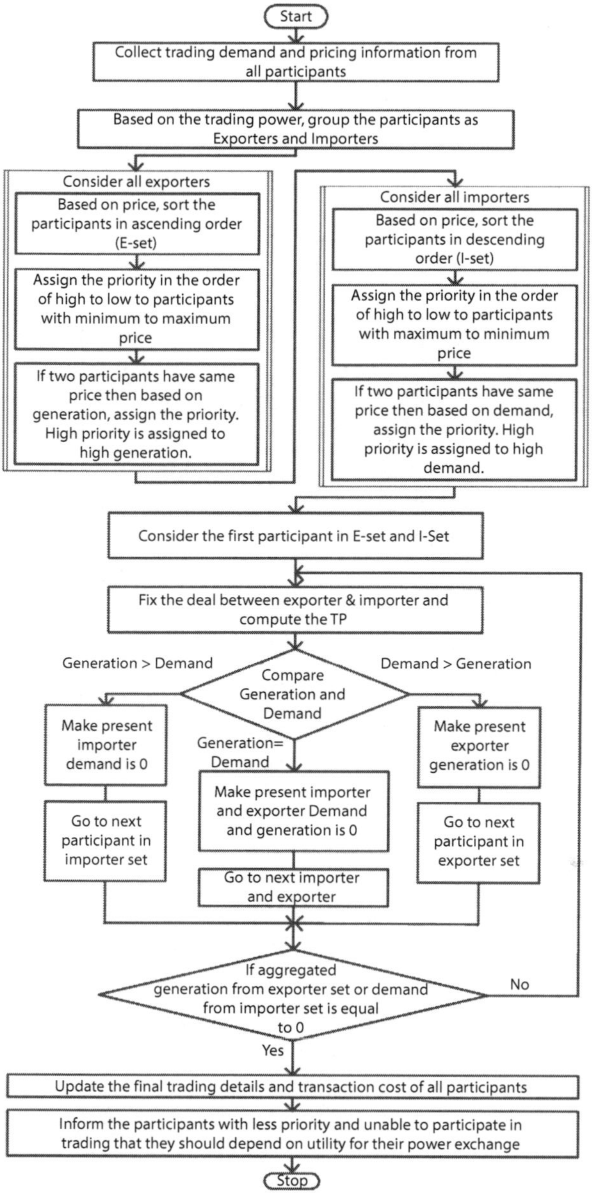

Figure 20.5 PAS marketplace-settling framework.

$$\mathfrak{B}^{t}_{P2P,i} = \sum_{\epsilon=1}^{\epsilon_i} \left(\eta^{t}_{i,\epsilon_i} \cdot \delta^{t}_{i,\epsilon} \cdot dt \right) \tag{20.21}$$

$$\mathfrak{B}^{t}_{P2P,a} = \sum_{\epsilon=1}^{\epsilon_a} \left(\eta^{t}_{i,\epsilon_a} \cdot \delta^{t}_{a,\epsilon} \cdot dt \right) \tag{20.22}$$

The proposed market approaches are developed on the basis of the participants' stated net demand. Participants' reported demand is entirely contingent on accurate forecasting of renewable energy resources and optimal operation of desirable loads. Nevertheless, owing to dynamics in end user's demand pattern, they are unable to maintain the reported demand at all trade hours. As a result, inaccurate demand quotes might impair the LEM's capability to exchange profitable demand in P2P energy market. To tackle this issue, a violation cost is incorporated into the participants' energy bills. The violation cost (Θ^{t}_{p}) can be calculated, as indicated in Equation 20.23. Further, deviations in stated demand may affect the participant's energy bill. Energy bill variation is termed as the deviation cost and it is computed using Equation 20.24.

$$\Theta^{t}_{p} = \left| \left(\delta^{t}_{A,p} - \delta^{t}_{Q,p} \right) \cdot \left(\frac{C^{t}_{S} + C^{t}_{B}}{2} \right) \cdot \vartheta^{t} \cdot dt \right| \tag{20.23}$$

$$\Omega^{t}_{p} = \begin{cases} \left(\delta^{t}_{A,p} - \delta^{t}_{Q,p} \right) \cdot \mathcal{M}^{t}_{imp} \cdot dt; & \text{if } 0 < \delta^{t}_{A,p} < \delta^{t}_{Q,p} \\ \left(\delta^{t}_{A,p} - \delta^{t}_{Q,p} \right) \cdot \mathcal{M}^{t}_{exp} \cdot dt; & \text{if } 0 > \delta^{t}_{A,p} > \delta^{t}_{Q,p} \\ 0 & else \end{cases} \tag{20.24}$$

The actual and stated net demands of the participant 'p' are correspondingly represented as $\delta^{t}_{A,p}$ and $\delta^{t}_{Q,p}$, respectively. ϑ^{t} represents the violation factor which will be decided by TEMO based on the community dynamics. Taking into account the electricity bills from the P2P and P2G markets, as well as the deviation price and penalty cost for violation, the total energy bill of the participant, p, for the trading period 't' is determined using Equation 20.25.

$$\mathfrak{B}^{t}_{p} = \mathfrak{B}^{t}_{P2G,p} + \mathfrak{B}^{t}_{P2P,p} + \Theta^{t}_{p} + \Omega^{t}_{p} \tag{20.25}$$

20.4 Simulation Study

For the purpose of comparing the recommended market methods and emphasizing the value of TEMS with respect to the reduction in the participant electricity bill, numerous case studies are used. Ten actively participating prosumers in TEMS make up the designated residential locality. There are many high-tech electrical and electronic appliances in residential dwellings today to make living easier [11]. Depending on the participant, different appliances have different operating times. Users are recommended to install renewable energy sources in the residential premises to lessen their reliance on the grid and increase their ability to save money on their electricity bills. Residential consumers strongly favor the usage of rooftop solar PV and small scale wind turbines renewable energy support. Additionally, users are fascinated with energy storage employing the battery because it can reduce their energy costs by meeting urgent demands during peak hours. Prosumers can scale their renewable energy sources and batteries by taking into account the space availability and low installation cost as RERs power generation impacts the drop in electricity bill [12]. The consistency of the P2P energy market may be seriously impacted by taking the longer time period for every transaction into account. Instead, taking the short time period into account can increase the energy market's consistency. However, the market clearing systems' processing complexity will rise, which could destroy the energy market. Additionally, the company has established time restrictions for the application of flexible loads. Given everything, the trading range frequency has been set at 15 min.

Participants in LEM must schedule appliance operation, properly taking into account utility dynamics, user wish and comfort, on-site RERs power production, and conventional power generation. Ideal demand characteristics and the amount of renewable energy that can be produced during the day for all participants in the area under consideration are depicted in Figure 20.6. The community power pattern is evaluated by taking into account individual participants' demand and generation, shown in Figure 20.7. The monthly variations in buying and selling rates of utility energy are shown in Figure 20.8. Local players' daily energy costs in the P2G and P2P market paradigms are shown in Figure 20.9. It is obvious that P2P participants have more economic gains than P2G participants. Additionally, the DAS-based market clearance P2P energy market success and failure possibilities are described in Table 20.1 and Figure 20.10. Double auction strategy (DAS) approach to successfully clear the energy market is considered in Case 1, as illustrated in Figure 20.10(a). However, in example 2, the DAS method was unable to determine the quantity and price of market clearance, as shown in Figure 20.10(b).

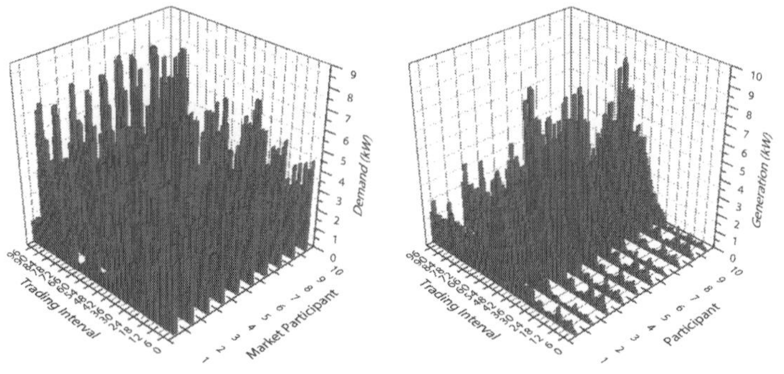

(**a**) Optimal demand (**b**) Available generation

Figure 20.6 Per day dynamics in participants energy demand and generation.

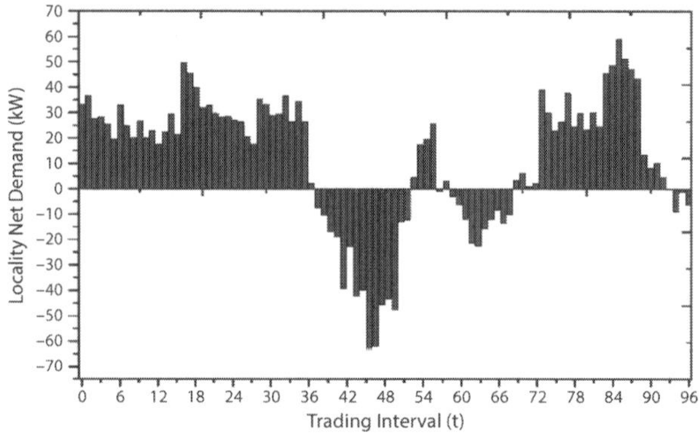

Figure 20.7 Net demand variation of participant's on daily basis.

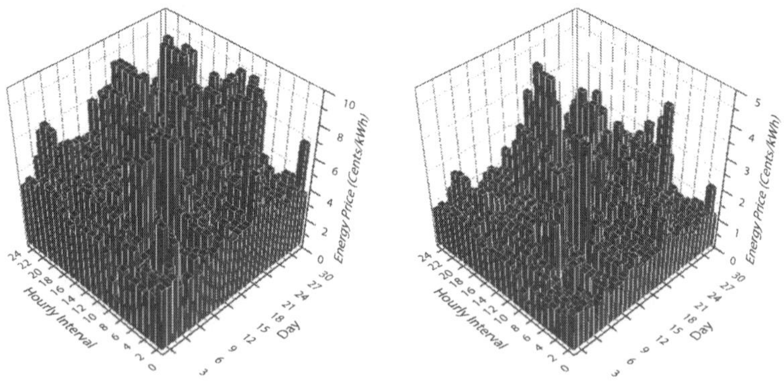

(**a**) Selling price (**b**) Buying price

Figure 20.8 Grid price variations.

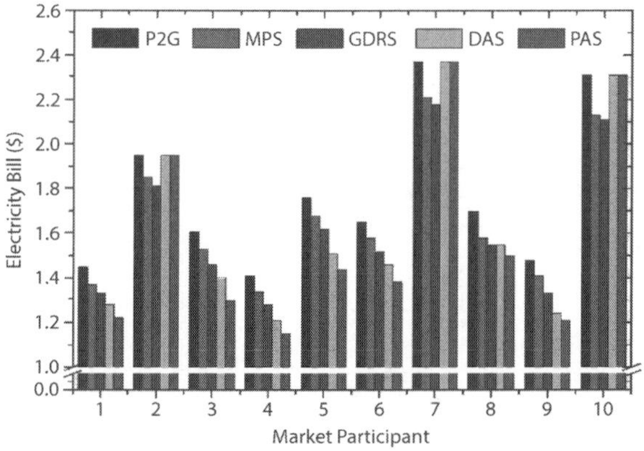

Figure 20.9 Per day electricity bill of participant for P2G and P2P schemes.

Table 20.1 Reported demand and cost of participants for LEM.

Participant ID (PID)	$\delta^t_{Q,p}$ (kW)	Study 1 Θ^t_p(c)	Study 2 Θ^t_p(c)
01	+1.00	4.60	2.70
02	+1.50	5.00	3.00
03	−1.00	3.00	4.80
04	−0.80	2.50	4.40
05	+1.20	4.40	2.50
06	+0.50	2.50	1.80
07	+1.30	3.50	2.00
08	−0.50	4.00	5.30
09	−1.10	3.50	5.00
10	−1.50	2.00	4.00

Monthly electricity expenses of the individual participants for various P2P market methods are shown in Table 20.2. For better comparison, monthly electricity bills of the participants when energy trading is not promoted as in case of P2G situation is also included in Table 20.2. The results validate all the participants will economically profit from the planned P2P energy trading. The simulation findings show that the trading process lowers everyone in the community's electricity bill as compared to members is uniquely dependent on the grid.

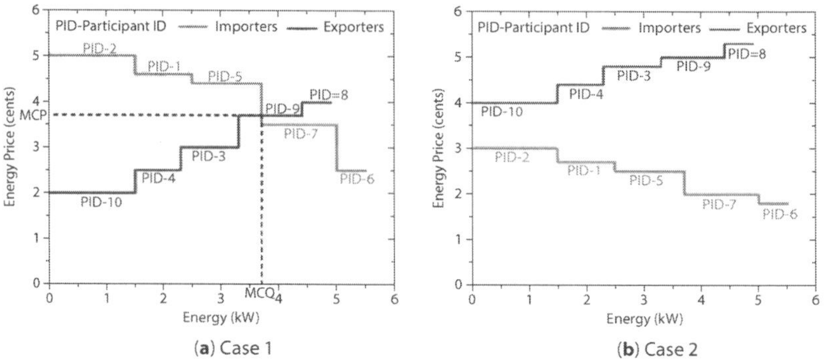

Figure 20.10 DAS-based energy market settlement.

Table 20.2 Energy bill of the participants per month.

PID	P2G	MPS		GDRS		DAS		PAS	
	$\mathfrak{B}_p$($)	$\mathfrak{B}_p$($)	$\mathcal{S}_p$(%)	$\mathfrak{B}_p$($)	$\mathcal{S}_p$(%)	$\mathfrak{B}_p$($)	$\mathcal{S}_p$(%)	$\mathfrak{B}_p$($)	$\mathcal{S}_p$(%)
1	136	128.23	5.72	121.6	10.59	116.32	14.48	115.46	15.11
2	137.75	129.74	5.82	123.34	10.47	117.6	14.63	116.56	15.39
3	117.11	103.71	11.45	105.4	10	99.76	14.82	96.88	17.28
4	119.94	108.39	9.63	108.02	9.94	103.11	14.04	97.91	18.37
5	119.79	106.22	11.33	106.9	10.77	100.97	15.72	99.39	17.03
6	120.3	111.65	7.2	106.41	11.55	102.35	14.93	97.94	18.59
7	140.05	132.67	5.27	123.64	11.72	116.66	16.71	113.08	19.26
8	124.04	113.79	8.27	111.76	9.91	106.11	14.46	104.32	15.9
9	118.3	110.59	6.52	103.59	12.44	100.64	14.93	97.53	17.56
10	139.71	131.7	5.74	124.78	10.69	120.36	13.86	115.77	17.14

$\mathfrak{B}_p$ – Participant p monthly energy bill
$\mathcal{S}n$ - Percentage savings when compared to energy bill in P2G scheme

Participants' earnings improve dramatically with active participation and accurate projection of expected net demand [13]. On the other hand, deviations in estimated demand create significant effect in the energy market. Consequently, in order to prevent players from changing the quoted trading demand, TEMO implemented a trade agreement breach cost. Tables 20.3 and 20.4 present the results of several market approaches for

Table 20.3 Participant penalty analysis for MPS and GDRS schemes.

Participant	$\delta^t_{Q,p}$ (kW)	$\delta^t_{A,p}$ (kW)	Θ^t_p (c)	MPS (cents)				GDRS (cents)			
				$\mathfrak{B}^t_{P2P,p}$	$\mathfrak{B}^t_{P2G,p}$	Ω^t_p	$\mathfrak{B}^t_p$	$\mathfrak{B}^t_{P2P,p}$	$\mathfrak{B}^t_{P2G,p}$	Ω^t_p	$\mathfrak{B}^t_p$
1	1.5	1.7	0.21	6.6	1.08	0	7.89	6.3	1.08	0	7.59
2	−1	−0.8	0.21	-3.5	0	0	−3.29	-3.1	0	0	−2.89
3	1.5	1.5	0	6.6	0	0	6.6	6.3	0	0	6.3
4	2	2.5	0.53	8.8	2.7	0	12.03	8.4	2.7	0	11.63
5	−1.5	0.5	2.1	−5.25	2.7	5.25	4.8	−4.65	2.7	4.35	4.5
6	2.5	1.5	1.05	11	0	−4.4	7.65	10.5	0	−4.2	7.35
7	0.5	0.8	0.32	2.2	1.62	0	4.14	2.1	1.62	0	4.04
8	-2	-2	0	−7	0	0	-7	-6.2	0	0	−6.2
9	−0.5	−1.8	1.37	−1.75	−2.08	0	-2.46	−1.55	−2.08	0	−2.26
10	1	1	0	4.4	0	0	4.4	4.2	0	0	4.2

Parameters of Utility: $C^t_S = 5.40$ c; $C^t_B = 1.60$ cents
Factor of TEMO: $\vartheta^t = 0.3$
LEM implementation using MPS: $\mathcal{M}^t_{imp} = 4.40$ cents; $\mathcal{M}^t_{exp} = 3.50$ cents
LEM implementation using GDRS: $\mathcal{M}^t_{imp} = 4.20$ cents; $\mathcal{M}^t_{exp} = 3.10$ cents

Table 20.4 Participants' deviations penalty investigation using DAS and PAS.

Participant	$\delta^t_{Q,p}$ (kW)	$P^t_{Q,p}$ (c)	$\delta^t_{A,p}$ (kW)	Θ^t_p (c)	DAS (cents)				PAS (cents)			
					$\mathfrak{B}^t_{P2P,p}$	$\mathfrak{B}^t_{P2G,p}$	Ω^t_p	$\mathfrak{B}^t_p$	$\mathfrak{B}^t_{P2P,p}$	$\mathfrak{B}^t_{P2G,p}$	Ω^t_p	$\mathfrak{B}^t_p$
1	1.5	3.2	1.7	0.21	4.8	1.08	0	6.09	1.6	1.08	0	2.89
2	−1	3.2	−0.8	0.21	−3.2	0	0	−2.99	−3.45	0	0	−3.24
3	1.5	2.9	1.5	0	0	8.1	0	8.1	0	8.1	0	8.1
4	2	4.5	2.5	0.53	6.4	2.7	0	9.63	6.66	2.7	0	9.89
5	−1.5	2.7	0.5	2.1	−4.8	2.7	4.8	4.8	−5.175	2.7	5.175	4.8
6	2.5	4.2	1.5	1.05	8	0	−3.2	5.85	8.675	0	−3.7	6.025
7	0.5	2.5	0.8	0.32	0	4.32	0	4.64	0	4.32	0	4.64
8	−2	2.1	−2	0	−6.4	0	0	−6.4	−6.66	0	0	−6.66
9	−0.5	2.4	−1.8	1.37	−1.6	−2.88	0	−3.11	−1.65	−2.88	0	−3.16
10	1	2	1	0	0	5.4	0	5.4	0	5.4	0	5.4

Parameters of Utility: $\Delta^t_S = 5.40$ cents; $\Delta^t_B = 1.60$ cents
Factor of TEMO: $\vartheta^t = 0.3$
LEM based on DAS: MCQ=5kW; MCC= 3.20 cents

the complete trading period with different deviations. Findings illustrate participants are fined according to the degree of deviation. In the energy market, false data quotation would also have a major negative impact on market clearance pattern including market clearance quantity and rate. To make up for the financial loss, participants are severely fined.

20.5 Conclusions

The proposed TEMO is used in this chapter together with various trading techniques to calculate the cost of every trading among the LEM players. Through the sale or purchase of surplus energy production or demand at a profit associated to utility rates, the offered strategies encourage prosumers and consumers to actively participate in LEM. The offered strategies are validated using a variety of case studies. The results of the simulation research demonstrate that all LEM participants profit significantly by executing the suggested trading schemes associated to trading power with the utility, i.e., P2G system. Mid-pricing strategy (MPS) method boosts the average savings to 6.19% in the monthly electricity expenses of the participants as compared to P2G strategy. However, GDRS method increases the average savings rate to 7.86%. Consequently, tested use case for dual auction solutions like DAS and PAS correspondingly report 11.31% and 13.72% of increase in average energy savings. Eventually, the importance of TEMSs towards the enhancement of end-user social welfare is highlighted by these findings. Additionally, the dependability and cyber security of the P2P energy market are provided by the penalty investigation for concerns of injecting fault data.

References

1. X. Fang, S. Misra, G. Xue and D. Yang, "Smart Grid — The New and Improved Power Grid: A Survey," in *IEEE Commun. Surv. Tutorials*, vol. 14, no. 4, pp. 944-980, Fourth Quarter 2012, doi: 10.1109/SURV.2011.101911.00087
2. B. P. Esther and K. S. Kumar, "A survey on residential demand side management architecture, approaches, optimization models and methods," *Renewable Sustain. Energy Rev.*, vol. 59, pp. 342–351, Jun. 2016, https://doi.org/10.1016/j.rser.2015.12.282
3. S. L. Arun and M. P. Selvan, "Intelligent residential energy management system for dynamic demand response in smart buildings," *IEEE Syst. J.*, vol. 12, no. 2, pp. 1329–1340, 2018. http://dx.doi.org/10.1049/iet-stg.2019.0131

4. Solar bonus report, Q. (2018) Department of Natural Resources, Mines and Energy, Oct. 2018, policy Guide. URL https://www.dnrm.qld.gov.au/__data/assets/pdf_file/0005/1379678/solar-bonus-scheme.pdf.
5. Liu N, Yu X, Wang C, Li C, Ma L, Lei J. Energy-sharing model with price-based demand response for microgrids of peer-to-peer prosumers. *IEEE Trans. Power Syst.,* 2017;32(5):3569–83. http://dx.doi.org/10.1109/TPWRS.2017.2649558
6. B. H. Rao, S. L. Arun, and M. P. Selvan, "Framework of locality electricity trading system for profitable peer-to-peer power transaction in locality electricity market," *IET Smart Grid,* vol. 3, no. 3, pp. 318–330, 2020. https://doi.org/10.1049/iet-stg.2019.0131.
7. P. Palensky and D. Dietrich, "Demand side management: Demand response, intelligent energy systems, and smart loads," *IEEE Trans. Ind. Informat.*, vol. 7, no. 3, pp. 381–388, Aug. 2011. DOI: 10.1109/TII.2011.2158841
8. Li J, Zhang C, Xu Z, Wang J, Zhao J, Zhang YA. Distributed transactive energy trading framework in distribution networks. *IEEE Trans. Power Syst.,* 2018;33(6):7215–27. http://dx.doi.org/10.1109/TPWRS.2018.2854649.
9. T. AlSkaif, J. L. Crespo-Vazquez, M. Sekuloski, G. van Leeuwen and J. P. S. Catalão, "Blockchain-Based Fully Peer-to-Peer Energy Trading Strategies for Residential Energy Systems," in *IEEE Trans. Ind. Inf.,* vol. 18, no. 1, pp. 231-241, Jan. 2022, doi: 10.1109/TII.2021.3077008.
10. Amin W, Huang Q, Umer K, Zhang Z, Afzal M, Khan AA, Ahmed SA. A motivational game-theoretic approach for peer-to-peer energy trading in islanded and grid-connected microgrid. *Int. J. Electr. Power Energy Syst.,* 2020; 123: 106307. http://dx.doi.org/10.1016/j.ijepes.2020.106307
11. Arun SL, Selvan MP. Dynamic demand response in smart buildings using an intelligent residential load management system. *IET Gener. Transm. Distrib.* 2017;11(17):4348–57. http://dx.doi.org/10.1049/iet-gtd.2016.1789.
12. M. M. R. Singaravel and S. A. Daniel, "Sizing of hybrid PMSG-PV system for battery charging of electric vehicles," *Front. Energy,* vol. 9, no. 1, pp. 68–74, Mar. 2015. https://doi.org/10.1007/s11708-015-0349-7
13. Hanumantha Rao Bokkisam, Ritesh Mohan Acharya, Selvan M. P. Framework of transactive energy market pool for community energy trading and demand response management using an auction-theoretic approach *Int. J. Electr. Power Energy Syst.,* Volume 137, May 2022, 107719 https://doi.org/10.1016/j.ijepes.2021.107719.

21

Recommending Medical Specialist and Detecting Point of Discomfort Using Computer Vision and Machine Learning

Shourya Gupta, Ritik Vashist, Ridhika Sahni, Kshitij Dwivedi, Sam Methuselah Penumala and Karmel Arockiasamy*

School of Computer Science and Engineering, Vellore Institute of Technology, Chennai, India

Abstract

In the last few years, the amount of information available on the internet related to healthcare has increased tremendously. Nowadays, with the advent of technology and ease of access to the Internet, people instantly Google if anything goes down, so is the same with the Health Care issues. As there are two sides to every coin, the abundance of information on the Internet has led to indecisiveness and people going around in circles. This has resulted in the need for an effective and reliable resource that can serve users' needs in the Health Care field. So, this project tends to handle all these issues with the reliability of the information at first. The objective of this project is to recommend the best- suited medical practitioner for the user. Detection of point of discomfort or pain is implemented with the help of Computer Vision. By using the Machine Learning technology, the system will ask a certain, customized set of questions to the user to match up with the probable medical issue. Thereby, the system will recommend the corresponding best-suited medical specialist to the user. This will help the user to save ample amount of money invested in many doctors before being referred to the right doctor. Furthermore, the website will provide report to both, the patient and the doctor beforehand which would help in better optimizing the consultation time. It will reduce shyness in patients, and will lessen the communication gap.

***Keywords*:** Health care, disease, medical practitioner, prediction, computer vision, machine learning, recommendation

**Corresponding author*: karmel.a@vit.ac.in

O.V. Gnana Swathika, K. Karthikeyan and Sanjeevikumar Padmanaban (eds.) Smart Grids as Cyber Physical Systems Volume 1: Artificial Intelligence, Cybersecurity, and Clean Energy for Next Generation Smart Grids, (337–348) © 2024 Scrivener Publishing LLC

21.1 Introduction

The Healthcare industry plays a vital role in our lives. With the advent of new tools and technologies like computer vision, machine learning, and data science, we need to modernize the way we look up at healthcare services. The majority of people today are not aware of exactly what kind of illness, pain, or discomfort they are suffering with. Also, the language barrier in India plays a major role in such a scenario. Even though there are numerous websites available on the Internet that concentrate on giving medical information to the users, the major drawback is that they are available in English only, thus are unable to be of any help to the people who do not even have basic knowledge of the medical domain. The UNESCO Global Education Monitoring report states that 35% of the world's illiterate population lives in India. [Arakkal, 2017 [1]]. So even if these websites have a simple and clear user interface, they lack in helping people having less knowledge or who are unaware of medical terms.

Currently, in this Covid scenario, people barely stepped out of their homes unless it was truly necessary. A wider gap has been created between doctors and patients in need. Though it has been quite shortened by online live consulting applications like Practo. On the contrary, most people are not aware of contacting the right medical practitioner for consultation for the pain or medical issue they are suffering from, which eventually leads to queuing up for general practitioners or physicians. Thus, increasing the time for the right treatment the patient requires in critical situations. Also, there are several negative effects of wait time on patients [Reichert, A. & Jacobs, R. 2018].

As we go to the hospital, we see a chain of doctors seeing a patient, which in turn, delays the process a lot. It also drains out the time and money out of our pockets involved in the hierarchy of medical practitioners. This chain has to be broken down to remove all the intermediaries.

Reliable information has a tough price to pay. Especially in the Healthcare sector, false/incomplete information can be fatal. Considering an example, a student has a lot of mouth ulcers, once he sets on his/her search on Google, she/he will find long articles about it on the Internet conflicting with each other at an instant while in agreement with the other. This situation lands a student into a fix of what needs to be done to tackle it, and there is a chance of obtaining a wrong, uniformed decision. So, our website tends to provide a solution to all of the above-mentioned problems and provides much more in one place. It provides a fast and free-of-cost environment. The rest of the chapter is structured as follows: section 21.2: Related Work, where the system used currently and the present situation will be discussed. Section

21.3: comparison table, explaining how different research papers solve the healthcare problems. Sections 21.5 and 21.6: conclusion and future work.

21.2 Related Work

Many research papers have been proposed in the field of healthcare. Although the literature covers a wide variety of fields, this review will focus on disease prediction and doctor recommendations. We have studied and analyzed a total of 13 research papers related to our project.

Reducing waiting times in mental health services has received more attention (Reichert *et al.*, 2018 [2]), as there is growing concern that prolonged treatment wait times result in negative health outcomes. The study (Grampurohit *et al.*, 2020, [3]) presents a comparative study of different machine learning models, decision tree, Random Forest classifier and Naive Bayes disease control for disease prediction. Through performance measurements, this research (Ferjani *et al.*, 2020 [4]) tries to identify patterns across various types of supervised machine learning (ML) models in disease diagnosis, support vector machine (SVM), random forest (RF), and linear regression (LR) were the most often utilized methods. For renal problems, SVM proved to be the most accurate. For cardiac disorders, LR was the most reliable. Because of its capacity to function well with huge datasets, RF produced the greatest results in the proper categorization of breast cancer. The CNN (Convolutional Neural Networks) model, on the other hand, proved to be the most reliable for predicting common disorders.

In the paper (Keniya *et al.*, 2020 [5]), the K-nearest neighbour (KNN) algorithm is studied to diagnose the disease early to ensure proper and appropriate treatment. The results predicted that the KNN model provided 93.5% accuracy. The paper (Taber *et al.*, 2014 [6]) aims to represent a detailed explanation and conceptual analysis of why people avoid medical care. The paper (Hamsagayathri *et al.*, 2021 [7]) presents a study of various ML algorithms for the diagnosis of various diseases such as heart disease and diabetes. SVM shows 94.60% accuracy of heart disease. The Naive Bayes provide 95% accuracy of diabetes diagnosis.

The paper (Chen *et al.*, 2011 [8]) presents the development of a decision support system to aid a doctor in medical prescription suggestions. The compute medication by many criteria decision-making approach is used to create an ontology of diabetes knowledge multi-criteria decision-making (MCDM). The primary experiment involves medical testing on diabetes patients to determine suggested medication, and the procedure has proven to be effective. The approach used in the research (Ju *et al.*, 2021 [9]) demonstrates greater recommendation accuracy and practicality in terms of department and doctor recommendation efficacy. The suggested online pre-diagnosis doctor recommendation algorithm combines ontology features and illness text mining. The model provides more accurate suggestion recommendations and enhances the patient experience. It demonstrates that, when picking an appropriate hospital department, a word vector- based approach that integrates ontology characteristics and illness text data mining performed well.

The study (Waqar *et al.*, 2019 [10]) demonstrates a hybrid doctor-recommender system that combines different recommendation methodologies for doctor recommendation: content basis, collaborative, and demographic filtering. Using an analytical hierarchical method, the program effectively transforms the specific features associated with doctors and their relative relevance into a nominal empirical model Antepartum haemorrhage (APH). This model is then utilized to provide doctor recommendations for patients. Examining the present state of the art in health recommender systems (Croon *et al.*, 2021 [11]). The products that were recommended were divided into four key categories: lifestyle, diet, general healthcare information, and particular health issues. Hybrid recommendation algorithms are used by the vast majority of health recommender systems.

The paper (Huang *et al.*, 2012 [12]) proposes a doctor's recommendation algorithm based on the doctor's performance model and the patient's preferences model. Development of a decision-support system to aid doctors in making better informed drug-selection decisions (Chen *et al.*, 2011 [13]). The values after normalization of the risk and benefit factors, respectively, can be found by sorting the range of values of the factors provided by the endocrine expert. This outcome can help doctors create prescriptions for patients, which has been shown effective by a diabetes doctors.

The research (Christensen *et al.*, 2018 [14]) focuses on combining natural language word embedding and network modelling techniques to study sound presentation of medical concepts using a weighted network matrix in the GloVe algorithm and demonstrating that embeddings improve neural network predictive network performance. To improve the level of compatibility between doctors and patients, researchers are looking at how to tailor and real-time recommend doctors for patients on a web-based appointment platform (Wen *et al.*, 2019 [15]). The suggested solution is shown to be sufficiently resilient to address the real-world issue because Adjust-EIB outperforms all other benchmark algorithms regardless of the E[X] condition.

Additionally, compared to SORT1 and SORT2-related algorithms, Adjust-EIB- and SORT-related algorithms perform generally consistently across a range of X conditions, demonstrating the robustness of the static method, SORT. The results show that the Adjust-EIB method consistently outperforms competing algorithms, especially when resources are scarce, indicating that it has the potential to significantly improve system performance.

21.3 Proposed System

The proposed system aims to deliver the only one of its kind website which identifies the user's point of care by detecting the affected body part, through computer vision, and thereon predicting the most probable disease along with the most appropriate medical practitioner to be consulted for treatment, with the help of decision tree classifier algorithm based on user selection of symptoms. As the majority of people are not aware of exactly what kind of illness, pain, or discomfort they are suffering from, we tend to provide them with a better perspective about it. Our solution works on identifying the pinpoint location where our user is experiencing pain or discomfort as they point their finger towards it, using computer vision. Further, the website asks a couple of questions customized to the affected area pointed by the user, to get a better understanding of the cause of the pain. Then, by using machine learning algorithm, *i.e.*, decision tree classifier, possible medical issue(s) the user might be suffering from are predicted. The system also predicts the best suited medical specialist for the disease diagnosed by the symptoms incurred. This is done by providing a document/report to the doctor/medical specialist (if selected) by the user and to the user as well, which contains the basic details, symptoms, disease predicted, and the other optional message, our user wants to convey to the doctor beforehand.

21.3.1 Computer Vision

The multidisciplinary field of science known as computer vision focuses on how computers can better comprehend digital photos or videos. It aims to comprehend and carry out tasks that the human visual system is capable of from an engineering perspective. The module as seen in Figure 21.1 is the identification of the pointed body part which is achieved with the help of computer vision. This is implemented by using OpenCV library, which uses MediaPipe API extensively. This API allows tracking of 33 distinct

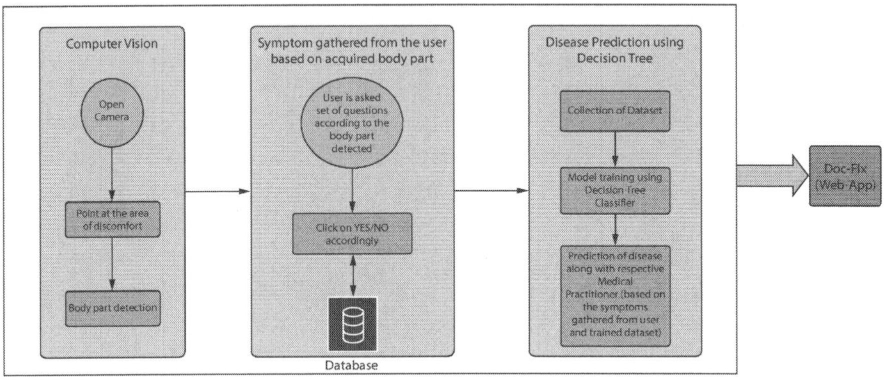

Figure 21.1 System design.

body parts on the human body in real-time, endowing the machine learning technology to cover a wide range of health problems.

21.3.2 Machine Learning

Machine learning is a study of computer algorithms that can be developed automatically through information and data usage. It is considered part of artificialintelligence.

21.3.2.1 Disease Prediction

For the Machine learning model training, the dataset used contains 163 diseases and 453 symptoms.

As shown in Figure 21.2, after gathering the symptoms from the user, the symptoms are matched with the trained model to get an accurate disease prediction. The dataset used gets trained using decision tree classifier that gave an accuracy of 90.54%.

21.3.2.2 Medical Practitioner Recommendation

Further, for the recommendation of medical practitioner as shown in Figure 21.2, each disease is mapped to the corresponding medical specialist in the classifier as per the previous trends. Thus, solving the problem of 'which doctor to consult?'. In total, there are 26 different medical practitioners on the website. Now, to further compress the information gained, the data is clustered with respect to different body parts.

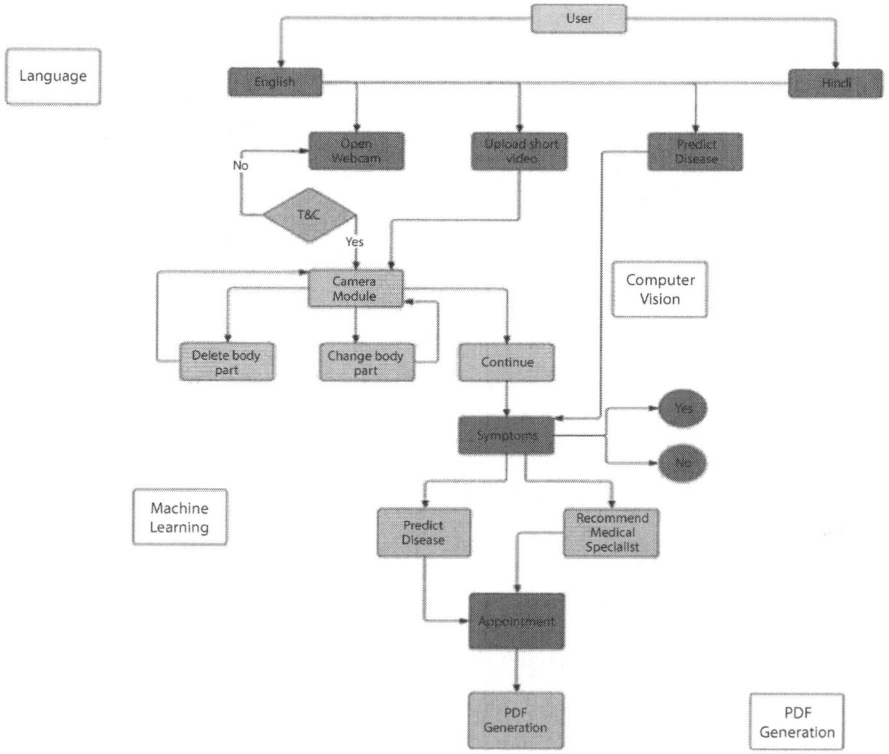

Figure 21.2 Implementation flow.

21.3.2.3 Report Generation

The website also has the feature of generating a report after filling out the form to make an appointment with the doctor. The generated report has the following content,

A. Name
B. Email
C. Contact No.
D. Date
E. Department
F. Doctor
G. Message
H. Symptoms

This will help the user as well as the doctor as it will give prior information regarding the disease, reducing the consultancy time, and hence, optimizing the appointments.

21.3.2.4 Language

India is a diversified country and a home of various cultures and languages. The website is bilingual English and Hindi, to serve better for Indian people. Based on several market studies as well as groundwork, the website is much needed in Hindi language.

21.4 Results

For our machine learning model training, we used a dataset that contains 163 diseases and 453 symptoms. Further, we mapped each disease to the corresponding medical specialist as per the previous trends. In total, we took 26 different medical practitioners for our website. Now, to further compress the information gained, we clustered our data in respect to different body parts. We considered 13 body parts for our work, and identification of body parts is done using OpenCV, which uses MediaPipe API extensively. In respect to our machine learning model, we trained the decision tree classifier on our dataset that gives us 90.54% accuracy. On the basis of several market studies as well as ground work, we made our website bilingual—English and Hindi.

21.5 Conclusion

The research work revolves around disease prediction and medical practitioner recommendation for the user in the medical field. It aims to deliver a one of its kind website which identifies the user's point of care by detecting the affected body part through computer vision, and thereon, predicting the most probable disease with the appropriate medical practitioner to be consulted for the treatment with the help of the decision tree classifier algorithm, based on user selection of symptoms. The dataset taken into account involved 163 diseases, 453 symptoms, 26 medical departments, and 13 body parts for prediction, which can be further populated after taking user input. The research work focused on eliminating the initial chain of doctors and to reduce the waiting time for patients for consultation by general physicians/doctors and at hospital outpatient department (OPD) by recommending the right medical practitioner in order to reduce financial burden and time wastage. It tends to bridge the communication gap between doctors and patients. The website is bilingual (English and Hindi), and in addition to predictions and recommendations, it generates a report to further optimize the appointment time. The main focus of our solution is for people with less medical knowledge, medical organizations who wish to eliminate corruptions in between the chain of doctors, and doctors who wish to optimize their consultation period.

21.6 Future Scope

a. Mobile application development can reach a larger audience and can assist a larger group of individuals.
b. The website can be converted into a software personalized for a particular medical institution so that doctors can benefit to better optimize their appointment time, and subsequently, cover more number of patients in a given time frame.
c. The hardware or an liquid-crystal display (LCD) display near the OPDs can be attached. Consider an example, in government institutions such as All India Institute of Medical Sciences (AIIMS), there are a huge number of patients from rural areas; the hardware installation can assist these individuals in better understanding their condition and selecting the appropriate medical practitioner at the same location of their medical visit.
d. Vaccination details or information regarding medical camps can be displayed on the website, allowing the user to call an allied hospital if necessary.
e. Voice-assisted instructions for disease detection and doctor can be incorporated, allowing illiterate people to benefit more, as most of them have difficulty in reading languages.
f. It can serve people all over the world by using natural language processing and offering the website in the majority of the languages spoken.
g. The feature of sorting can be implemented based on patient ratings and comments to display the best and most suited doctor based on their preferences.
h. Multiple family members' information can be entered to expand the database and provide more data points.
i. A reminder for future hospital visits can be added to notify the patient well in advance.

References

1. Arakkal, I. 2017. India has 35 percent of the world's illiterate, says UNESCO. Deccan Chronicle. https://www.deccanchronicle.com/nation/current-affairs/061217/india-has-35-per-cent-of-worlds-illiterate-says- unesco.html
2. Reichert, A. & Jacobs, R. 2018. The impact of waiting time on patient outcomes: Evidence from early intervention in psychosis services in England, *Health Econ.*, Volume 27, Issue 11.
3. Grampurohit, S. & Sagarnal, C. 2020 Disease Prediction using Machine Learning Algorithms, *International Conference for Emerging Technology (INCET)* Belgaum, India, https://doi.org/10.1109/INCET49848.2020.9154130
4. Ferjani, M. 16 December 2020 Disease Prediction Using Machine Learning, Preprints and early-stage research may not have been peer-reviewed yet. http://dx.doi.org/10.13140/RG.2.2.18279.47521
5. Keniya, R. & Khakharia, A. *et al.* July 2020 Disease prediction from various symptoms using machine learning, Preprints and early-stage research may not have been peer-reviewed yet. https://dx.doi.org/10.2139/ssrn.3661426
6. Taber, J. M. & Leyva, B. & Persoskie, A. Nov 12, 2014, Why do People Avoid Medical Care? A Qualitative Study Using National Data, *J. Gen. Intern. Med.*, https://dx.doi.org/10.1007%2Fs11606-014- 3089-1
7. Hamsagayathri, P. & Vigneshwaran, S. 2021 Symptoms Based Disease Prediction Using Machine Learning Techniques, *2021 Third International Conference on Intelligent Communication Technologies and Virtual Mobile Networks (ICICV)*, https://doi.org/10.1109/ICICV50876.2021.9388603
8. Chen, R. C. & Chiu, J. V. & Bau, C. T. 10-13 July, 2011 The recommendation of medicines based on multiple criteria decision making and domain ontology- an example of anti-medicines diabetic, *International Conference on Machine Learning and Cybernetics,* Guilin, https://doi.org/10.1109/ICMLC.2011.6016682
9. Ju, C. & Zhang, S. 2021, Doctor Recommendation Model for Pre- Diagnosis Online in China: Integrating Ontology Characteristics and Disease Text Mining, *IEEE the 6th International Conference on Big Data Analytics,* https://doi.org/10.1109/ICBDA51983.2021.9402991
10. Waqar, M. & Majeed, N. & Dawood, H. *et al.* 25 May 2019, An adaptive doctor- recommender system, *Behav. Inf. Technol.*, 38(4):1-15, http://dx.doi.org/10.1080/0144929X.2019.1625441
11. Croon, R. D. & Houdt, L. V. & Htun, N. N. *et al.* 29.6.2021, Health Recommender Systems: Systematic Review, *J. Med. Internet Res.*, ISSN 1438-8871, https://doi.org/10.2196/18035
12. Feng, Y. & Liu, P. & Pan, Q. *et al.* 2012 A Doctor Recommendation algorithm based on doctor performances and patient preferences, *International Conference on Wavelet Active Media Technology and Information Processing (ICWAMTIP),* https://doi.org/10.1109/ICWAMTIP.2012.6413447

13. Ching, R. & Chiu, J. Y. & Batj, C. T. *et al.* 2011 The recommendation of medicines based on multiple criteria decision making and domain ontology- An example of anti-diabetic medicines, *International Conference on Machine Learning and Cybernetics,* https://doi.org/10.1109/ICMLC.2011.6016682
14. Christensen, T. & Frandsen, A & Glazier, S *et al.* 2018 Machine Learning Methods for Disease Prediction with Claims Data, *IEEE International Conference on Healthcare Informatics (ICHI),* https://doi.org/10.1109/ICHI.2018.00108
15. Wen, H. & Song, J. & Pan, X. *et al.* 2019 Physician Recommendation on Healthcare Appointment Platforms Considering Patient Choice, *IEEE Trans. Autom. Sci. Eng.*, (Volume: 17, Issue: 2, April 2020, https://doi.org/10.1109/TASE.2019.2950724

22

Reliability Assessment and Reliability Improvement of System by High Renewable Penetration

Satyaki Biswas, Sadasiva Behera and Nalin. B. Dev Choudhury*

Dept. of Electrical, NIT Silchar Assam, India

Abstract

The power/electricity generation sector is shifting toward renewable energy sources over conventional fossil fuels all over the world, due to climate change and the energy resource crisis, and reliability is a very crucial consideration in the generation expansion of a power system, especially with high penetration of renewable energy because of their intermittent nature. This chapter presents the reliability assessment and reliability improvement with solar and wind energy penetration in the system. Some standard reliability indices like loss of load probability (LOLP), loss of load expectations (LOLE), and expected energy not supplied (EENS) are used as parameters to estimate reliability. The objectives of the proposed method are categorized into two different sections, first, we estimate the reliability of the system with and without renewable energy sources (RES) integration and compared it. Secondly, the reliability of the system is estimated without an RES integration and with solar and wind integration, and individually and concertedly, are compared. Markov chain Monte Carlo simulation is used to calculate the availability and unavailability of the energy sources in the system. An index was used to calculate the credible capacity of wind, which gives an improved outlook of the fluctuation due to the impact of renewables on the reliability of the overall system.

Keywords: Renewable energy penetration, reliability estimation, Markov chain Monte Carlo

**Corresponding author*: nalin@ee.nits.ac.in

O.V. Gnana Swathika, K. Karthikeyan and Sanjeevikumar Padmanaban (eds.) Smart Grids as Cyber Physical Systems Volume 1: Artificial Intelligence, Cybersecurity, and Clean Energy for Next Generation Smart Grids, (349–368) © 2024 Scrivener Publishing LLC

22.1 Introduction

Electricity is one of the most inalienable parts of our day-to-day life. The usage of electric energy commercially started with light bulbs, and since then, has evolved rapidly, which has shaped our current society. This advancement in technology, as well as progressiveness among humans over the past decades regarding environmental awareness and finding a balance between industrialization and nature, has brought us to a landmark crossroads in our lives. The unsustainable ways of conventional energy production along with its impending ecological impacts are propelling us towards renewable energy sources (RES) [1, 2]. The integration of RES with our current power infrastructure is to be dealt with thoughtfully to fulfill our energy demands as well as taking into account customer satisfaction in addition to planning future expansion strategies of the system [3]. The context of power adequacy and security is becoming a significant subject matter in today's world owing to the limited resources available to us. Such a stream of study concerned with adequacy and security falls under reliability engineering. For hundreds of years, the existing generation, transmission, and retailing systems have been servicing the needed demand. But since the customer loads vary at every instance oftime, it is of paramount importance to keep up with these ever-changing demands in order to maintain the power balance between generation and load plus losses. A significant mismatch between the two could lead to blackouts, contingencies, or huge losses [4]. Therefore, reliability studies in the power sector provide us an insight into many subjects such as component failures, generation capacity evaluation, and security measures, to name a few [5].

There are numerous techniques for assessing adequacy in the system. While the inclusion of the fluctuating RES is often entangled with voltage fluctuations and other side effects in the system, this study investigates the benefits of the inclusion of RES from the perspective of generation reliability. Starting from looking into the traditional analytical techniques used to evaluate the system along with the newer approach to the evaluation of indices proposed by various authors has been studied. Further, a thorough analysis has been done to study the effects on the reliability of the system under various cases such as increment of peak load, different levels of renewable penetration, and in a different configuration using the conventional IEEE indices for comparative studies. The established indicators for reliability set by IEEE examine the system at basic levels, thus setting a benchmark. But with the advancement in statistical tools and computer simulations, many newly developed techniques are able to adapt to the fluctuating nature of renewables. One such method proposed in the literature is based on covariance between the wind power outputs and its weights, giving a better estimation of the impact of the intermittent nature of renewables on the system's reliability. This has been explored in this chapter alongside the traditional indices for comparisons. Although the erratic nature of these supplies is undesirable, the effects of this intermittency have to be reflected in the measured reliability indices ofthe system to have a practical inference on the reliability of the system. This chapter is arranged in the following sections in a coherent manner to make a thorough analysis. Section 22.1 deals with traditional reliability indices and the mathematical formulations behind them, section 22.2 is about the particular methodology used to derive the indices, followed by section 22.3, which talks about the new indices proposed by the authors Chao Yan *et al.*, which allow for the inclusion of the fluctuating nature of the renewables by the use of modified index system. Finally, the results are put forth by evaluating the above scenarios on the RTS-79 system with wind and solar generation added.

In 1979, a report was published by the IEEE task force of application of probability methods subcommittee, which introduced a network model to test the reliability of power systems. This system is comprised of 32 generation units, 17 loads, and 24 nodes. This model is famously known as IEEE RTS-79, which is usually used to estimate the reliability of a system [6]. Roy Billington and RN Allan [7] published an article to discuss the philosophical aspects and basic concepts of power system reliability. The authors also discussed the conformity among different parts of the system and the balance between generation, transmission, and distribution. Zhaohong Bie *et al.* [8] introduced a non-sequential hybrid Monte Carlo simulation to evaluate the reliability of active distribution, considering various operation modes and different conditions including single and multiple contingencies. S.Suleman *et al.* [9] introduced an analytical method to model the outputs and independent outages of a large wind farm to estimate the reliability of the system, considering the variability of wind speed, random force outages of the turbine, and correlation between turbine output. This method is validated in the IEEE reliability test system by using Monte Carlo simulation. Ayodele *et al.* [10] introduced a novel approach to techno-economic analysis and as well as to calculate generation reliability of a hybrid-mini-grid system and validated the obtained result by HOMER software. Capacity outage probability table (COPT) was utilized to validate the reliability and to calculate loss of load probability (LOLP) and loss of load expectation (LOLE). Hamed *et al.* [11] proposed a method to minimize the calculation time of smart grid reliability assessment, by reducing the developing load state reduction strategies.

To validate the proposed method, it was applied to IEEE 69 bus system. And indices like LOLP, LOLE, and expected energy not supplied (EENS) were calculated to measure the reliability. T. Adefarati and R.C. Bansal [12] proposed a method to develop a more economical microgrid system with better reliability. This microgrid is integrated with diesel generators, energy storage systems, and renewable energy sources (RERs) like wind turbine generators and photovoltaics. Indices like LOLP, LOLE, and EENS are used to assess the reliability of the system. Abedalsalam *et al.* [13] introduced a decentralized microgrid control architecture which is ensuring higher reliability over the centralized control architecture. To validate this comparison between a single controller and a decentralized controller, the Markov Chain model was used. Xiyung Yang *et al.* [14] introduced a wind power probabilistic interval prediction model and a novel reliability assessment approach. Egill Tommason and Lennart söder [15] proposed a method to estimate the capacity credit of resources in a multi-area and to predict the strategically optimal distribution of "reserves" among areas of the "multi-area power system". The authors proved that this method is economical by using Monte Carlo simulation and sensitivity analysis. The conventional reliability indices like LOLP, LOLE and loss of energy expectation (LOEE) are suitable for conventional generators but usually fail to detect the stochastic nature of RES-based distributed energy resources (DERs). Priyanka Paliwal *et al.* [16] proposed new probabilistic reliability indices to estimate the impact of penetration of DERs on system reliability. Atef Hasanpour and Emad Roghanian [17] proposed a two-stage model scheduling the maintenance of generation units of the power system. The game theory is used to solve the non-cooperative generation companies' maintenance scheduling problem when the generations are stochastic in nature and the reliability is measured by indices like LOLP, LOLE, and EENS to predict optimal maintenance schedule to satisfy the generation company (GENCOs) and independent system operator (ISO). Sara Ostovar *et al.* [18] developed a two-state Markov model for reliability assessment of a hybrid power system integrated with photovoltaic and energy storage system. Failure mode and effect analysis is used to consider the failure of hybrid system components. Radhanon Diewvilai and Kulyos Audomvongseree [19] developed a method for optimal generation expansion planning integrated with energy storage system, renewable energy sources, and annual hourly power balance constraints. The authors introduced a less complex method to break down and solve a multi-period mixed-integer linear program (MILP) generation expansion planning problem with realistic hurdles. Reliability indices like LOLE were calculated to estimate the impacts of integrating ESS and RES on system reliability. Though this proposed method has some limitations, like the transmission sector is neglected, and even constraints like available fuel options, different renewable energy profiles, and regional load profiles, are not considered. Jiayi Zeng *et al.* [20] proposed a multi-state reliability assessment model for concentrated solar plant (CSP). The author simplified the eight-state model to a four-state Markov model by combining the states which show similar partial function failure. To evaluate the fruitfulness of this proposed method, this method was tested on the modified IEEE RTS system. Abir Muhtadi *et al.* [21] discussed the importance of flywheel energy storage system (FESS), which is the potential to increase the system reliability of a DER-based microgrid. The authors also conducted a comparative analysis of the impact on the DER-based microgrid reliability, while integrated with battery energy storage systems (BESS) and while integrated with FESS, considering a various range of rotational speed and corresponding losses; which are demonstrated on the IEEE RTS-79 system. Reliability is a very important parameter for a power system, especially in the time of expansion of the generation sector and transmission sector. Usually the reliability indices like LOLP and LOLE determine the adequacy of the bulk power system in near future. Ying-Yi Hong *et al.* [22] proposed a novel scenario-based reliability estimation technique that is especially beneficial for large complex power systems with high penetration of uniformly allocated renewable energy sources. The proposed method is less computational time consuming because this method considers the scenarios over the chronological cases in Monte Carlo simulation. However, this is proved in this paper that a high penetration of renewables and energy storage systems can reduce the LOLP and LOLE; which indicates the increment of the system reliability. In the power system planning sector, capacity credit is a very frequently used and recognized index to quantify wind energy reliability. The standard characterization of this term has been called by Chao yan *et al.* [23] as unrepresentative and insufficient to properly address the effect on renewables on the reliability of system, and thus, a new index system is introduced, which can be easily defined/constructed from the original IEEE definitions to quantify the effect on system reliability due to renewable energy fluctuations. To check the effectiveness of this index system, various renewable energy sources are integrated into a modified IEEE RTS 79 model To address the research gap, our paper aims to take into consideration the "fluctuating behavior" of the renewable energy system, while defining the entire system reliability. Techniques like Markov chain Monte Carlo (MCMC) can be used to describe overall system reliability, also other probabilistic distributions such as lognormal can be explored while modeling renewable resources. The current literature provides a plethora of techniques that can be utilized for carrying out reliability studies. But a comprehensive study and comparison entailing the standard indices as well as the recently proposed modifications are yet to be logged. Many statistical approaches ranging from sequential, time-varying and discrete Bayesian tools exist which can be executed in a different manner to approach the overall system reliability.

22.2 Methodology

22.2.1 Standard IEEE-Reliability Indices

The term "adequacy" implies the sufficiency of the generation system to satisfy the demand at all points of time disregarding any other effects. It only zeros in supply-demand demand balance whereas the voltage limits and stability would fall under the security analysis realm. The conventional way to evaluate the generation side reliability in the volves calculation of basic indices such as LOLP, LOLE, and EENS.

The generator adequacy indices are concerned with the total installed generation being able to meet the total load. Usually. this is carried out by the evaluation of a COPT, which is a base for generation model. The load can be designated using hourly, weekly, daily or annual peak load curves or else using load duration curve. The basic indices (*e.g.*, LOLP, LOLE) communicate the probability of loss of load, duration or load loss or else its frequency and not the amount of lost load, for this severity indices are used *e.g.*, EENS.

22.2.2 Capacity Outage Probability Table (COPT)

Let there be "N" number of generators available, all with the same outage rate characteristics and failure time characteristics. Then the probability of "r" units failing out of the total "N" generators, with unavailability of unit being "U" and availability "A" is nothing but a Bernoulli variable, defined as follows:

$$\mathrm{P}\,(r) = C^{n} U^{r} A^{n-r} \tag{22.1}$$

This procedure gives a straightforward evaluation of probabilities of capacity outage states in a step- wise fashion starting from the total capacity of the whole generation system. This forms the base of calculation of COPT table but can be mathematically intensive when the number of units increase drastically leading to large number of total combinations. Thus, a way around this cumbersome process is a recursive method.

The recursive algorithm is used for a multi-state generation units' calculation which is formulated as follows:

Given the forced outage rate to be "*U*," the cumulative probability of a particular state of capacity outage of XMW for an added unit of capacity C MW and

$$P(X) = (1 - U)P'(X) + (U)P'(X - C) \tag{22.2}$$

Where P'(X) = cumulative probability of the capacity outage state of X MW before the unit is added and P(X) = cumulative probability of the capacity outage state of X MW after the unit is added.

This table allows one to find out the traditional IEEE indices the analytical way, the main indices which are under consideration are LOLP, LOLE, and EENS.

22.3 Modified IEEE Indices

These traditional indices give the power system engineer the benchmark for operation and planning decisions. But over the years, as access to data and computer technology has increased, and also the power grid is transforming into a greener technology, the integration of renewables is imperative to the environment. But this non-conventional source of energy comes at the cost of stochastic behavior and hence, a continuity is supply by renewables alone is impossible.

On the other hand, its integration to the grid is the need of the hour, thus, many authors have suggested specific techniques which account for such fluctuating nature of renewables. For example, the widely used renewable resources such as wind and solar are indexed using wind credible capacity and photovoltaic (PV) system credible capacity, respectively. Most techniques deal with this problem using probabilistic techniques to evaluate the capacity credit or statistical factors to evaluate the relation between power output and the weather. But the authors in Reference [1] took a different route by instead using the covariance between wind power outputs and the load, *i.e.*, their weights to define a novel way to calculate credible capacity of wind as well as formulated new indices from the standard ones which give an improved outlook of the impact of renewables on the reliability on the overall system.

Let "*T*" be the total time period for estimation, then for *time* $t \in T$ and *generation* $g \in G$, *Rt* is the reliability function, *Lt* is the load level at time, *Cg* is generation capacity and *Pwind, t* is the wind turbine output at time, *t*, and *Cc*, credible capacity, or equivalent load carrying capacity (ELCC) which considers reliability levels to be constant over the scenario, then the following relation hold:

$$\sum_{t\in T} R_t \left\{ P_{wind}, t + \sum_{g\in G} C_g, L_t + C_c \right\} = \sum_{t\in T} R_t \left\{ \sum_{g\in G} C_g, L_t \right\} \tag{22.3}$$

This can be expressed in a combined format as:

Where $E(.)$ *is the Expected value*

Let κ be the weighting parameter for various R_t functions and P_{wind} output values, then another way of representing the same is:

$$C_c = E(P^{wind}) + \frac{C_{ov}(K, P_{wind})}{E(K)} \tag{22.4}$$

$E(P^{wind})$ is expected value of wind power production, *i.e.*, mean value of the specified time segment (expected value of a discrete or even continuous random variable is always equal to its mean)

$\frac{C_{ov}(K, P_{wind})}{E(K)}$ practically indicates the relativity among wind output and load. This is the fluctuation term. Which uses covariance between probabilistic terms *κ* and *Pwi*, divided by the expected weights.

The other novel index system indices for adequacy calculations by given by authors in Reference [23] is summarized as follows:

$$\text{Fluctuation System Minute (FSM)} = \frac{EENS_{re} - EENS_0}{P_{load}} \tag{22.5}$$

Where, P_{load} is the nominal capacity of the RES added.

Where *EENS0* is the expected energy not supplied in the base case without renewable generation.

$$\text{Loss of Load Probability Improvement} = \frac{LOLP_{re} - LOLP_0}{LOLP_0} \tag{22.6}$$

It indicates by how much LOLP improves when renewable is added to the system than when it's not, expressed in parentage.

The premise of this process is to allow users to draw samples from a distribution that is not straightforward to obtain. Thus, rather than sampling the equation involving posterior, we instead sample from the pre-defined non-normalized distribution and use it to find the statistical estimates of mean, standard deviation, variance, etc. This particular algorithm emulates the "Markov Chain" in the sense with which it obtains its samples, whereas the "Monte Carlo" is due to the sampling process. The uniqueness of this state-of the art algorithm arises from its ability to solve problems which are deterministic by principle using the randomness of sampling [7, 14, 22, 24].

The context in which this project uses MCMC is to find the overall reliability in terms of generator availability (*i.e.*, operational state) or not available (*i.e.*, non-operational) at a given instance of time, or else in derated state where the generator does not run at its full capacity. As the IEEE-RTS 79' system data provides us with individual capacity levels and the individual unit availability data which we can get using crude Monte Carlo simulation, we then proceed further to find over system reliability by finding the conditional probability that the generator is operational/available given a certain level of generator capacity at a given instance of time. By taking into account the derated state, the system is thus modeled as a three-state model rather than the popular two-state model, which accounts for working and failure states only.

The generator side reliability is evaluated using the COPT table, indices such as LOLP, LOLE, and EENS can thus be derived once the tabulation is established. For a large system, the COPT table is developed using the recursive algorithm rather than binomial distribution due to its laborious calculations.

22.4 Results and Analysis

The IEEE RTS-79 is equipped with 24 buses and a total of 32 generating units with a maximum generation levelof 3405 MW, the types of generators include diesel, hydro, coal, and nuclear each unit with its own forced outage rate, defined along with mean time to repair (MTTR) and mean time to failure (MTTF) in hours, the transmission system comprises of 38 lines which are at two different voltages level: 138kV and 230kV. The comprehensive dataset for RTS-79' is available. Considering a wind power value of 150 MW and solar power of 166 MW, the turbine and solar farm failure data from References [14] and [15]. The failure data will be used to get the indices using the analytical methodfor evaluating the probability of the capacity outage method.

The annual peak load that the system realizes is 2850 MW while a minimum being 876.48 MW. The reliability test system (RTS) system is modified and integrated with non-conventional sources as well to study the reliability of the system. The generator side reliability is first computed to find out basic indices such as LOLP, LOLE, and EENS. In order to do this, the analytical method of evaluating capacity outage probabilities is used here. Stemming from the rudimentary concept of binomial distribution, the capacity outage for each level of generation starting from 0 MW to the maximum of 3405 MW increased at each step by 1 MW, we calculate its corresponding probability of outage.

Here in this case, if we apply the Bernoulli concept then

$$P\,(success) = FOR \; whereas \; p\,(failure) = no \; outage \; i.e.\; (1 - FOR) \tag{22.7}$$

This method stands for a modest two-state system whereas for a multi-state system, a recursive approach is taken up as explained previously in section 22.2.

Case 1: Indices for RTS-79 system with no renewables added

The results of generation side standard IEEE reliability indices are first and foremost analyzed for the RTS system alone as indicated in Figure 22.1 and then we move on to see the effect of each renewable source added in the system and finally with both wind and solar in the system.

Now, this table will be utilized to calculate indices for the peak load of 2850 MW in RTS, for which first the amount of reserve is required to be calculated, which is essentially the difference between maximum generation

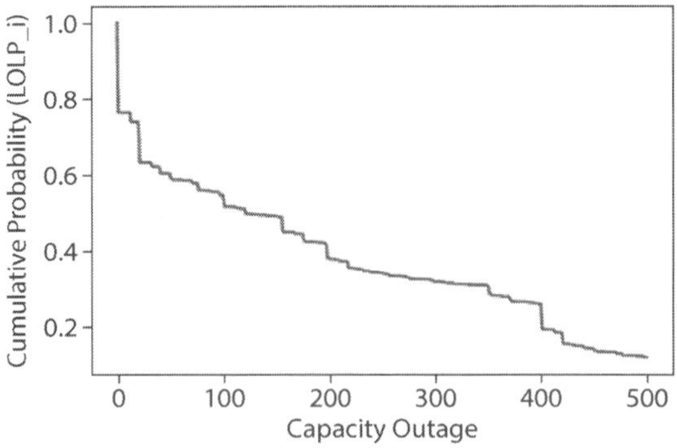

Figure 22.1 Cumulative probability *vs.* capacity for RTS-79' system.

capacity of the system and the peak load, this amounts to 3405 – 2850 = 555 *MW* of the reserve. From the tabulation above, the exact probability for when the outage capacity becomes equal to this reserve is nothing but the LOLP of the system, when multiplied by total time (in hours/year) gives the LOLE, and the exact probability at which the outage is exceeding or becomes equal to is the point which will give us EENS when multiplied with the total no. of hours.

Case 2: Indices for RTS-79 system with renewables (wind and solar) added

This process is then repeated after the integration of wind farm of 150 MW capacity and solar generator of 166-MW capacity with the RTS generator data, after which we utilize their respective COPT tables to find out generator side reliability indices. As per the data taken from Reference [39], the wind farm is made up of 50 turbines of 3 MW each, adding up to a total of 150 MW, each with an FOR of 0.04. Thus, here the wind system is added as a wind farm similar to Reference [40]. The FOR value for solar/PV systems is taken as 0.03 as seen from various references such as Reference [40] and [41], and the capacity is added is 166 MW. The results for each case are summarized in Table 22.1.

It can be observed from Table 22.2 that as generation increases, reliability increases since one can now be able to meet the demand easily, *i.e.*, there is more generation available and hence lesser loss of load. Thus, these basic indices have an inverse relation to reliability. The barchart for different configurations are as indicated in Figures 22.2–22.4.

All the indices at the average off-peak load level decrease from the base values by almost 99%, which shows that as a load to be serviced decreases, given the generation is held at the same level then, reliability increases since the loss of load indices reduce significantly.

Table 22.1 Indices for RTS-79 system at peak and off-peak load.

Load level	LOLP	LOLE (hours/year)	EENS (KWh/year)
Peak load	0.000955313	8.345613307	1283.639706
Off-peak load	3.97E–06	0.034677931	1.93608454

Table 22.2 Reliability indices for various RTS configurations.

System	LOLP	LOLE (hour/year)	EENS (kWh/year)
RTS	0.000955312	8.345613	1283.6397705
RTS + Wind	0.000439800	3.842100	552.33807465
RTS + Solar	0.000417412	3.646515	489.30816251
RTS + Wind + Solar	0.00014675	1.282075	169.273242325

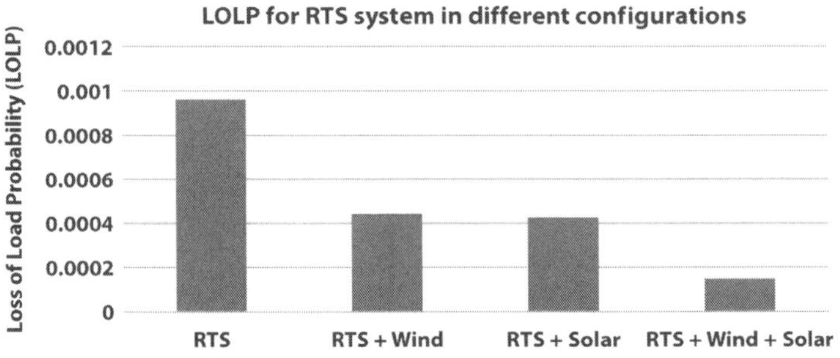

Figure 22.2 Bar charts depicting effects on LOLP indifferent RTS configurations.

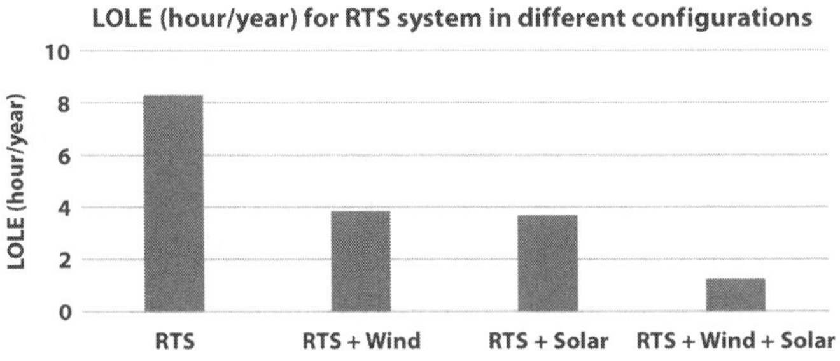

Figure 22.3 Bar charts depicting effects on LOLE (hour/year) in different RTS configurations.

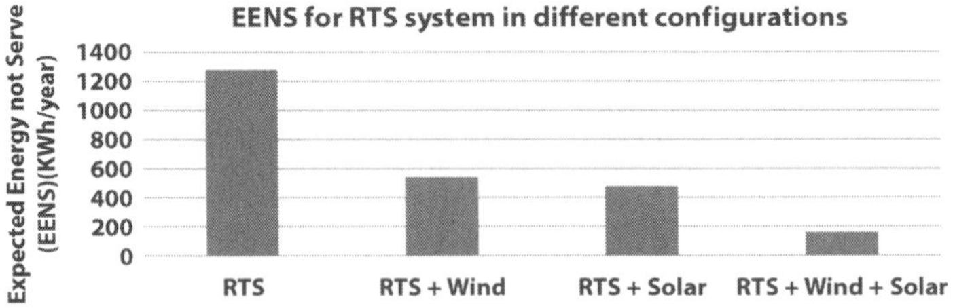

Figure 22.4 Bar charts depicting effects on EENS (KWh/year) in different RTS configurations.

Case 3: Indices for RTS-79 system at increasing levels of peak load (with solar and wind added)
With the amount of generation kept as the same, only the peak load is increased, the effects of this system are observed as shown in Table 22.3. As peak load increases, one anticipates that there are more chances of load loss since the generation is still at the base level, which is solidified by these results, thus, as these indices are increasing in value, theprobability of not meeting the demand is increasing hence the reliability decreases.

Case 4: Indices for RTS-79 system at increasing penetration levels of solar and wind generation in the system
Another criterion of analyzing the system indices is the increasing penetration of wind and solar power in the RTS system. LOLP, LOLE and EENS for this case are as indicated in Figures 22.5–22.7 respectively. The capacity penetration level of renewable sin system is given by:

$$Renewable\ Capacity\ Penetration = \frac{Installed\ renewable\ capacity\ (MW)}{Peak\ Load\ (MW)} \tag{22.8}$$

Table 22.3 Indices for RTS-79 system at increasing levels of peak load.

Peak load level in RTS with renewables (MW)	LOLP	LOLE (hours/ year)	EENS (KWh/ year)
2850	0.000146758	1.28208	169.273
2950	0.000296053	2.58632	352.135
3050	0.000509097	4.44747	724.197

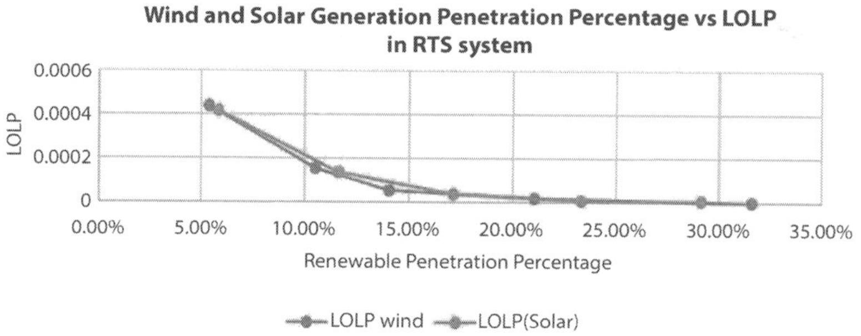

Figure 22.5 LOLP of RTS-79 system with increasing penetration levels of solar and wind generation.

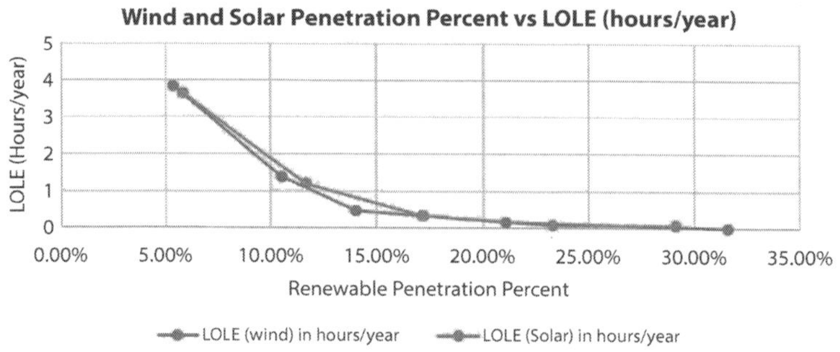

Figure 22.6 LOLE (hours/year) of RTS-79 system with increasing penetration levels of solar and wind generation.

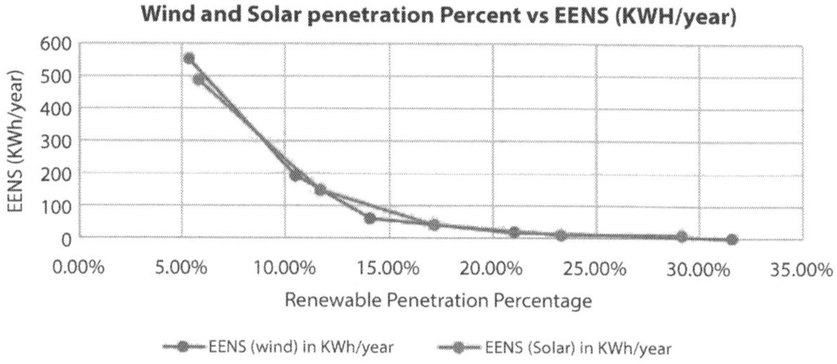

Figure 22.7 EENS (KWh/year) of RTS-79 system with increasing penetration levels of solar and wind generation.

Up until now, the system has been just looked at from the point of standard reliability indices, but as mentioned by Chao Yan *et al.* (2016) [21], the impact of renewables can only be justified by the inclusion of credible capacity, which is broadly used by researchers, especially for wind generators. The credible capacity is nothing but the essence of how much amount can the added wind generation be a substitute for the existing generation system. But this basic evaluation method is modified by the author by adding a fluctuation term in *Cc* calculation. This term signifies the ratio of covariance between weighing parameters and wind power outputs to the expectation of weights. In the case of discrete or even continuous random variables, it is mathematically known that the expectation of a random variable in statistics is nothing but the weighted arithmetic mean, given by

$$E(X) = \sum_{for\ all\ x} x \times P(x) \tag{22.9}$$

where x is the random variable and $P\ (x)$ is the individual probability.

Similarly for a continuous random variable:

$$E(X)\int_{-\infty}^{+\infty} xf(x)dx \tag{22.10}$$

Hence, weights are nothing but probabilities of their occurrence. Using this, the expectation of wind power outputis calculated as 74.99, the expectation of weights is 0.00657. The covariance among these two quantities is 0. 4572. Thus, the C_c comes out to be 143.99 MW.

Observations made from Chao Yan *et al.* as indicated in Table 22.4, index system:

The new novel indices as modeled by authors are easy to compute, but it fails to address why fluctuation still brings adverse effects when the relevance between the load and wind is not considered. The load and wind/solar or any other renewable to be considered as "beneficial" to the system can occur at a particular penetration level which can vary from system to system and place. The authors verify their result by adding two separate 300-MW and 200-MW wind energy generators, without considering

Table 22.4 Modified reliability indices for RTS-79 system with renewables proposed by Chao Yan.

Chao Yan defined index system	Value
Fluctuations system minute (FSM)	0.07647 min
Expected energy not served benefit (EENSB)	1.45305/h
Loss of load probability system improvement (LOLPI)	4.06%
Credible capacity of wind by the inclusion of "fluctuation term"	143.99 MW

how solar energy resources would affect the system and how the new indices would turn out in that case. The availability reliability of a component expresses the component's availability probability at an instance of time. This is computed from the uptime to total time period ratio. In order to define this for all generators of the modified RTS system, we use a classic MCS technique to simulate the availability (A) and unavailability (U) as shown in Figure 22.8. The idea behind MCMC is that it allows one to sample from such a distribution which would otherwise be impossible to do so due to its non-normalized characteristics, this will thus help the user to construct the likelihood distribution as shown in Figure 22.9. While we can

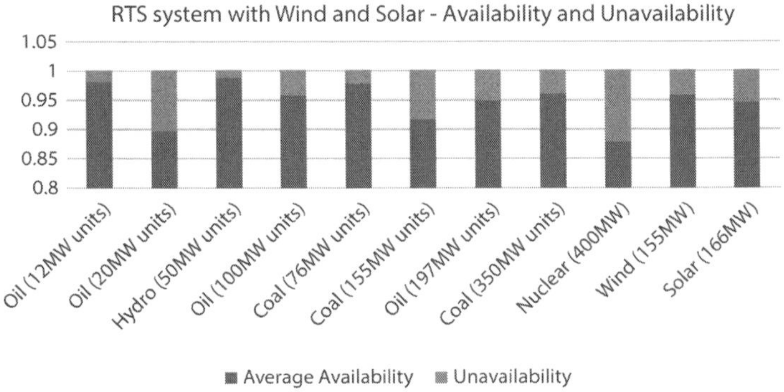

Figure 22.8 Individual availability and unavailability factors for the modified RTS- 79' system using MCS.

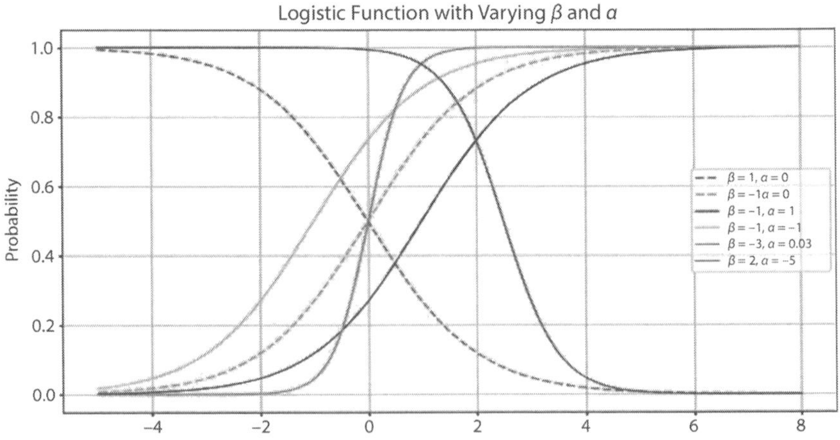

Figure 22.9 Logistic function to model the states in MCM.

predict visually that the logistic function would fit out data points to give us the end result, at this point, the parameters characterizing this function, namely α and β, are currently known. Hence, to get around this, we generate a large number of samples, in the range of thousands and more, to get an approximate likelihood distribution. This part of sampling "N" number of times to get the end result which tends to converge to a steady-state over time is the "Monte Carlo" aspect of MCMC. Practically any probability distribution function that would give you the number of parameters to be estimated, it can be a normal distribution, gamma distribution, lognormal, etc. The most commonly and widely used one is the normal distribution as it is easy to simulate. Using this as a base distribution for sampling, the user needs to give an initial guess as input to the normal distribution specifying its mean (μ) and the standard deviation (r). It is to be noted that μ can take any values whether be it positive or negative but r will always be positive.

22.5 Conclusion

The chapter presents the evaluation of generator-side reliability indices with the integration of renewables such as solar power and wind power generation. The system is first and foremost examined by standardly defined IEEE reliability indices under various configurations and conditions such as increasing peak load, increasing renewable penetration, and in off-peak hours. Then the criticism made against these traditional indices by authors Chao Yan *et al.* has been taken into account and their modifications of the original index system, such as the inclusion of covariance indicating the stochastic nature of wind power have been used to evaluate a modified RTS-79 bus system. The novel terms are stated to be more communicative of the sporadic nature of renewables in reliability analysis and hence, these indices are also used for examining the system and validations are performed by the IEEE RTS system.

Furthermore, the generator-forced outage rates are used to evaluate the availability and unavailability of the various generators using the crude MCS technique, this is then used as a base to shape a Markov chain which takes into account the derated nature of operation of generating units. While most of the literature present uses the pervasive and extensively applied MCMC technique for wind power interval techniques, demand response, load uncertainty, etc., using the technique mostly for time-varying models, this thesis applies this method to a discrete model to determine the unit's state of operation as a deterministic variable.

References

1. E. Du *et al.*, "Operation of a high renewable penetrated power system with CSP plants: A look-aheadstochastic unit commitment model," *IEEE Trans. Power Syst.*, vol. 34, no. 1, pp. 140–151, 2019, doi:10.1109/TPWRS.2018.2866486.
2. L. Polleux, G. Guerassimoff, J. P. Marmorat, J. Sandoval-Moreno, and T. Schuhler, "An overview of the challenges of solar power integration in isolated industrial microgrids with reliability constraints," *Renew. Sustain. Energy Rev.*, vol. 155, no.December 2021, p. 111955, 2022, doi: 10.1016/j.rser.2021.111955.
3. N. Alshammari and J. Asumadu, "Optimum unit sizing of hybrid renewable energy system utilizing harmony search, Jaya and particle swarm optimization algorithms," *Sustain. Cities Soc.*, vol. 60, no. May, p. 102255, 2020, doi: 10.1016/j.scs.2020.102255.
4. J. Dong, F. Gao, X. Guan, Q. Zhai, and J. Wu, "Storage-Reserve Sizing with Qualified Reliability for Connected High Renewable Penetration Micro-Grid," *IEEE Trans. Sustain. Energy*, vol. 7, no. 2, pp. 732–743, 2016, doi: 10.1109/TSTE.2015.2498599.
5. M. Firouzi, A. Samimi, and A. Salami, "Reliability Evaluation of a Composite Power System in the Presence of Renewable Generations," *Reliab. Eng. Syst. Saf.*, vol. 222, no. January, p. 108396, 2022, doi: 10.1016/j.ress.2022.108396.
6. P. M. Subcommittee, "wks / year," no. 6, pp. 2047–2054, 1979.
7. R. N. Allan, "Power-system reliability in perspective the economics of reliability should be compared," no. March, pp. 231–236, 1984.
8. Z. Bie, P. Zhang, G. Li, B. Hua, M. Meehan, and X. Wang, "Reliability evaluation of active distribution systems including microgrids," *IEEE Trans. Power Syst.*, vol. 27, no. 4, pp. 2342–2350, 2012, doi: 10.1109/TPWRS.2012.2202695.
9. S. Sulaeman, S. Member, and M. Benidris, "Wind Variability and Turbine Forced Outages," vol. 8, no. 2, pp. 629–637, 2017.
10. A. B. Esan, A. F. Agbetuyi, O. Oghorada, K. Ogbeide, A. A. Awelewa, and A. E. Afolabi, "Reliability assessments of an islanded hybrid PV-diesel-battery system for a typical rural community in Nigeria," *Heliyon*, vol. 5, no. 5, p. e01632, 2019, doi: 10.1016/j.heliyon.2019.e01632.
11. H. Hashemi-Dezaki, A. M. Hariri, and M. A. Hejazi, "Impacts of load modeling on generalized analytical reliability assessment of smart grid under various penetration levels of wind/solar/non- renewable distributed generations," *Sustain. Energy, Grids Networks*, vol. 20, p. 100246, 2019, doi: 10.1016/j.segan.2019.100246.
12. T. Adefarati and R. C. Bansal, "Reliability, economic and environmental analysis of a microgrid system in the presence of renewable energy resources," *Appl. Energy*, vol. 236, no. November 2018, pp. 1089– 1114, 2019, doi: 10.1016/j.apenergy.2018.12.050.

13. A. Bani-Ahmed, M. Rashidi, A. Nasiri, and H. Hosseini, "Reliability Analysis of a Decentralized Microgrid Control Architecture," *IEEE Trans. Smart Grid*, vol. 10, no. 4, pp. 3910–3918, 2019, doi: 10.1109/TSG.2018.2843527.
14. O. Erdinc, A. Tascikaraoglu, N. G. Paterakis, I. Dursun, M. C. Sinim, and J. P. S. Catalao, "Comprehensive Optimization Model for Sizing and Siting of DG Units, EV Charging Stations, and Energy Storage Systems," *IEEE Trans. Smart Grid*, vol. 9, no. 4, pp. 3871–3882, 2018, doi: 10.1109/TSG.2017.2777738.
15. E. Tómasson and L. Söder, "Coordinated optimal strategic demand reserve procurement in multi-area power systems," *Appl. Energy*, vol. 270, no. May, p. 114984, 2020, doi: 10.1016/j.apenergy.2020.114984.
16. P. Paliwal, N. P. Patidar, and R. K. Nema, "Probabilistic indices for analysing the impact of penetration of distributed energy resources on system reliability," *IET Renew. Power Gener.*, vol. 14, no. 12, pp. 2154–2165, 2020, doi: 10.1049/iet-rpg.2019.1214.
17. A. Hassanpour and E. Roghanian, "A two-stage stochastic programming approach for non-cooperative generation maintenance scheduling model design," *Int. J. Electr. Power Energy Syst.*, vol. 126, no. PA, p. 106584, 2021, doi: 10.1016/j.ijepes.2020.106584.
18. S. Ostovar, A. Esmaeili-Nezhad, M. Moeini-Aghtaie, and M. Fotuhi-Firuzabad, "Reliability assessment of distribution system with the integration of photovoltaic and energy storage systems," *Sustain. Energy, Grids Networks*, vol. 28, p. 100554, 2021, doi: 10.1016/j.segan.2021.100554.
19. R. Diewvilai and K. Audomvongseree, "Generation expansion planning with energy storage systems considering renewable energy generation profiles and full-year hourly power balance constraints," *Energies*, vol. 14, no. 18, 2021, doi: 10.3390/en14185733.
20. J. Zeng *et al.*, "A multi-state reliability evaluation model of CSP plants considering partial function failure," *Electr. Power Syst. Res.*, vol. 199, no. June, 2021, doi: 10.1016/j.epsr.2021.107396.
21. A. Muhtadi, D. Pandit, N. Nguyen, S. Elsaiah, and J. Mitra, "FESS for reliability improvement of DER- based microgrids: Comparison of distributed storage alternatives," *SEST 2021 - 4th Int. Conf. Smart Energy Syst. Technol.*, 2021, doi: 10.1109/SEST50973.2021.9543108.
22. Y. Y. Hong, C. I. Wu, T. H. Hsiao, and C. S. Lin, "Reliability of a Power System with High Penetration of Renewables: A Scenario-Based Study," *IEEE Access*, vol. 9, pp. 78050–78059, 2021, doi: 10.1109/ACCESS.2021.3083793.
23. C. Yan, Z. Bie, Y. Ling, G. Li, F. Xiong, and J. Wen, "A comprehensive reliability assessment index system for regional grid with a large renewable energy penetration," *Asia-Pacific Power Energy Eng. Conf. APPEEC*, vol. 2016-Decem, pp. 619–622, 2016, doi: 10.1109/APPEEC.2016.7779578.
24. R. Billinton and A. Sankarakrishnan, "Comparison of Monte Carlo simulation techniques for composite power system reliability assessment," *IEEE WESCANEX Commun. Power, Comput.*, vol. 1, no. 95, pp. 145–150, 1995, doi: 10.1109/wescan.1995.4939.

23

Distance Measurement Using Ultrasonic Sensor

Rajesh Babu Damala[1], Rajesh Kumar Patnaik[2] and Praveen Korla*

[1]Department of Electrical and Electronic Engineering, EEE Department at GMRIT, Rajam, Andhrapradesh, India
[2]Department of EEE, Centurion University of Technology and Management, Paralakhemundi, Odisha, India
[3]Department of EEE, GMR Institute of Technology, Rajam, Andhrapradesh, India

Abstract
This chapter describes an ultrasonic sensor that can gauge how far certain motor vehicle points are from the ground. The sensor works by tracking how long an ultrasonic pulse takes to travel after being reflected by the ground. A threshold comparator can quickly identify reflected pulses that are generated using a limited optimization method. A sub-wavelength detection may be made using a method like this that takes response of frequency into consideration. At rest or at moderate speeds, experimental testing at a frequency of 40 kHz. The sensor is nevertheless functional at speeds of maximum 30 m/s with error of 1 mm. The sensor can self-adapt to diverse conditions to provide optimum results and is made up of inexpensive components, making it appropriate for first-car equipment in many situations.

Keywords: Ultrasonic sensor, Arduino UNO, distance measurement, LCD display, jumper wires

23.1 Introduction

Ultrasonic sensors offer a low-cost, simple-to-implement, and reliable method for measuring distance and detecting the presence of objects.

**Corresponding author*: 20341A0266@gmrit.edu.in

O.V. Gnana Swathika, K. Karthikeyan and Sanjeevikumar Padmanaban (eds.) Smart Grids as Cyber Physical Systems Volume 1: Artificial Intelligence, Cybersecurity, and Clean Energy for Next Generation Smart Grids, (369–376) © 2024 Scrivener Publishing LLC

Due to its durability and great performance, most industrial sensors generate and detect ultrasonic waves using piezo ceramic components. They primarily function at a very specific resonance frequency that requires modest stimulating energies. This causes very narrowband bursts whose flight periods are recorded in relation to the separation between the sensor and the item. The distance to the item is estimated under the assumption that the sound speed is known. When the amplitude of the incoming signal increases beyond a predetermined threshold, an echo is considered to have been heard. This method of echo reception is sometimes known as simple amplitude detection. When adopting this strategy, there are a few factors to keep in mind. Ultrasonic signals experience significant attenuation when in flight [1]. The regular design of the reflecting item has an impact on the echo amplitude; edges only reflect a small portion of the incoming signal energy, whereas correctly- aligned planes reflects the maximum of it. Reference [2] discusses in depth the various reflection qualities of various objects without knowing in advance the characteristics of the reflecting item, the sensor cannot account for this effect and may consequently cause systematic inaccuracies that rely on the shape of the object. This issue can be resolved by giving the sensor knowledge of the object's reflecting characteristics beforehand and only using object kinds that are clearly specified. The direction of axis of the sensor has the largest signal amplitude and sensor sensitivity. As a result, objects that produce powerful echoes are identified across a considerably larger region, where as other methods cannot do.

23.2 Associated Hardware Component Details

As their name suggests, ultrasonic sensors measure distance using ultrasonic waves. An ultrasonic pulse is emitted by the sensor head and returned to it by the object. Time period between emission and reception is used by ultrasonic sensors to find out the distance of targeted object. For both emission and reception, ultrasonic sensor shown in Figure 23.1 employs a single ultrasonic component i.e. optical sensor uses a transmitter and a receiver.

Jumper Wires

Cables were merely cables with connectors one on each end that allowed you to join two ports with no soldering. The jumper cables as shown in Figure 23.2 are widely used in breadboards and prototype devices to make it easy to join a circuits as required and pretty easy.

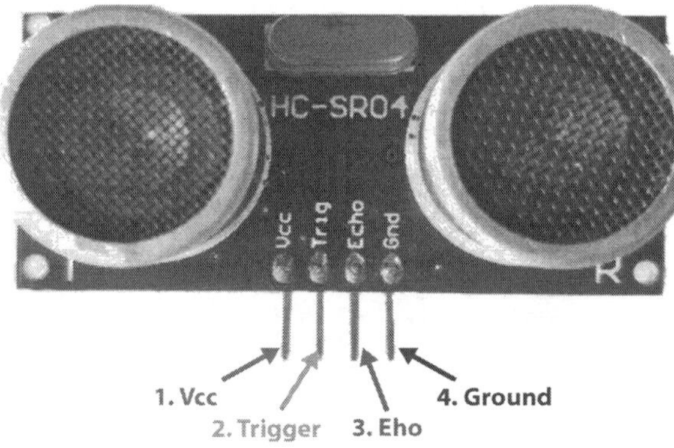

Figure 23.1 Ultrasonic sensor.

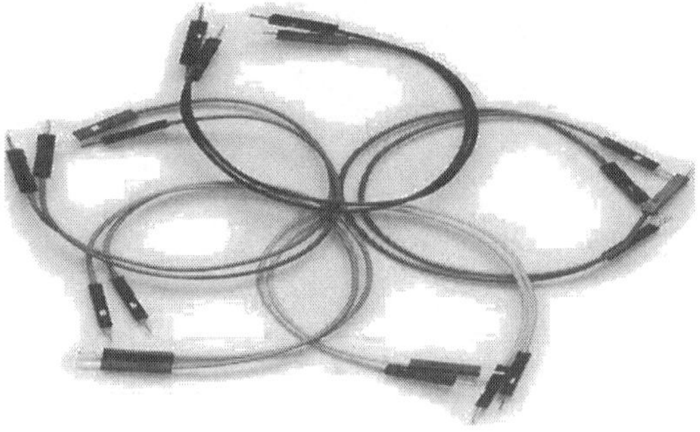

Figure 23.2 Jumper wires.

16×2 LCD Module with I2C

Its 16 columns and 2 rows give the 162 liquid crystal display (LCD) module its name as shown in Figure 23.3. There are many possible combinations, including 8-1, 8-2, 10-1, 16-1, etc. However, the 162 LCD is the most popular, therefore, we are employing it in this instance. I2C stands for inter-integrated communication protocol (I-square-C, or IIC). Normally, this is done to facilitate communication between a single master and several slaves. We may lessen the connections while utilizing I2C, which is one of its finest features (wiring). If you utilize a standard LCD panel, you require a total of 12 connections. When using an I2C LCD display, just four connections are required. You can see the advantage of the I2C protocol by looking at the aforementioned example. The two-line protocol is another name for the I2C protocol. If you correctly viewed the image above, you would have noticed a black adapter. The PCF8574T IC chip is used by that

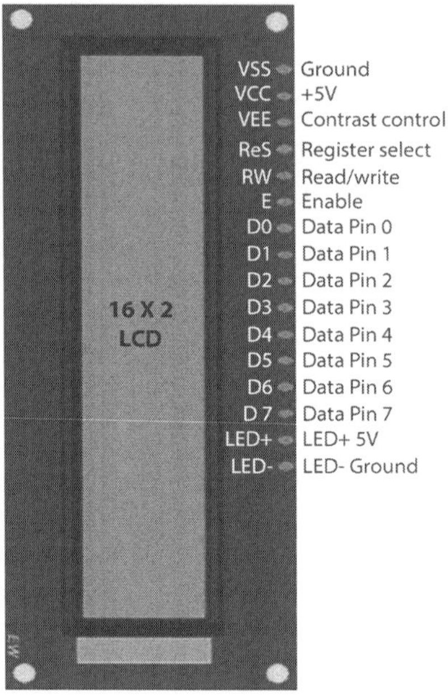

Figure 23.3 16×2 LCD module with I2C LCD display.

black adapter to convert parallel data from I2C serial data to for the LCD display. A potentiometer, as you can see in Figure 23.3, is blue and is used to control the display's brightness.

Arduino

Arduino UNO is an 8-bit ATmega328P microcontroller with a controller board as shown in Figure 23.4. It has supplementary parts such a voltage regulator, serial communication, crystal oscillator, and others that help the ATmega328P microprocessor. The Arduino UNO has six analogue input pins, a USB port, a power barrel connector, 14 digital input/output pins, an ICSP header, and a reset button. Each pin operates at 5 V and has an inbuilt pull-up resistor that ranges from 20 to 50 K Ohm that is unplugged by default. It can also supply a maximum of 40 mA of current to the receiver.

Battery

One or more electrochemical cells with external connections make form a battery as shown in Figure 23.5, a type of power source that can be used to run devices like flashlights and telephones. When a battery provides power, the negative terminal serves as the anode the positive terminal serves as the cathode. The positive port will receive electrons from the negative terminal via an outside electric circuit.

Figure 23.4 Schematic diagram of Ardunio UNO.

Figure 23.5 Battery of 9 W for DC power supply.

Potentiometer

A potentiometer as shown in Figure 23.6 which is a three-terminal adjustable resistor that the value of resistance could be adjusted physically. A wiper which is a slipping contact that slides on the resistive component is attached to the third terminal. The wiper point determines the potentiometer's output voltage. Another name for a potentiometer is a pot. In addition to numerous other applications, this kind of pot is frequently employed as volume control in audio system. The potentiometer can be designed from a several of resources, including as conductive plastic, metal film, wire wound potentiometers, carbon composition, and cermet.

Breadboard

A breadboard as shown in Figure 23.7 is a rectangular piece of plastic with many small holes on it. One can easily make circuits such as this one with an LED, battery, and resistor, and switch using these holes.

Figure 23.6 Potentiometer.

Figure 23.7 Breadboard.

23.3 Circuit Diagram

The pin connection can be seen from Figure 23.8. For simple understanding, different pin connections are shown in different colors.

23.4 Methodology

Ultrasonic wave measurement system with an Arduino interface in the 20 Hz to 20 kHz range is the range that humans can hear. We can use the ultrasonic sensor HC-SR04 to make use of these frequency range waves. When this sensor is connected to an Arduino sensing and control system, a precise measurement of distance can be measured using novel methods. This technique is effective for taking accurate measurements of small distances. Ultrasonic sensors are used to determine how far an obstruction is from the sensor.

Calculating the distance is possible once the sound speed is known as indicated in Figure 23.9. The LCD display shows the measured distance.

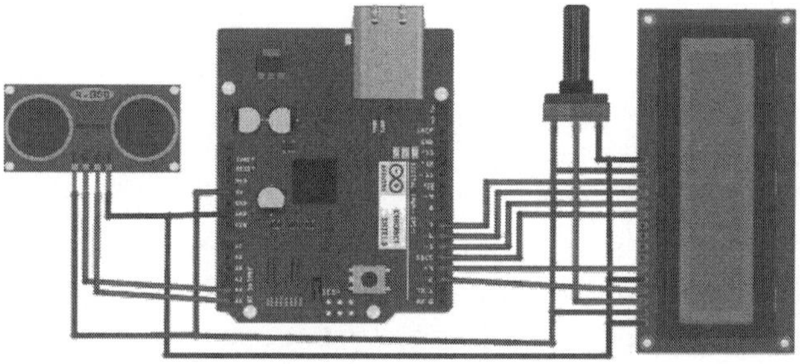

Figure 23.8 Schematic circuit diagram for distance measurement.

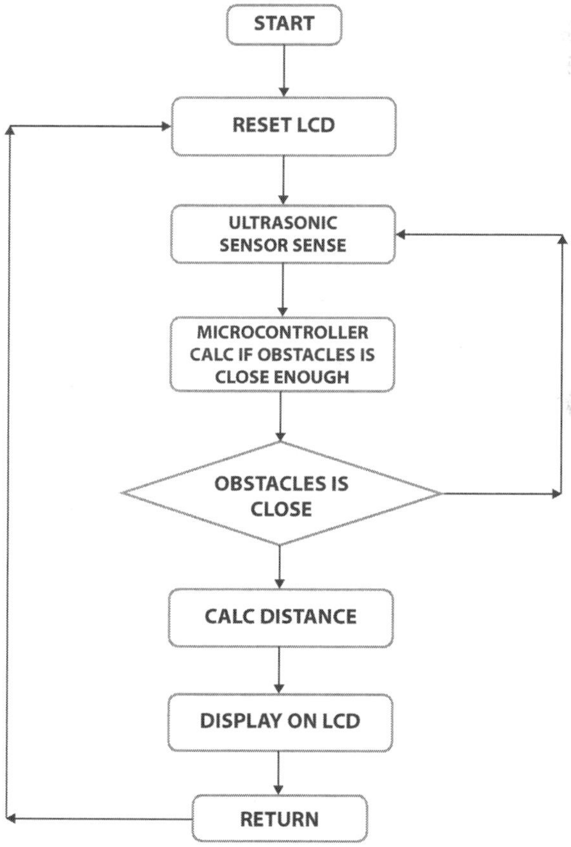

Figure 23.9 Flow chart for distance measurement.

23.5 Conclusions

Precise distance measurement device is developed with an ultrasonic sensor and an Arduino. This consists of a transmitter and receiver parts of an ultrasonic module unit with high frequency waves. If any object comes within the range of these sensors, reflection takes place. The time taken by the reflected waves by the object can be calculated by the microphone in receiver part of sensor. The distance between two points is calculated as Distance = time*speed. The human auditory range can be used to accurately assess distance. New prototyping hardware and compatibility, as well as interface with other consumer flood shielding and TV/smartphones. Mining equipment may be required where necessary. Already compatible with many major simulation software packages such as lab view and MATLAB.

References

1. D. Wobschall, M. Zeng and B. Srinivasaraghavan, "An Ultrasonic/Optical Pulse Sensor for Precise Distance Measurements," *2005 Sensors for Industry Conference*, 2005, pp. 31-34, doi: 10.1109/SICON.2005.257865.
2. Lianjun Zhang and Lifang Zhao, "Research of ultrasonic distance measurement system based on DSP," *2011 International Conference on Computer Science and Service System (CSSS)*, 2011, pp. 2455-2458, doi: 10.1109/CSSS.2011.5974489.

Index

Also of Interest

Other Books by the Same Editors

SMART GRIDS FOR SMART CITIES VOLUME 1, Edited by O.V. Gnana Swathika, K. Karthikeyan, and Sanjeevikumar Padmanaban, ISBN: 9781119872078. Written and edited by a team of experts in the field, this first volume in a two-volume set focuses on an interdisciplinary perspective on the financial, environmental, and other benefits of smart grid technologies and solutions for smart cities.

SMART GRIDS FOR SMART CITIES VOLUME 2: Real-Time Applications in Smart Cities, Edited by O.V. Gnana Swathika, K. Karthikeyan, and Sanjeevikumar Padmanaban, ISBN: 9781394215874. Written and edited by a team of experts in the field, this second volume in a two-volume set focuses on an interdisciplinary perspective on the financial, environmental, and other benefits of smart grid technologies and solutions for smart cities.

INTEGRATED GREEN ENERGY SOLUTIONS VOLUME 1, Edited by Milind Shrinivas Dangate, W. S. Sampath, O. V. Gnana Swathika, and Sanjeevikumar Padmanaban, ISBN: 9781119847434. This first volume in a two-volume set presents the state of the art for the concepts, practical applications, and future of renewable energy and how to move closer to true sustainability.

INTEGRATED GREEN ENERGY SOLUTIONS VOLUME 2, Edited by Milind Shrinivas Dangate, W. S. Sampath, O. V. Gnana Swathika, and Sanjeevikumar Padmanaban, ISBN: 9781394193660. This second volume in a two-volume set continues to present the state of the art for the concepts, practical applications, and future of renewable energy and how to move closer to true sustainability.

Check out these other related titles from Scrivener Publishing

SMART GRIDS AND INTERNET OF THINGS, Edited by Sanjeevikumar Padmanaban, Jens Bo Holm-Nielsen, Rajesh Kumar Dhanaraj, Malathy Sathyamoorthy, and Balamurugan Balusamy, ISBN: 9781119812449. Written and edited by a team of international professionals, this ground-breaking new volume covers the latest technologies in automation, tracking, energy distribution and consumption of Internet of Things (IoT) devices with smart grids.

SMART GRIDS AND GREEN ENERGY SYSTEMS, Edited by A. Chitra, V. Indragandhi and W. Razia Sultana, ISBN: 9781119872030. Presenting the concepts and advances of smart grids within the context of "green" energy systems, this volume, written and edited by a global team of experts, goes into the practical applications that can be utilized across multiple disciplines and industries, for both the engineer and the student.

SMART GRIDS AND MICROGRIDS: Concepts and Applications, Edited by P. Prajof, S. Mohan Krishna, J. L. Febin Daya, Umashankar Subramaniam, and P. V. Brijesh, ISBN: 9781119760559. Written and edited by a team of experts in the field, this is the most comprehensive and up to date study of smart grids and microgrids for engineers, scientists, students, and other professionals.

MICROGRID TECHNOLOGIES, Edited by C. Sharmeela, P. Sivaraman, P. Sanjeevikumar, and Jens Bo Holm-Nielsen, ISBN 9781119710790. Covering the concepts and fundamentals of microgrid technologies, this volume, written and edited by a global team of experts, also goes into the practical applications that can be utilized across multiple industries, for both the engineer and the student.

INTEGRATION OF RENEWABLE ENERGY SOURCES WITH SMART GRIDS, Edited by A. Mahaboob Subahani, M. Kathiresh and G. R. Kanagachidambaresan, ISBN: 9781119750420.
Provides comprehensive coverage of renewable energy and its integration with smart grid technologies.

Encyclopedia of Renewable Energy, by James G. Speight, ISBN 9781119363675. Written by a highly respected engineer and prolific author in the energy sector, this is the single most comprehensive, thorough, and up to date reference work on renewable energy.

Green Energy: Solar Energy, Photovoltaics, and Smart Cities, edited by Suman Lata Tripathi and Sanjeevikumar Padmanaban, ISBN 9781119760764. Covering the concepts and fundamentals of green energy, this volume, written and edited by a global team of experts, also goes into the practical applications that can be utilized across multiple industries, for both the engineer and the student.

DESIGN AND DEVELOPMENT OF EFFICIENT ENERGY SYSTEMS, edited by Suman Lata Tripathi, Dushyant Kumar Singh, Sanjeevikumar Padmanaban, and P. Raja, ISBN 9781119761631. Covering the concepts and fundamentals of efficient energy systems, this volume, written and edited by a global team of experts, also goes into the practical applications that can be utilized across multiple industries, for both the engineer and the student.

INTELLIGENT RENEWABLE ENERGY SYSTEMS: Integrating Artificial Intelligence Techniques and Optimization Algorithms, edited by Neeraj Priyadarshi, Akash Kumar Bhoi, Sanjeevikumar Padmanaban, S. Balamurugan, and Jens Bo Holm-Nielsen, ISBN 9781119786276. This collection of papers on artificial intelligence and other methods for improving renewable energy systems, written by industry experts, is a reflection of the state of the art, a must-have for engineers, maintenance personnel, students, and anyone else wanting to stay abreast with current energy systems concepts and technology.

SMART CHARGING SOLUTIONS FOR HYBRID AND ELECTRIC VEHICLES, edited by Sulabh Sachan, Sanjeevikumar Padmanaban, and Sanchari Deb, ISBN 9781119768951. Written and edited by a team of experts in the field, this is the most comprehensive and up to date study of smart charging solutions for hybrid and electric vehicles for engineers, scientists, students, and other professionals.